Basic Concepts

Today we live in a predominantly electrical world. While this statement may sound strange at first, a moment's reflection will indicate its inherent truth. The two primary areas of electrotechnology that permeate essentially every aspect of our lives are power and information. Without them, life as we know it would undergo stupendous changes. And yet, we have learned to generate, convert, transmit, and utilize these technologies for the enhancement of the whole human race.

Electrotechnology is a driving force in the changes that are occurring in every engineering discipline. For example, surveying is now done with lasers and electronic range finders, and automobiles employ electronic dashboards and electronic ignition systems. Industrial processes that range from chemical refineries and metal foundries to waste water treatment plants use (1) electronic sensors to obtain information about the process, (2) instrumentation systems to gather the information, and (3) computer control systems to process the information and generate electronic commands to actuators, which correct and control the process.

Fundamental to electrotechnology is the area of circuit analysis. A thorough knowledge of this subject provides an understanding of such things as cause and effect, amplification and attenuation, feedback and control, and stability and oscillation. Of critical importance is the fact that these principles can be applied, not only to engineering systems, but also to economic and social systems. Thus, the ramifications of circuit analysis are immense, and a solid understanding of this subject is well worth the effort expended to obtain it.

1.1 SYSTEM OF UNITS

The system of units we employ is the international system of units, the Système International des Unités, which is normally referred to as the SI standard system. This system, which is composed of the basic units meter (m), kilogram (kg), second (s), ampere (A), degree kelvin (°K), and candela (cd), is defined in all modern physics texts and therefore will not be defined here. However, we will discuss the units in some detail as we encounter them in our subsequent analyses.

The standard prefixes that are employed in SI are shown in Fig. 1.1. Note the decimal relationship between these prefixes. These standard prefixes are employed throughout our study of electric circuits.

10^{-12} 10^{-9} 10^{-6} 10^{-3} 1 10^{3} 10^{6} 10^{9} 10^{12}

pico (p)   nano (n)   micro (μ)   milli (m)   kilo (k)   mega (M)   giga (G)   tera (T)

Figure 1.1 Standard SI prefixes.

Circuit technology has changed drastically over the years. For example, in the early 1960s the space on a circuit board occupied by the base of a single vacuum tube was about the size of a quarter (25-cent coin). Today that same space could be occupied by an Intel Pentium integrated circuit chip containing 3.1 million transistors. These chips are the engine for a host of electronic equipment.

1.2 BASIC QUANTITIES

Before we begin our analysis of electronic circuits, we must define terms that we will employ. However, in this chapter and throughout the book our definitions and explanations will be as simple as possible in order to foster an understanding of the use of the material. No attempt will be made to give complete definitions of many of the quantities because such definitions are not only unnecessary at this level but are often confusing. Although most of us have an intuitive concept of what is meant by a circuit, we will simply refer to an *electric circuit* as an interconnection of electrical components.

It is very important at the outset that the reader understand the basic strategy that we will employ in our analysis of electric circuits. This strategy is outlined in Fig. 1.2, and we will be concerned only with the portion of the diagram to the right of the dashed line. In subsequent courses or further study the reader will learn how to model physical devices such as electronic components. Our procedure here will be to employ linear models for our circuit components, then define the variables that are used in the appropriate circuit equations to yield solution values, and finally, interpret the solution values for the variables in order to determine what is actually happening in the physical circuit. The variables may be time varying or constant depending on the nature of the physical parameters they represent.

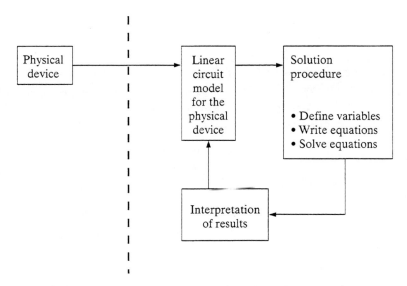

Figure 1.2 Basic strategy employed in circuit analysis.

The most elementary quantity in an analysis of electric circuits is the electric *charge*. We know from basic physics that the nature of charge is based on concepts of atomic theory. We view the atom as a fundamental building block of matter that is composed of a positively charged nucleus surrounded by negatively charged electrons. In the metric system, charge is measured in coulombs (C). The charge on an electron is negative and equal in magnitude to 1.602×10^{-19} C. However, our interest in electric charge is centered around its motion, since charge in motion results in an energy transfer. Of particular interest to us are those situations in which the motion is confined to a definite closed path.

An electric circuit is essentially a pipeline that facilitates the transfer of charge from one point to another. The time rate of change of charge constitutes an electric *current*. Mathematically, the relationship is expressed as

$$i(t) = \frac{dq(t)}{dt} \quad \text{or} \quad q(t) = \int_{-\infty}^{t} i(x)\,dx \qquad \textbf{1.1}$$

where i and q represent current and charge, respectively (lowercase letters represent time dependency and capital letters are reserved for constant quantities). The basic unit of current is the ampere (A) and 1 ampere is 1 coulomb per second.

Although we know that current flow in metallic conductors results from electron motion, the conventional current flow, which is universally adopted, represents the movement of positive charges. It is important that the reader think of current flow as the movement of positive charge regardless of the physical phenomena that take place. The symbolism that will be used to represent current flow is shown in Fig. 1.3. The variable representing the current in the wire in Fig. 1.3a was defined as I_1 flowing in the wire from left to right in the figure. A set of equations was written for the circuit and a solution value of 2 A was obtained; that is, $I_1 = 2$ A. This means that the physical current in the wire is flowing from left to right, in the direction of our variable, and is 2 A. $I_1 = 2$ A in Fig. 1.3a indicates that at any point in the wire shown, 2 C of charge pass from left to right each second. The same procedure was followed for the wire shown in Fig. 1.3b and the variable

I_2 was defined in the same manner. A set of equations was written and a solution value of −3 A was obtained. An interpretation of this result is that the physical current in the wire is from right to left, opposite to our reference direction for I_1; its value is 3 A. $I_2 = -3$ A in Fig. 1.3b indicates that at any point in the wire shown, 3 C of charge pass from right to left each second. Therefore, it is important to specify not only the magnitude of the variable representing the current, but also its direction. After a solution is obtained for the current variable, the actual physical current is known.

$I_1 = 2$ A

$I_2 = -3$ A

Circuit 1

Circuit 2

Figure 1.3 Conventional current flow: (a) positive current flow; (b) negative current flow.

(a)

(b)

There are two types of current that we encounter often in our daily lives, alternating current (ac) and direct current (dc), which are shown as a function of time in Fig. 1.4. *Alternating current* is the common current found in every household, used to run the refrigerator, stove, washing machine, and so on. Batteries, which are used in automobiles or flashlights, are one source of *direct current*. In addition to these two types of currents, which have a wide variety of uses, we can generate many other types of currents. We will examine some of these other types later in the book. In the meantime, it is interesting to note that the magnitude of currents in elements familiar to us ranges from soup to nuts, as shown in Fig. 1.5.

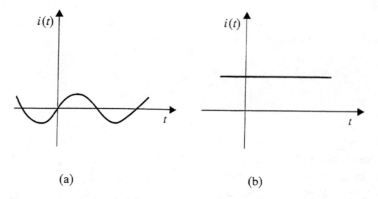

(a)

(b)

Figure 1.4 Two common types of current: (a) alternating current (ac); (b) direct current (dc).

We have indicated that charges in motion yield an energy transfer. Now we define the *voltage* (also called the *electromotive force* or *potential*) between two points in a circuit as the difference in energy level of a unit charge located at each of the two points. Work or energy, $w(t)$ or W, is measured in joules (J); 1 joule is 1 newton meter (N·m). Hence, voltage [$v(t)$ or V] is measured in volts (V) and 1 volt is 1 joule per coulomb; that is, 1 volt = 1 joule per coulomb = 1 newton meter per coulomb.

$$V = \frac{W}{q}$$

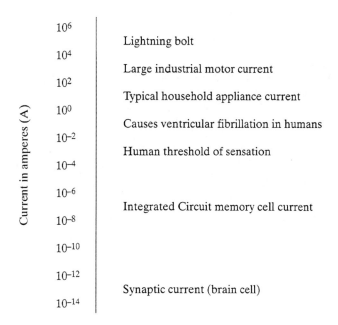

Current in amperes (A)

10^6	
10^4	Lightning bolt
10^2	Large industrial motor current
10^0	Typical household appliance current
10^{-2}	Causes ventricular fibrillation in humans
10^{-4}	Human threshold of sensation
10^{-6}	
10^{-8}	Integrated Circuit memory cell current
10^{-10}	
10^{-12}	
10^{-14}	Synaptic current (brain cell)

Figure 1.5 Typical current magnitudes.

If a unit positive charge is moved between two points, the energy required to move it is the difference in energy level between the two points and is the defined voltage. It is extremely important that the variables that are used to represent voltage between two points be defined in such a way that the solution will let us interpret which point is at the higher potential with respect to the other.

In Fig. 1.6a the variable that represents the voltage between points A and B has been defined as V_1, and it is assumed that point A is at a higher potential than point B, as indicated by the $+$ and $-$ signs associated with the variable and defined in the figure. The $+$ and $-$ signs define a reference direction for V_1. The equations for the circuit were written and a solution value of 2 V was obtained; that is $V_1 = 2$ V, as shown in the figure. The physical interpretation of this is that the difference in potential of points A and B is 2 V and point A is at the higher potential. If a unit positive charge is moved from point A through the circuit to point B, it will give up energy to the circuit and have 2 J less energy when it reaches point B. If a unit positive charge is moved from point B to point A, extra energy must be added to the charge by the circuit, and hence the charge will end up with 2 J more energy at point A than it started with at point B.

DRILL

D1.2 Determine the energy required to move 120 C of charge from point B to point A in the network in Fig. 1.6a.

ANSWER: 240 J

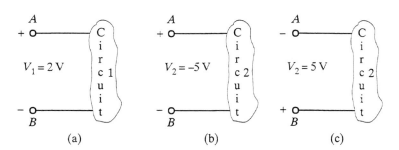

Figure 1.6 Voltage representations.

The same procedure was followed for the circuit in Fig. 1.6b, and the variable V_2 was defined in the same manner. A set of equations was written and a solution value of -5 V was obtained; that is, $V_2 = -5$ V. The physical interpretation of $V_2 = -5$ V is that the potential between points A and B is 5 V and point B is at the higher potential. If the circuit in Fig. 1.6b was reworked with the variable for the voltage defined as shown in Fig. 1.6c, the solution would have been $V_2 = 5$ V. The physical interpretation of this is that the difference in potential of points A and B is 5 V, with point B at the higher potential.

Note that it is important to define a variable with a reference direction so that the answer can be interpreted to give the physical condition in the circuit. We will find that it is not possible in many cases to define the variable so that the answer is positive, and we will also find that it is not necessary to do so. A negative number for a given variable gives exactly the same information as a positive number for a new variable that is the same as the old variable, except that it has an opposite reference direction. Hence, when we define either current or voltage, it is absolutely necessary that we specify both magnitude and direction. Therefore, it is incomplete to say that the voltage between two points is 10 V or the current in a line is 2 A, since only the magnitude and not the direction for the variables has been defined.

The range of magnitudes for voltage, equivalent to that for currents in Fig. 1.5, is shown in Fig. 1.7. Once again, note that this range spans many orders of magnitude.

Figure 1.7 Typical voltage magnitudes.

At this point we have presented the conventions that we employ in our discussions of current and voltage. *Energy* is yet another important term of basic significance. Figure 1.8 illustrates the voltage–current relationships for energy transfer. In this figure, the block representing a circuit element has been extracted from a larger circuit for examination.

In Fig. 1.8a, energy is being supplied *to* the element by whatever is attached to the terminals. Note that 2 A, that is, 2 C, of charge are moving from point A to point B through the element each second. Each coulomb loses 3 J of energy as it passes through the element from point A to point B. Therefore, the element is absorbing 6 J of energy per second. Note that when the element is *absorbing* energy, a positive current enters the positive terminal. In Fig. 1.8b energy is being supplied *by* the element to whatever is connected to terminals A-B. In this case, note that when the element is *supplying* energy, a positive current enters the negative terminal and leaves via the positive terminal. In this convention a negative current in one direction is equivalent to a positive current in the opposite direction, and vice versa. Similarly, a negative voltage in one direction is equivalent to a positive voltage in the opposite direction.

DRILL

D1.3   Five joules of energy are absorbed by 1 C of charge when moved from point B to point A in 1 s. Find the voltage between points A and B.

ANSWER: 5 V

Figure 1.8 Voltage-current relationships for energy absorbed (a) and energy supplied (b).

(a)   (b)

EXAMPLE 1.1

Suppose that your car will not start. To determine if the battery is faulty, you turn on the light switch and find that the lights are very dim, indicating a weak battery. You borrow a friend's car and a set of jumper cables. However, how do you connect his car's battery to yours? What do you want his battery to do?

SOLUTION   Essentially, his car's battery must supply energy to yours, and therefore it should be connected in the manner shown in Fig. 1.9. Note that the positive current leaves the positive terminal of the good battery (supplying energy) and enters the positive terminal of the weak battery (absorbing energy). Note that the same connections are used when charging a battery.

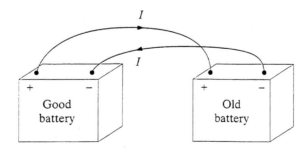

Figure 1.9 Diagram for Example 1.1.

In practical applications there are often considerations other than simply the electrical relations (e.g., safety). Such is the case with jump starting an automobile.

Automobile batteries produce explosive gases that can be ignited accidentally, causing severe physical injury. Be safe—follow the procedure described in your auto owner's manual.   ❏

We have defined voltage in joules per coulomb as the energy required to move a positive charge of 1 C through an element. If we assume that we are dealing with a differential amount of charge and energy, then

$$v = \frac{dw}{dq} \qquad\qquad \textbf{1.2}$$

Multiplying this quantity by the current in the element yields

$$vi = \frac{dw}{dq}\left(\frac{dq}{dt}\right) = \frac{dw}{dt} = p \qquad\qquad \textbf{1.3}$$

which is the time rate of change of energy or power measured in joules per second, or watts (W). Since, in general, both v and i are functions of time, p is also a time-varying quantity. Therefore, the change in energy from time t_1 to time t_2 can be found by integrating Eq. (1.3); that is,

$$w = \int_{t_1}^{t_2} p \, dt = \int_{t_1}^{t_2} vi \, dt \qquad\qquad \textbf{1.4}$$

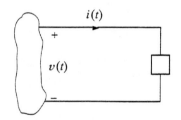

$i(t)$

$v(t)$

Figure 1.10 Sign convention for power.

The passive sign convention is used to determine whether power is being absorbed or supplied.

At this point, let us summarize our sign convention for power. To determine the sign of any of the quantities involved, the variables for the current and voltage should be arranged as shown in Fig. 1.10. The variable for the voltage $v(t)$ is defined as the voltage across the element with the positive reference at the same terminal that the current variable $i(t)$ is entering. This convention is called the *passive sign convention* and will be so noted in the remainder of this book. The product of v and i, with their attendant signs, will determine the magnitude and sign of the power. If the sign of the power is positive, power is being absorbed by the element; if the sign is negative, power is being supplied by the element.

EXAMPLE 1.2

We wish to determine the power absorbed by, or supplied by, the elements in Fig. 1.8.

SOLUTION   In Fig. 1.8a, $P = VI = (3\,\text{V})(2\,\text{A}) = 6\,\text{W}$ is absorbed by the element. In Fig. 1.8b, $P = VI = (3\,\text{V})(-2\,\text{A}) = -6\,\text{W}$ is absorbed by the element, or $+6\,\text{W}$ is supplied by the element.   ❏

EXAMPLE 1.3

Given the two diagrams shown in Fig. 1.11, determine whether the element is absorbing or supplying power and how much.

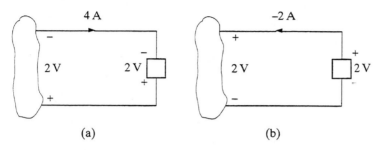

(a)   (b)

Figure 1.11 Elements for Example 1.3.

SOLUTION In Fig. 1.11a the power is $P = (2\,\text{V})(-4\,\text{A}) = -8\,\text{W}$. Therefore, the element is supplying power. In Fig. 1.11b, the power is $P = (2\,\text{V})(2\,\text{A}) = 4\,\text{W}$. Therefore, the element is absorbing power.   ❏

EXTENSION EXERCISE

E1.1 Determine the amount of power absorbed or supplied by the elements in Fig. E1.1.

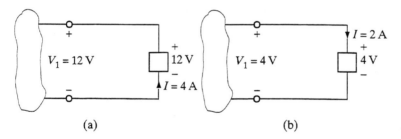

(a)   (b)

Figure E1.1

ANSWER: (a) $P = -48\,\text{W}$; (b) $P = 8\,\text{W}$.

EXAMPLE 1.4

We wish to determine the unknown voltage or current in Fig. 1.12.

SOLUTION In Fig. 1.12a, a power of $-20\,\text{W}$ indicates that the element is delivering power. Therefore, the current enters the negative terminal (terminal A), and from Eq. (1.3) the voltage is 4 V.

Figure 1.12 Elements for Example 1.4.

In Fig 1.12b, a power of +40 W indicates that the element is absorbing power and, therefore, the current should enter the positive terminal B. The current thus has a value of −8 A, as shown in the figure. □

EXTENSION EXERCISE

E1.2 Determine the unknown variables in Fig. E1.2.

Figure E1.2

ANSWER: (a) $V_1 = -20$ V; (b) $I = -5$ A.

Finally, it is important to note that these electrical networks satisfy the principle of conservation of energy. For our present purposes this means that the power that is supplied in a network is exactly equal to the power absorbed.

1.3 CIRCUIT ELEMENTS

It is important to realize at the outset that in our discussions we will be concerned with the behavior of a circuit element, and we will describe that behavior by a mathematical model. Thus, when we refer to a particular circuit element, we mean the *mathematical model* that describes its behavior.

Thus far we have defined voltage, current, and power. In the remainder of this chapter we will define both independent and dependent current and voltage sources. Although we will assume ideal elements, we will try to indicate the shortcomings of these assumptions as we proceed with the discussion.

In general, the elements we will define are terminal devices that are completely characterized by the current through the element and/or the voltage across it. These elements, which we will employ in constructing electric circuits, will be broadly classified as being either active or passive. The distinction between these two classifications depends essentially upon one thing—whether they supply or absorb energy. As the words themselves imply, an *active* element is capable of generating energy and a *passive* element cannot generate energy.

However, we will show later that some passive elements are capable of storing energy. Typical active elements are batteries, generators, and transistor models. The three common passive elements are resistors, capacitors, and inductors.

In Chapter 2 we will launch an examination of passive elements by discussing the resistor in detail. However, before proceeding with that element, we first present some very important active elements.

1. Independent voltage source
2. Independent current source
3. Two dependent voltage sources
4. Two dependent current sources

Independent Sources

An *independent voltage source* is a two-terminal element that maintains a specified voltage between its terminals *regardless of the current through it*. The general symbol for an independent source, a circle, is shown in Fig. 1.13a. As the figure indicates, terminal A is

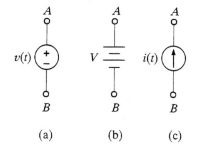

Figure 1.13 Symbols for (a) independent voltage source; (b) constant voltage source; (c) independent current source.

$v(t)$ volts positive with respect to terminal B. The word *positive* may be somewhat misleading. What is meant in this case is that $v(t)$ is referenced positive at terminal A and that the physical voltage across the device must be interpreted from the numerical value of $v(t)$. That is, if $v(t)$ at $t = 2$ s is -10 V, point B is at a higher potential than point A at $t = 2$ s. The symbol $v(t)$ is normally employed for time-varying voltages. However, if the voltage is time invariant (i.e., constant), the symbol shown in Fig. 1.13b is sometimes used. This symbol, which is used to represent a battery, illustrates that terminal A is V volts positive with respect to terminal *B*, where the long line on the top and the short line on the bottom indicate the positive and negative terminals, respectively, and thus the polarity of the element.

In contrast to the independent voltage source, the *independent current source* is a two-terminal element that maintains a specified current *regardless of the voltage across its terminals*. The general symbol for an independent current source is shown in Fig. 1.13c, where $i(t)$ is the specified current and the arrow indicates the positive direction of current flow.

Although sources come in a variety of forms, it is informative to note that there are modern high-precision instruments on the market that are capable of functioning as sources. The unit shown in Fig. 1.14 is such an instrument. This very versatile device can function as either a voltage source or a current source. It has a number of other important features, and they will be specifically addressed in the following chapter.

Figure 1.14 Keithley Model 2400 Source Meter. (Courtesy of Keithley Instruments, Inc.)

In their normal mode of operation, independent sources supply power to the remainder of the circuit. However, they may also be connected into a circuit in such a way that they absorb power. A simple example of this latter case is a battery charging circuit such as that shown in Example 1.1.

It is important that we pause here to interject a comment concerning a shortcoming of the models. In general, mathematical models approximate actual physical systems only under a certain range of conditions. Rarely does a model accurately represent a physical system under every set of conditions. To illustrate this point, consider the model for the voltage source in Fig. 1.13a. We assume that the voltage source delivers *v* volts regardless of what is connected to its terminals. Theoretically, we could adjust the external circuit so that an infinite amount of current would flow, and therefore the voltage source would deliver an infinite amount of power. This is, of course, physically impossible. A similar argument could be made for the independent current source. Hence, the reader is cautioned to keep in mind that models have limitations and are thus valid representations of physical systems only under certain conditions.

DRILL

D1.4 Determine the power supplied by the 12-V source.

ANSWER: 24 W

EXAMPLE 1.5

Determine the power absorbed or supplied by the elements in the network in Fig. 1.15.

Figure 1.15 Network for Example 1.5.

SOLUTION   The current flow is out of the positive terminal of the 24-V source, and therefore this element is supplying (2)(24) = 48 W of power. The current is into the positive terminals of elements 1 and 2, and therefore elements 1 and 2 are absorbing (2)(6) = 12 W and (2)(18) = 36 W, respectively. Note that the power supplied is equal to the power absorbed.   ❏

EXTENSION EXERCISE

E1.3   Find the power that is absorbed or supplied by the elements in Fig. E1.3.

Figure E1.3

ANSWER: Current source supplies 36 W, element 1 absorbs 54 W, and elements 2 supplies 18 W.

Dependent Sources

In contrast to the independent sources, which produce a particular voltage or current completely unaffected by what is happening in the remainder of the circuit, dependent sources generate a voltage or current that is determined by a voltage or current at a specified location in the circuit. These sources are very important because they are an integral part of the mathematical models used to describe the behavior of many electronic circuit elements.

For example, both metal-oxide-semiconductor field-effect transistors (MOSFETs) and bipolar transistors, both of which are commonly found in a host of electronic equipment, are modeled with dependent sources, and therefore the analysis of electronic circuits involves the use of these controlled elements.

In contrast to the circle used to represent independent sources, a diamond is used to represent a dependent or controlled source. Figure 1.16 illustrates the four types of dependent sources. The input terminals on the left represent the voltage or current that controls the dependent source, and the output terminals on the right represent the output current or voltage of the controlled source. Note that in Figs. 1.16a and d the quantities μ and β are dimensionless constants because we are transforming voltage to voltage and current to current. This is not the case in Figs. 1.16b and c; hence when we employ these elements a short time later, we must describe the units of the factors r and g.

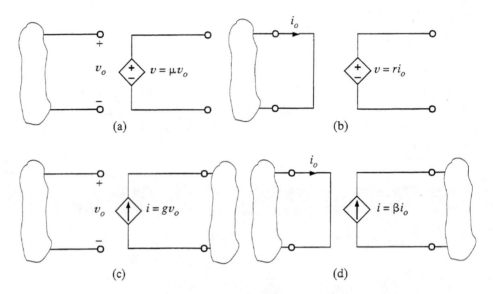

Figure 1.16 Four different types of dependent sources.

EXAMPLE 1.6

Given the two networks shown in Fig. 1.17, we wish to determine the outputs.

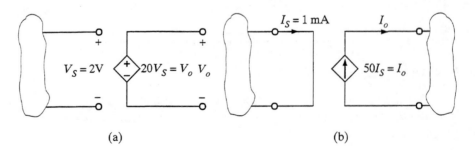

Figure 1.17 Circuits for Example 1.6.

SOLUTION   In Fig. 1.17a the output voltage is $V_o = \mu V_S$ or $V_o = 20 V_S = (20)(2\,\text{V}) = 40\,\text{V}$. Note that the output voltage has been amplified from 2 V at the input terminals to 40 V at the output terminals; that is, the circuit is an amplifier with an amplification factor of 20.

In Fig. 1.17b, the output current is $I_o = \beta I_S = (50)(1 \text{ mA}) = 50 \text{ mA}$; that is, the circuit has a current gain of 50, meaning that the output current is 50 times greater than the input current. ❑

EXTENSION EXERCISE

E1.4 Determine the power supplied by the dependent sources in Fig. E1.4.

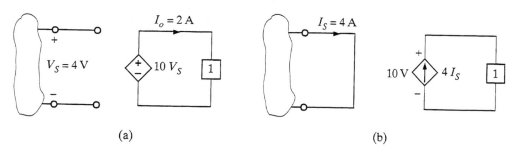

(a)   (b)

Figure E1.4

ANSWER: (a) Power supplied = 80 W; (b) power supplied = 160 W.

EXAMPLE 1.7

Let us compute the power that is absorbed or supplied by the elements in the network in Fig. 1.18.

Figure 1.18 Circuit for Example 1.7.

SOLUTION Using the sign convention for power illustrated in Fig. 1.10, we find the following:

$$P_{36\,V} = (36)(-4) = -144 \text{ W}$$
$$P_1 = (12)(4) = 48 \text{ W}$$
$$P_2 = (24)(2) = 48 \text{ W}$$
$$P_{DS} = (1I_x)(-2) = (4)(-2) = -8 \text{ W}$$
$$P_3 = (28)(2) = 56 \text{ W}$$

Note that the 36-V source and the dependent source are supplying power to the network and the remaining elements are absorbing power. Furthermore, energy is conserved, since the power supplied to the network is identically equal to the power absorbed by the network.   ❏

EXAMPLE 1.8

Let us find the current I_o in the network in Fig. 1.19.

Figure 1.19 Circuit used in Example 1.8.

SOLUTION   First, we must determine the power absorbed or supplied by each element in the network. Using the sign convention for power, we find

$$P_{2\,A} = (6)(-2) = -12 \text{ W}$$
$$P_1 = (6)(I_o) = 6I_o \text{ W}$$
$$P_2 = (12)(-9) = -108 \text{ W}$$
$$P_3 = (10)(-3) = -30 \text{ W}$$
$$P_{4\,V} = (4)(-8) = -32 \text{ W}$$
$$P_{DS} = (8I_x)(11) = (16)(11) = 176 \text{ W}$$

Since energy must be conserved,

$$-12 + 6I_o - 108 - 30 - 32 + 176 = 0$$

or

$$6I_o + 176 = 12 + 108 + 30 + 32$$

Hence,

$$I_o = 1 \text{ A} \qquad ❏$$

EXTENSION EXERCISE

E1.5   Find the power that is absorbed or supplied by the circuit elements in the network in Fig. E1.5.

Figure E1.5

ANSWER: $P_{24V} = 96$ W supplied, $P_1 = 32$ W absorbed, $P_{4I_x} = 64$ W absorbed.

As we conclude this chapter, a number of important comments are in order. The reader must thoroughly understand our approach as outlined in Fig. 1.2 because in the following chapters we will simply define and refer to voltages and currents at specific locations within a circuit. Before any equations are written, readers should think carefully about what kinds of answers they expect. For example, if all the sources are constant, all the voltages and currents in the circuit will be constant, and therefore capital letters should be used for all the variables representing the voltages and currents. It is also useful for readers to think ahead about how variables are related to one another and define the variables so that a minimum number of minus signs are used. Although this is not important in the final solution, as indicated earlier, it does force readers to think about the problem from beginning to end before attacking it so that the solution will not come as a surprise.

1.4 SUMMARY

The standard prefixes employed

$$p = 10^{-12} \qquad k = 10^3$$
$$n = 10^{-9} \qquad M = 10^6$$
$$\mu = 10^{-6} \qquad G = 10^9$$
$$m = 10^{-3} \qquad T = 10^{12}$$

The relationships between current and charge

$$i(t) = \frac{dq(t)}{dt} \qquad \text{or} \qquad q(t) = \int_{-\infty}^{t} i(x)\, dx$$

The relationships among power, energy, current, and voltage are

$$p = \frac{dw}{dt} = vi$$

$$W = \int_{t_1}^{t_2} p \, dt = \int_{t_1}^{t_2} vi \, dt$$

The passive sign convention   The passive sign convention states that if the voltage and current associated with an element are as shown in Fig. 1.10, the product of v and i, with their attendant signs, determines the magnitude and sign of the power. If the sign is positive, power is being absorbed by the element, and if the sign is negative, the element is supplying power.

Independent and dependent sources   An ideal independent voltage (current) source is a two-terminal element that maintains a specified voltage (current) between its terminals regardless of the current (voltage) through (across) the element. Dependent or controlled sources generate a voltage or current that is determined by a voltage or current at a specified location in the circuit.

Conservation of energy   The electric circuits under investigation satisfy the conservation of energy.

PROBLEMS

Section 1.2

Use the relationship between current and charge.	**1.1** In an electric conductor, a charge of 300 C passes any point in a 5-s interval. Determine the current in the conductor.
Similar to P1.1	**1.2** The current in a conductor is 1.5 A. How many coulombs of charge pass any point in a time interval of 1.5 min?
Similar to P1.1	**1.3** Determine the number of coulombs of charge produced by a 12-A battery charger in an hour.
Similar to P1.1	**1.4** Determine the time required for a 12-A battery charger to deliver a charge of 10,000 C.
Similar to P1.1	**1.5** A lightning bolt carrying 20,000 A lasts for 70 μs. If the lightning strikes a tractor, determine the charge deposited on the tractor if the tires are assumed to be perfect insulators
$V = W/q$	**1.6** A 10^8-V lightning bolt carrying 20,000 A lasts for 10 μs. If the lightning strikes a tree and burns it, determine the energy deposited on the tree.
Similar to P1.6	**1.7** Determine the energy required to move 1000 C of charge through 12 V.
Similar to P1.6	**1.8** If a 12-V battery supplies 10 A, find the amount of energy delivered.

1.9 If a 12-V battery delivers 100 J in 5 s, find (a) the amount of charge delivered and (b) the current produced.

Similar to P1.6 and P1.1

1.10 Five coulombs of charge pass through the element in Fig. P1.10 from point A, to point B. If the energy absorbed by the element is 120 J, determine the voltage across the element.

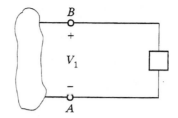

Figure P1.10

Similar to P1.6

1.11 A charge of 50 C passes through the element in Fig. P1.10 from point A to B. If $V_1 = 90$ V, determine the power required by this element.

Use the relationship between $P, W, I,$ and V

1.12 Determine the amount of power absorbed or supplied by the element in Fig. P1.12. if

(a) $V_1 = 9$ V and $I = 2$ A.

(b) $V_1 = 9$ V and $I = -3$ A.

(c) $V_1 = -12$ V and $I = 2$ A.

(d) $V_1 = -12$ V and $I = -3$ A.

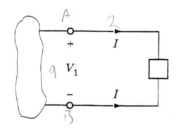

Figure P1.12

Similar to P1.11

1.13 Determine the magnitude and direction of the voltage across the elements in Fig. P1.13.

(a)

(b)

Similar to P1.11

(c)

(d)

Figure P1.13

1.14 Determine the power supplied to the elements in Fig. P1.14.

Similar to P1.11. Sum the power for each element.

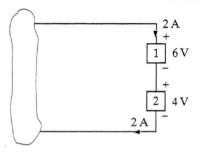

Figure P1.14

1.15 Determine the power supplied to the elements in Fig. P1.15.

Similar to P1.14

Figure P1.15

1.16 Two elements are connected in series, as shown in Fig. P1.16. Element 1 supplies 24 W of power. Is element 2 absorbing or supplying power, and how much?

1. Find the current.
2. Find the power.

Figure P1.16

Section 1.3

Important note: The values used in the remaining problems of Section 1.3 are not arbitary. They have been selected to satisfy the basic laws of circuit analysis that will be studied in the following chapters.

1.17 Determine the power that is absorbed or supplied by the circuit elements in Fig. P1.17.

Similar to P1.14

(a)   (b)

Figure P1.17

1.18 Find the power that is absorbed or supplied by the circuit elements in Fig. P1.18.

| (a) | (b) |

Similar to P1.14

Figure P1.18

1.19 Find the power that is absorbed or supplied by the network elements in Fig. P1.19.

| (a) | (b) |

Similar to P1.14

Figure P1.19

Advanced Problems

1.20 The charge entering an element is shown in Fig. P1.20. Find the current in the element in the time interval $0 \leq t \leq 0.5$ s. [*Hint:* The equation for $q(t)$ is $q(t) = 1 + (1/0.5)t, t \geq 0$.]

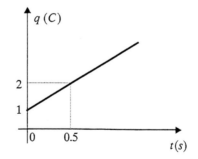

Figure P1.20

1.21 The current that enters an element is shown in Fig. 1.21. Find the charge that enters the element in the time interval $0 < t < 20$ s.

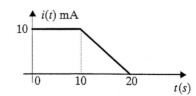

Figure P1.21

1.22 Is the source V_S in the network in Fig. P1.22 absorbing or supplying power, and how much?

Figure P1.22

1.23 Find V_x in the network in Fig. P1.23.

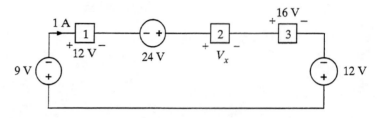

Figure P1.23

1.24 Find I_x in the network in Fig. P1.24

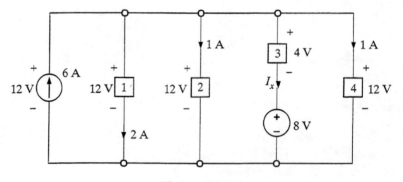

Figure P1.24

1.25 Find I_s such that the power absorbed by the two elements in Fig. P1.25 is 24 W.

Figure P1.25

1.26 Find V_x in the network Fig. P1.26.

Figure P1.26

1.27 Find I_o in the network in Fig. P1.27.

Figure P1.27

1.28 The charge entering the positive terminal of an element is $q(t) = -30e^{-4t}$ mC. If the voltage across the element is $120e^{-2t}$ V, determine the energy delivered to the element in the time interval $0 < t < 50$ ms.

1.29 The charge entering the positive terminal of an element is given by the expression $q(t) = -12e^{-2t}$ mC. The power delivered to the element is $p(t) = 2.4e^{-3t}$ W. Compute the current in the element, the voltage across the element, and the energy delivered to the element in the time interval $0 < t < 100$ ms.

1.30 The voltage across an element is $12e^{-2t}$ V. The current entering the positive terminal of the element is $2e^{-2t}$ A. Find the energy absorbed by the element in 1.5 s starting from $t = 0$.

Resistive Circuits

2

In this chapter we introduce the basic concepts and laws that are fundamental to circuit analysis. These laws are Ohm's law, Kirchhoff's current law (KCL), and Kirchhoff's voltage law (KVL). We cannot overemphasize the importance of these three laws because they will be used extensively throughout our entire study of circuit analysis. The reader who masters their use quickly will not only find the material in this text easy to learn, but will be well positioned to grasp subsequent topics in the field of electrical engineering.

As a general rule, most of our activities will be confined to analysis; that is, to the determination of a specific voltage, current, or power somewhere in a network. The techniques we introduce have wide application in circuit analysis, even though we will discuss them within the framework of simple networks.

Our approach here is to begin with the simplest passive element, the resistor, and the mathematical relationship that exists between the voltage across it and the current through it, as specified by Ohm's law. As we build our confidence and proficiency by successfully analyzing some elementary circuits, we will introduce other techniques, such as voltage division and current division, that will accelerate our work.

We will introduce circuits containing dependent sources, which are used to model active devices such as transistors. Thus, our study of circuit analysis provides a natural introduction to many topics in the area of electronics.

The devices and circuits used in dc measurements are introduced, and thus we learn the fundamentals of measuring such quantities as voltage, current, and resistance.

Finally, we present some real-world applications that begin to indicate the usefulness of circuit analysis, and then we briefly introduce the topic of circuit design in an elementary fashion. In future chapters these topics will be revisited often to present some fascinating examples that describe problems we encounter in our everyday lives.

2.1 OHM'S LAW

Ohm's law is named for the German physicist Georg Simon Ohm, who is credited with establishing the voltage–current relationship for resistance. As a result of his pioneering work, the unit of resistance bears his name.

Ohm's law states that the voltage across a resistance is directly proportional to the current flowing through it. The resistance, measured in ohms, is the constant of proportionality between the voltage and current.

A circuit element whose electrical characteristic is primarily resistive is called a resistor and is represented by the symbol shown in Fig. 2.1a. A resistor is a physical device that can be purchased in certain standard values in an electronic parts store. These resistors, which find use in a variety of electrical applications, are normally carbon composition or wire-wound. In addition, resistors can be fabricated using thick oxide or thin metal films for use in hybrid circuits, or they can be diffused in semiconductor integrated circuits. Some typical resistors are shown in Fig. 2.1b.

The mathematical relationship of Ohm's law is illustrated by the equation

$$v(t) = R \times i(t) \quad \text{where } R \geqq 0 \qquad\qquad \textbf{2.1}$$

> The passive sign convention will be employed in conjunction with Ohm's law.

Figure 2.1 (a) Symbol for a resistor; (b) some practical devices. (1), (2), and (3) are high-power resistors. (4) and (5) are variable resistors. (6) is a high-precision resistor. (7)–(12) are fixed resistors with different power ratings.

DRILL

D2.1   Determine $v(t)$ in the two circuits

(a)

(b)

ANSWER: (a) 10 V
             (b) −10 V

or equivalently, by the voltage–current characteristic shown in Fig. 2.2a. Note carefully the relationship between the polarity of the voltage and the direction of the current. In addition, note that we have tacitly assumed that the resistor has a constant value and therefore that the voltage–current characteristic is linear.

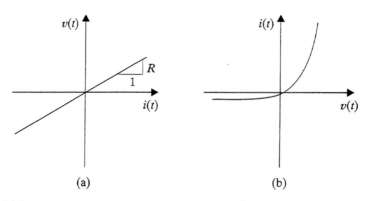

(a)   (b)

Figure 2.2 Graphical representation of the voltage–current relationship for (a) a linear resistor and (b) a diode.

The symbol Ω is used to represent ohms, and therefore,

$$1\,\Omega = 1\,\text{V/A}$$

Although in our analysis we will always assume that the resistors are *linear* and are thus described by a straight-line characteristic that passes through the origin, it is important that readers realize that some very useful and practical elements do exist that exhibit a *nonlinear* resistance characteristic; that is, the voltage–current relationship is not a straight line. Diodes, which are used extensively in electric circuits, are examples of nonlinear resistors. A typical diode characteristic is shown in Fig. 2.2b.

Since a resistor is a passive element, the proper current–voltage relationship is illustrated in Fig. 2.1a. The power supplied to the terminals is absorbed by the resistor. Note that the charge moves from the higher to the lower potential as it passes through the resistor and the energy absorbed is dissipated by the resistor in the form of heat. As indicated in Chapter 1, the rate of energy dissipation is the instantaneous power, and therefore

$$p(t) = v(t)i(t) \qquad\qquad \textbf{2.2}$$

which, using Eq. (2.1), can be written as

$$p(t) = Ri^2(t) = \frac{v^2(t)}{R} \qquad\qquad \textbf{2.3}$$

This equation illustrates that the power is a nonlinear function of either current or voltage and that it is always a positive quantity.

Conductance, represented by the symbol G, is another quantity with wide application in circuit analysis. By definition, conductance is the inverse of resistance; that is,

$$G = \frac{1}{R} \qquad \text{2.4}$$

The unit of conductance is the siemens, and the relationship between units is

$$1\,S = 1\,A/V$$

Using Eq. (2.4), we can write two additional expressions,

$$i(t) = Gv(t) \qquad \text{2.5}$$

and

$$p(t) = \frac{i^2(t)}{G} = Gv^2(t) \qquad \text{2.6}$$

Equation (2.5) is another expression of Ohm's law.

Two specific values of resistance, and therefore conductance, are very important: $R = 0$ and $R = \infty$.

In examining the two cases, consider the network in Fig. 2.3a. The variable resistance symbol is used to describe a resistor such as the volume control on a radio or television set. As the resistance is decreased and becomes smaller and smaller, we finally reach a point where the resistance is zero and the circuit is reduced to that shown in Fig. 2.3b; that is, the resistance can be replaced by a short circuit. On the other hand, if the resistance is increased and becomes larger and larger, we finally reach a point where it is essentially infinite and the resistance can be replaced by an open circuit, as shown in Fig. 2.3c. Note that in the case of a short circuit where $R = 0$,

$$v(t) = Ri(t)$$
$$= 0$$

D2.2   Determine $i(t)$ in the following two circuits:

(a)

(b)

ANSWER: (a) 8 A

(b) −2 A

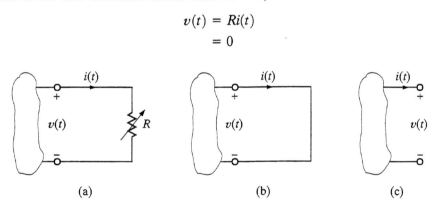

(a)   (b)   (c)

Figure 2.3 Short-circuit and open-circuit descriptions.

Therefore, $v(t) = 0$, although the current could theoretically be any value. In the open-circuit case where $R = \infty$,

$$i(t) = v(t)/R$$
$$= 0$$

Therefore, the current is zero regardless of the value of the voltage across the open terminals.

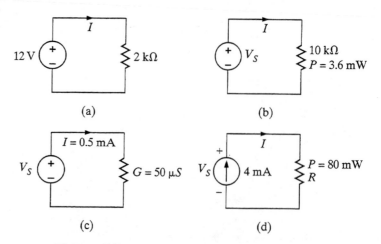

Figure 2.4 Circuits for Examples 2.1 to 2.4.

EXAMPLE 2.1

In the circuit in Fig. 2.4a, determine the current and the power absorbed by the resistor.

SOLUTION Using Eq. (2.1), we find the current to be

$$I = V/R = 12/2k = 6 \, \text{mA}$$

Note that because many of the resistors employed in our analysis are in kΩ, we will use k in the equations in place of 1000. The power absorbed by the resistor is given by Eq. (2.2) or (2.3) as

$$P = VI = (12)(6 \times 10^{-3}) = 0.072 \, \text{W}$$
$$= I^2R = (6 \times 10^{-3})^2(2k) = 0.072 \, \text{W}$$
$$= V^2/R = (12)^2/2k = 0.072 \, \text{W} \qquad \square$$

EXAMPLE 2.2

The power absorbed by the 10 kΩ resistor in Fig. 2.4b is 3.6 mW. Determine the voltage and the current in the circuit.

SOLUTION Using the power relationship, we can determine either of the unknowns.

$$V_S^2/R = P$$
$$V_S^2 = (3.6 \times 10^{-3})(10k)$$
$$V_S = 6 \, \text{V}$$

and

$$I^2R = P$$
$$I^2 = (3.6 \times 10^{-3})/10k$$
$$I = 0.6 \, \text{mA}$$

Furthermore, once V_S is determined, I could be obtained by Ohm's law and likewise once I is known, then Ohm's law could be used to derive the value of V_S. Note carefully that the equations for power involve the terms I^2 and V_S^2. Therefore, $I = -0.6$ mA and $V_S = -6$ V also satisfy the mathematical equations and, in this case, the direction of *both* the voltage and current is reversed. ❏

EXAMPLE 2.3

Given the circuit in Fig. 2.4c, we wish to find the value of the voltage source and the power absorbed by the resistance.

SOLUTION   The voltage is

$$V_S = I/G = \left(0.5 \times 10^{-3}\right)/\left(50 \times 10^{-6}\right) = 10 \text{ V}$$

The power absorbed is then

$$P = I^2/G = \left(0.5 \times 10^{-3}\right)^2/\left(50 \times 10^{-6}\right) = 5 \text{ mW}$$

Or we could simply note that

$$R = 1/G = 20 \text{ k}\Omega$$

and therefore

$$V_S = IR = \left(0.5 \times 10^{-3}\right)(20\text{k}) = 10 \text{ V}$$

and the power could be determined using $P = I^2 R = V_S^2/R = V_S I$. ❏

EXAMPLE 2.4

Given the network in Fig. 2.4d, we wish to find R and V_S.

SOLUTION   Using the power relationship, we find that

$$R = P/I^2 = \left(80 \times 10^{-3}\right)/\left(4 \times 10^{-3}\right)^2 = 5 \text{ k}\Omega$$

The voltage can now be derived using Ohm's law as

$$V_S = IR = \left(4 \times 10^{-3}\right)(5\text{k}) = 20 \text{ V}$$

The voltage could also be obtained from the remaining power relationships in Eqs. (2.2) and (2.3). ❏

Before leaving this initial discussion of circuits containing sources and a single resistor, it is important to note a phenomenon that we will find to be true in circuits containing many sources and resistors. The presence of a voltage source between a pair of terminals tells us precisely what the voltage is between the two terminals regardless of what is happening in the balance of the network. What we do not know is the current in the voltage source. We must apply circuit analysis to the entire network to determine this current. Likewise, the presence of a current source connected between two terminals specifies the exact value of the current through the source between the terminals. What we do not know is the value of the voltage across the current source. This value must be calculated by applying circuit analysis to the entire network. Furthermore, it is worth emphasizing that when applying Ohm's law, the relationship $V = IR$ specifies a relationship between the voltage *directly across* a resistor R and that current that is *present in* this resistor. Ohm's law does not apply when the voltage is present in one part of the network and the current exists in another. This is a common mistake made by students who try to apply $V = IR$ to a resistor R in the middle of the network while using a V at some other location in the network.

EXTENSION EXERCISES

E2.1   Given the circuits in Fig. E2.1, find (a) the current I and the power absorbed by the resistor in Fig. E2.1a, and (b) the voltage across the current source and the power supplied by the source in Fig. E2.1b.

Figure E2.1   (a)   (b)

ANSWER: (a) $I = 0.3 \text{ mA}, P = 3.6 \text{ mW}$, (b) $V_S = 3.6 \text{ V}, P = 2.16 \text{ mW}$.

E2.2   Given the circuits in Fig. E2.2, find (a) R and V_S in the circuit in Fig. E2.2a, and (b) find I and R in the circuit in Fig. E2.2b.

(a)   (b)

Figure E2.2

ANSWER: (a) $R = 10 \text{ k}\Omega, V_S = 4 \text{ V}$, (b) $I = 20.8 \text{ mA}, R = 576 \text{ }\Omega$.

2.2 KIRCHHOFF'S LAWS

The previous circuits that we have considered have all contained a single resistor and were analyzed using Ohm's law. At this point we begin to expand our capabilities to handle more complicated networks that result from an interconnection of two or more of these simple elements. We will assume that the interconnection is performed by electrical conductors (wires) that have zero resistance; that is, perfect conductors. Because the wires have zero resistance, the energy in the circuit is in essence lumped in each element, and we employ the term *lumped-parameter circuit* to describe the network.

To aid us in our discussion, we will define a number of terms that will be employed throughout our analysis. As will be our approach throughout this text, we will use examples to illustrate the concepts and define the appropriate terms. For example, the circuit shown in Fig. 2.5 will be used to describe the terms *node*, *loop*, and *branch*. A *node* is simply a point of connection of two or more circuit elements. The reader is cautioned to note that although one node can be spread out with perfect conductors, it is still only one node. For example, node 5 consists of the entire bottom connector of the circuit. In other words, if we start at some point in the circuit and move along perfect conductors in any direction until we encounter a circuit element, the total path we cover represents a single node. Therefore, we can assume that a node is one end of a circuit element together with all the perfect conductors that are attached to it. Examining the circuit, we note that there are numerous paths through it. A *loop* is simply any *closed path* through the circuit in which no node is encountered more than once. For example, starting from node 1, one loop would contain the elements R_1, v_2, R_4, and i_1; another loop would contain R_2, v_1, v_2, R_4, and i_1; and so on. However, the path R_1, v_1, R_5, v_2, R_3, and i_1 is not a loop because we have encountered node 3 twice. Finally, a *branch* is a portion of a circuit containing only a single element and the nodes at each end of the element. The circuit in Fig. 2.5 contains eight branches.

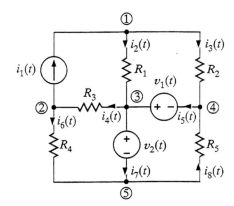

Figure 2.5 Circuit used to illustrate KCL.

Given the previous definitions, we are now in a position to consider Kirchhoff's laws, named after German scientist Gustav Robert Kirchhoff. These two laws are quite simple but extremely important. We will not attempt to prove them because the proofs are beyond our current level of understanding. However, we will demonstrate their usefulness and attempt to make the reader proficient in their use. The first law is *Kirchhoff's current law*, which states that *the algebraic sum of the currents entering any node is zero*. In mathematical form the law appears as

KCL is an extremely important and useful law.

$$\sum_{j=1}^{N} i_j(t) = 0 \qquad\qquad 2.7$$

where $i_j(t)$ is the jth current entering the node through branch j and N is the number of branches connected to the node. To understand the use of this law, consider node 3 shown in Fig. 2.5. Applying Kirchhoff's current law to this node yields

$$i_2(t) - i_4(t) + i_5(t) - i_7(t) = 0$$

D2.3   Write the KCL equation for the following node:

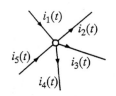

ANSWER:
$$i_1(t) - i_2(t) - i_3(t)$$
$$- i_4(t) + i_5(t) = 0$$

We have assumed that the algebraic signs of the currents entering the node are positive and, therefore, that the signs of the currents leaving the node are negative.

If we multiply the foregoing equation by −1, we obtain the expression

$$-i_2(t) + i_4(t) - i_5(t) + i_7(t) = 0$$

which simply states that *the algebraic sum of the currents leaving a node is zero.* Alternatively, we can write the equation as

$$i_2(t) + i_5(t) = i_4(t) + i_7(t)$$

which states that *the sum of the currents entering a node is equal to the sum of the currents leaving the node.* Both of these italicized expressions are alternative forms of Kirchhoff's current law.

Once again it must be emphasized that the latter statement means that the sum of the *variables* that have been defined entering the node is equal to the sum of the *variables* that have been defined leaving the node, not the actual currents. For example, $i_j(t)$ may be defined entering the node, but if its actual value is negative, there will be positive charge leaving the node.

Note carefully that Kirchhoff's current law states that the *algebraic* sum of the currents either entering or leaving a node must be zero. We now begin to see why we stated in Chapter 1 that it is critically important to specify both the magnitude and the direction of a current.

EXAMPLE 2.5

Let us write KCL for every node in the network in Fig. 2.5 assuming that the currents leaving the node are positive.

SOLUTION   The KCL equations for nodes 1 through 5 are

$$-i_1(t) + i_2(t) + i_3(t) = 0$$
$$i_1(t) - i_4(t) + i_6(t) = 0$$
$$-i_2(t) + i_4(t) - i_5(t) + i_7(t) = 0$$
$$-i_3(t) + i_5(t) - i_8(t) = 0$$
$$-i_6(t) - i_7(t) + i_8(t) = 0$$

Note carefully that if we add the first four equations we obtain the fifth equation. What does this tell us? Recall that this means that this set of equations is not linearly independent. We can show that the first four equations are, however, linearly independent. Store this idea in memory because it will become very important when we learn how to write the equations necessary to solve for all the currents and voltages in a network in the following chapter.   ❏

EXAMPLE 2.6

The network in Fig. 2.5 is represented by the topological diagram shown in Fig. 2.6. We wish to find the unknown currents in the network.

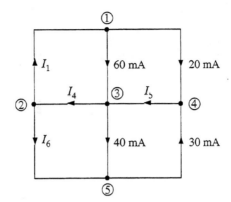

Figure 2.6 Topological diagram for the circuit in Fig. 2.5.

SOLUTION   Assuming the currents leaving the node are positive, the KCL equations for nodes 1 through 4 are

$$-I_1 + 0.06 + 0.02 = 0$$
$$I_1 - I_4 + I_6 = 0$$
$$-0.06 + I_4 - I_5 + 0.04 = 0$$
$$-0.02 + I_5 - 0.03 = 0$$

The first equation yields I_1 and the last equation yields I_5. Knowing I_5 we can immediately obtain I_4 from the third equation. Then the values of I_1 and I_4 yield the value of I_6 from the second equation. The results are $I_1 = 80$ mA, $I_4 = 70$ mA, $I_5 = 50$ mA, and $I_6 = -10$ mA.

As indicated earlier, dependent or controlled sources are very important because we encounter them when analyzing circuits containing active elements such as transistors. The following example presents a circuit containing a current-controlled current source.   ❑

EXAMPLE 2.7

Let us write the KCL equations for the circuit shown in Fig. 2.7.

Figure 2.7 Circuit containing a dependent current source.

SOLUTION The KCL equations for nodes 1 through 4 follow.

$$i_1(t) + i_2(t) - i_5(t) = 0$$
$$-i_2(t) + i_3(t) - 50i_2(t) = 0$$
$$-i_1(t) + 50i_2(t) + i_4(t) = 0$$
$$i_5(t) - i_3(t) - i_4(t) = 0$$

If we added the first three equations, we would obtain the negative of the fourth. What does this tell us about the set of equations?   ❑

Finally, it is possible to generalize Kirchhoff's current law to include a closed surface. By a closed surface we mean some set of elements completely contained within the surface that are interconnected. Since the current entering each element within the surface is equal to that leaving the element (i.e., the element stores no net charge), it follows that the current entering an interconnection of elements is equal to that leaving the interconnection. Therefore, Kirchhoff's current law can also be stated as follows: *The algebraic sum of the currents entering any closed surface is zero.*

EXAMPLE 2.8

Let us find I_4 and I_1 in the network represented by the topological diagram in Fig. 2.6.

SOLUTION This diagram is redrawn in Fig. 2.8, and node 1 is enclosed in surface 1 and nodes 3 and 4 are enclosed in surface 2. A quick review of the previous example indicates that we derived a value for I_4 from the value of I_5. However, I_5 is now completely enclosed in surface 2. If we apply KCL to surface 2, assuming the currents out of the surface are positive, we obtain

$$I_4 - 0.06 - 0.02 - 0.03 + 0.04 = 0$$

or

$$I_4 = 70 \, \text{mA}$$

which we obtained without any knowledge of I_5. Likewise for surface 1, what goes in must come out and, therefore, $I_1 = 80$ mA. The reader is encouraged to cut the

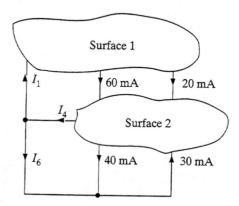

Figure 2.8 Diagram used to demonstrate KCL for a surface.

network in Fig. 2.6 into two pieces in any fashion and show that KCL is always satisfied at the boundaries.   ❏

EXTENSION EXERCISES

E2.3   Given the networks in Fig. E2.3, find (a) I_1 in Fig. E2.3a and (b) I_T in Fig. E2.3b.

Figure E2.3

ANSWER: (a) $I = -50$ mA, (b) $I = 70$ mA.

E2.4   Find (a) I_1 in the network in Fig. E2.4a and (b) I_1 and I_2 in the circuit in Fig. E2.4b.

Figure E2.4

ANSWER: (a) $I_1 = 6$ mA, (b) $I_1 = 8$ mA and $I_2 = 5$ mA.

E2.5   Find the current i_x in the circuits in Fig. E2.5.

Figure E2.5

ANSWER: (a) $i_x = 4$ mA, (b) $i_x = 12$ mA.

Kirchhoff's second law, called *Kirchhoff's voltage law*, states that *the algebraic sum of the voltages around any loop is zero*. As was the case with Kirchhoff's current law, we will defer the proof of this law and concentrate on understanding how to apply it. Once again the reader is cautioned to remember that we are dealing only with lumped-parameter circuits. These circuits are conservative, meaning that the work required to move a unit charge around any loop is zero.

Recall that in Kirchhoff's current law, the algebraic sign was required to keep track of whether the currents were entering or leaving a node. In Kirchhoff's voltage law the algebraic sign is used to keep track of the voltage polarity. In other words, as we traverse the circuit, it is necessary to sum to zero the increases and decreases in energy level. Therefore, it is important we keep track of whether the energy level is increasing or decreasing as we go through each element.

EXAMPLE 2.9

Consider the circuit shown in Fig. 2.9. If V_{R_1} and V_{R_2} are known quantities, let us find V_{R_3}.

Figure 2.9 Circuit used to illustrate KVL.

SOLUTION In applying KVL, we must traverse the circuit and sum to zero the increases and decreases in energy level. Since the network is a single loop, we have only one closed path. We will adopt a policy of considering an increase in energy level as negative and a decrease in energy level as positive. Using this policy and starting at point *a* in the network and traversing it in a clockwise direction, we obtain the equation

$$+V_{R_1} - 5 + V_{R_2} - 15 + V_{R_3} - 30 = 0$$

which can be written as

$$+V_{R_1} + V_{R_2} + V_{R_3} = 5 + 15 + 30$$

$$= 50$$

Now suppose that V_{R_1} and V_{R_2} are known to be 18 V and 12 V, respectively. Then $V_{R_3} = 20$ V.

Consider the network in Fig. 2.10.

Figure 2.10 Circuit used to explain KVL.

Let us demonstrate that only two of the three possible loop equations are linearly independent.

SOLUTION   Note that this network has three closed paths: the left loop, right loop, and outer loop. Applying our policy for writing KVL equations and traversing the left loop starting at point a, we obtain

$$V_{R_1} + V_{R_4} - 16 - 24 = 0$$

The corresponding equation for the right loop starting at point b is

$$V_{R_2} + V_{R_3} + 8 + 16 - V_{R_4} = 0$$

The equation for the outer loop starting at point a is

$$V_{R_1} + V_{R_2} + V_{R_3} + 8 - 24 = 0$$

Note that if we add the first two equations, we obtain the third equation. Therefore, as we indicated in Example 2.5, the three equations are not linearly independent. Once again, we will address this issue in the next chapter and demonstrate that we need only the first two equations to solve for the voltages in the circuit.   ❑

Finally, we employ the convention V_{ab} to indicate the voltage of point a with respect to point b: that is, the variable for the voltage between point a and point b, with point a considered positive relative to point b. Since the potential is measured between two points, it is convenient to use an arrow between the two points with the head of the arrow located at the positive node. Note that the double-subscript notation, the + and − notation, and the single-headed arrow notation are all the same if the head of the arrow is pointing toward the positive terminal and the first subscript in the double-subscript notation. All of these equivalent forms for labeling voltages are shown in Fig. 2.11.

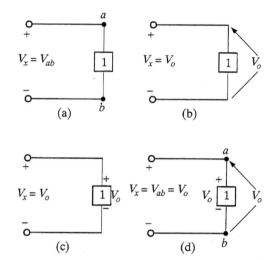

Figure 2.11 Equivalent forms for labeling voltage.

EXAMPLE 2.11

Consider the network in Fig. 2.12a. Let us apply KVL to determine the voltage between two points. Specifically, in terms of the double-subscript notation, let us find V_{ae} and V_{ec}.

Figure 2.12 Network used in Example 2.11.

SOLUTION The circuit is redrawn in Fig. 2.12b. Since points a and e as well as e and c are not physically close, the arrow notation is very useful. Our approach to determining the unknown voltage is to apply KVL with the unknown voltage in the closed path. Therefore, to determine V_{ae} we can use the path $aefa$ or $abcdea$. The equations for the two paths in which V_{ae} is the only unknown are

$$V_{ae} + 10 - 24 = 0$$

and

$$16 - 12 + 4 + 6 - V_{ae} = 0$$

Note that both equations yield $V_{ae} = 14$ V. Even before calculating V_{ae}, we could calculate V_{ec} using the path *cdec* or *cefabc*. However, since V_{ae} is now known, we can also use the path *ceabc*. KVL for each of these paths is

$$4 + 6 + V_{ec} = 0$$
$$-V_{ec} - V_{ae} + 16 - 12 = 0$$

and

$$-V_{ec} + 10 - 24 + 16 - 12 = 0$$

Each of these equations yields $V_{ec} = -10$ V.   ❑

In general, the mathematical representation of Kirchhoff's voltage law is

> KVL is an extremely important and useful law.

$$\sum_{j=1}^{N} v_j(t) = 0 \qquad\qquad 2.8$$

where $v_j(t)$ is the voltage across the *j*th branch (with the proper reference direction) in a loop containing N voltages. This expression is analogous to Eq. (2.7) for Kirchhoff's current law.

EXAMPLE 2.12

Given the network in Fig. 2.13 containing a dependent source, let us write the KVL equations for the two closed paths *abda* and *bcdb*.

Figure 2.13 Network containing a dependent source.

SOLUTION   The two KVL equations are

$$V_{R_1} + V_{R_2} - V_S = 0$$
$$20V_{R_1} + V_{R_3} - V_{R_2} = 0 \qquad ❑$$

EXTENSION EXERCISES

E2.6   In the circuit in Fig. E2.6, find V_{ad}, V_{ac} and V_{bd}.

ANSWER: $V_{ad} = 12$ V, $V_{ac} = 10$ V, and $V_{bd} = 6$ V.

Figure E2.6

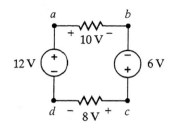

Figure E2.7

E2.7   In the network in Fig. E2.7, find V_{ac} and V_{db}.

ANSWER: $V_{ac} = 4\,\text{V}$, and $V_{db} = -2\,\text{V}$.

E2.8   Find V_{ad} and V_{eb} in the network in Fig. E2.8.

Figure E2.8

ANSWER: $V_{ad} = 26\,\text{V}$, $V_{eb} = 10\,\text{V}$.

E2.9   Find V_{bd} in the circuit in Fig. E2.9.

Figure E2.9

ANSWER: $V_{bd} = 11\,\text{V}$.

Before proceeding with the analysis of simple circuits, it is extremely important that we emphasize a subtle but very critical point. Ohm's law as defined by the equation $V = IR$ refers to the relationship between the voltage and current as defined in Fig. 2.14a. If the direction of either the current or the voltage, but not both, is reversed, the relationship between the current and the voltage would be $V = -IR$. In a similar manner, given the circuit in Fig. 2.14b, if the polarity of the voltage between the terminals A and B is specified as shown, then the direction of the current I is from point B through R to

Figure 2.14   Circuits used to explain Ohm's law.

point A. Likewise, in Fig. 2.14c, if the direction of the current is specified as shown, then the polarity of the voltage must be such that point D is at a higher potential than point C and, therefore, the arrow representing the voltage V is from point C to point D.

2.3 SINGLE-LOOP CIRCUITS

Voltage Division

At this point we can begin to apply the laws we have presented earlier to the analysis of simple circuits. To begin, we examine what is perhaps the simplest circuit—a single closed path, or loop, of elements. The elements of a single loop carry the same current and, therefore, are said to be in *series*. However, we will apply Kirchhoff's voltage law and Ohm's law to the circuit to determine various quantities in the circuit.

Our approach will be to begin with a simple circuit and then generalize the analysis to more complicated ones. The circuit shown in Fig. 2.15 will serve as a basis for discussion. This circuit consists of an independent voltage source that is in series with two resistors. We have assumed that the current flows in a clockwise direction. If this assumption is correct, the solution of the equations that yields the current will produce a positive value. If the current is actually flowing in the opposite direction, the value of the current variable will simply be negative, indicating that the current is flowing in a direction opposite to that assumed. We have also made voltage polarity assignments for v_{R_1} and v_{R_2}. These assignments have been made using the convention employed in our discussion of Ohm's law and our choice for the direction of $i(t)$; that is, the convention shown in Fig. 2.14a.

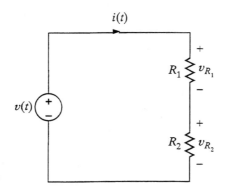

Figure 2.15 Single-loop circuit.

Applying Kirchhoff's voltage law to this circuit yields

$$-v(t) + v_{R_1} + v_{R_2} = 0$$

or

$$v(t) = v_{R_1} + v_{R_2}$$

However, from Ohm's law we know that

$$v_{R_1} = R_1 i(t)$$
$$v_{R_2} = R_2 i(t)$$

Therefore,

$$v(t) = R_1 i(t) + R_2 i(t)$$

Solving the equation for $i(t)$ yields

$$i(t) = \frac{v(t)}{R_1 + R_2}$$ **2.9**

Knowing the current, we can now apply Ohm's law to determine the voltage across each resistor:

$$
\begin{aligned}
v_{R_1} &= R_1 i(t) \\
&= R_1 \left[\frac{v(t)}{R_1 + R_2} \right] \\
&= \frac{R_1}{R_1 + R_2} v(t)
\end{aligned}
$$ **2.10**

Similarly,

The manner in which voltage divides between two series resistors

$$v_{R_2} = \frac{R_2}{R_1 + R_2} v(t)$$ **2.11**

Although simple, Eqs. (2.10) and (2.11) are very important because they describe the operation of what is called a *voltage divider*. In other words, *the source voltage $v(t)$ is divided between the resistors R_1 and R_2 in direct proportion to their resistances.*

In essence, if we are interested in the voltage across the resistor R_1, we bypass the calculation of the current $i(t)$ and simply multiply the input voltage $v(t)$ by the ratio $R_1(R_1 + R_2)$. As illustrated in Eq. (2.10), we are using the current in the calculation, but not explicitly.

Note that the equations satisfy Kirchhoff's voltage law, since

$$-v(t) + \frac{R_1}{R_1 + R_2} v(t) + \frac{R_2}{R_1 + R_2} v(t) = 0$$

EXAMPLE 2.13

Consider the circuit shown in Fig. 2.16. The circuit is identical to Fig. 2.15 except that R_1 is a variable resistor such as the volume control for a radio or television set. Suppose that $V_S = 9\,\text{V}$, $R_1 = 90\,\text{k}\Omega$ and $R_2 = 30\,\text{k}\Omega$.

Figure 2.16 Voltage-divider circuit.

Let us examine the change in both the voltage across R_2 and the power absorbed in this resistor as R_1 is changed from 90 kΩ to 15 kΩ.

SOLUTION   Since this is a voltage-divider circuit, the voltage V_2 can be obtained directly as

$$
\begin{aligned}
V_2 &= \left[\frac{R_2}{R_1 + R_2}\right]V_S \\
&= \left[\frac{30\text{k}}{90\text{k} + 30\text{k}}\right](9) \\
&= 2.25 \text{ V}
\end{aligned}
$$

Now suppose that the variable resistor is changed from 90 kΩ to 15 kΩ. Then

$$
\begin{aligned}
V_2 &= \left[\frac{30\text{k}}{30\text{k} + 15\text{k}}\right]9 \\
&= 6 \text{ V}
\end{aligned}
$$

The direct voltage-divider calculation is equivalent to determining the current I and then using Ohm's law to find V_2. Note that the larger voltage is across the larger resistance. This voltage-divider concept and the simple circuit we have employed to describe it are very useful because, as will be shown later, more complicated circuits can be reduced to this form.

Finally, let us determine the instantaneous power absorbed by the resistor R_2 under the two conditions $R_1 = 90$ kΩ and $R_1 = 15$ kΩ. For the case $R_1 = 90$ kΩ, the power absorbed by R_2 is

$$
P_2 = I^2 R_2 = \left(\frac{9}{120\text{k}}\right)^2 (30\text{k})
$$
$$
= 0.169 \text{ mW}
$$

In the second case

$$
P_2 = \left(\frac{9}{45\text{k}}\right)^2 (30\text{k})
$$
$$
= 1.2 \text{ mW}
$$

The current in the first case is 75 μA, and in the second case it is 200 μA. Since the power absorbed is a function of the square of the current, the power absorbed in the two cases is quite different.   ❑

Let us now demonstrate the practical utility of this simple voltage-divider network.

EXAMPLE 2.14

Consider the circuit in Fig. 2.17a, which is an approximation of a high-voltage dc transmission facility. We have assumed that the bottom portion of the transmission line is a perfect conductor and will justify this assumption in the next chapter. The load can be represented by a resistor of value 183.5 Ω. Therefore, the equivalent circuit of this network is shown in Fig. 2.17b.

(a)   (b)

Figure 2.17 A high-voltage dc transmission facility.

Let us determine both the power delivered to the load and the power losses in the line.

SOLUTION   Using voltage division, the load voltage is

$$V_{\text{load}} = \left[\frac{183.5}{183.5 + 16.92} \right] 400k$$

$$= 366.24 \text{ kV}$$

The input power is 800 MW and the power transmitted to the load is

$$P_{\text{load}} = I^2 R_{\text{load}}$$

$$= 734 \text{ MW}$$

Therefore, the power loss in the transmission line is

$$P_{\text{line}} = P_{\text{in}} - P_{\text{load}} = I^2 R_{\text{line}}$$

$$= 66 \text{ MW}$$

Since $P = VI$, suppose now that the utility company supplied power at 200 kV and 4 kA. What effect would this have on our transmission network? Without making a single calculation, we know that because power is proportional to the square of the current, there would be a large increase in the power loss in the line and, therefore, the efficiency of the facility would decrease substantially. That is why, in general, we transmit power at high voltage and low current.   ❏

Multiple Source/Resistor Networks

At this point we wish to extend our analysis to include a multiplicity of voltage sources and resistors. For example, consider the circuit shown in Fig. 2.18a. Here we have assumed that the current flows in a clockwise direction, and we have defined the variable $i(t)$ accordingly. This may or may not be the case, depending on the value of the various voltage sources. Kirchhoff's voltage law for this circuit is

$$+v_{R_1} + v_2(t) - v_3(t) + v_{R_2} + v_4(t) + v_5(t) - v_1(t) = 0$$

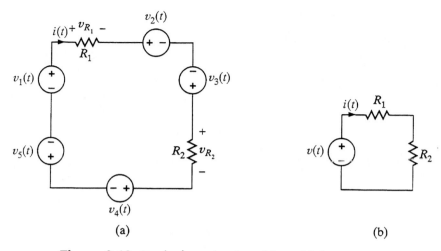

Figure 2.18 Equivalent circuits with multiple sources.

or, using Ohm's law,

$$(R_1 + R_2)i(t) = v_1(t) - v_2(t) + v_3(t) - v_4(t) - v_5(t)$$

which can be written as

$$(R_1 + R_2)i(t) = v(t)$$

where

$$v(t) = v_1(t) + v_3(t) - [v_2(t) + v_4(t) + v_5(t)]$$

so that under the preceding definitions, Fig. 2.18a is equivalent to Fig. 2.18b. In other words, the sum of several voltage sources in series can be replaced by one source whose value is the algebraic sum of the individual sources. This analysis can, of course, be generalized to a circuit with N series sources.

Now consider the circuit with N resistors in series, as shown in Fig. 2.19a. Applying Kirchhoff's voltage law to this circuit yields

Figure 2.19 Equivalent circuits.

$$v(t) = v_{R_1} + v_{R_2} + \cdots + v_{R_N}$$
$$= R_1 i(t) + R_2 i(t) + \cdots + R_N i(t)$$

and therefore,

$$v(t) = R_S i(t) \tag{2.12}$$

where

$$R_S = R_1 + R_2 + \cdots + R_N \tag{2.13}$$

and hence,

$$i(t) = \frac{v(t)}{R_S} \tag{2.14}$$

Note also that for any resistor R_i in the circuit, the voltage across R_i is given by the expression

$$v_{R_i} = \frac{R_i}{R_S} v(t) \tag{2.15}$$

which is the voltage-division property for multiple resistors in series.

Equation (2.13) illustrates that *the equivalent resistance of N resistors in series is simply the sum of the individual resistances.* Thus, using Eq. (2.13), we can draw the circuit in Fig. 2.19b as an equivalent circuit for the one in Fig. 2.19a.

EXAMPLE 2.15

Given the circuit in Fig. 2.20, let us find I, V_{bd}, and the power absorbed by the 30-kΩ resistor.

Figure 2.20 Circuit used in Example 2.15.

SOLUTION   KVL for the network yields the equation

$$10kI + 20kI + 12 + 30kI - 6 = 0$$
$$60kI = -6$$
$$I = -0.1 \text{ mA}$$

Therefore, the magnitude of the current is 0.1 mA, but its direction is opposite to that assumed.

The voltage V_{bd} can be calculated using either of the closed paths *abdea* or *bcdb*. The equations for both cases are

$$10kI + V_{bd} + 30kI - 6 = 0$$

and

$$20kI + 12 - V_{bd} = 0$$

Using $I = -0.1$ mA in either equation yields $V_{bd} = 10$ V. Finally, the power absorbed by the 30 kΩ resistor is

$$P = I^2 R = 0.3 \text{ mW}$$

EXAMPLE 2.16

A dc transmission facility is modeled by the approximate circuit shown in Fig. 2.21. If the load voltage is known to be $V_{\text{load}} = 458.3$ kV, we wish to find the voltage at the sending end of the line and the power loss in the line.

Figure 2.21 Circuit used in Example 2.16.

SOLUTION   Knowing the load voltage and load resistance, we can obtain the line current using Ohm's law:

$$I_{\text{L}} = 458.3\text{k}/220$$

$$= 2.083 \text{ kA}$$

The voltage drop across the line is

$$V_{\text{line}} = (I_L)(R_{\text{line}})$$

$$= 41.66 \text{ kV}$$

Now, using KVL,

$$V_S = V_{\text{line}} + V_{\text{load}}$$

$$= 500 \text{ kV}$$

Note that since the network is simply a voltage-divider circuit, we could obtain V_S immediately from our knowledge of R_{line}, R_{load}, and V_{load}. That is,

$$V_{\text{load}} = \left[\frac{R_{\text{load}}}{R_{\text{load}} + R_{\text{line}}} \right] V_S$$

and V_S is the only unknown in this equation.

The power absorbed by the line is

$$P_{\text{line}} = I_L^2 R_{\text{line}}$$
$$= 86.79 \text{ MW}$$

❏

EXTENSION EXERCISES

E2.10   Find I and V_{bd} in the circuit in Fig. E2.10.

Figure E2.10

ANSWER: $I = -0.05$ mA and $V_{bd} = 10$ V.

E2.11   In the network in Fig. E2.11, if V_{ad} is 3 V, find V_S.

Figure E2.11

ANSWER: $V_S = 9$ V.

2.4 SINGLE-NODE-PAIR CIRCUITS

Current Division

An important circuit is the single-node-pair circuit. In this case the elements have the same voltage across them and, therefore, are in *parallel*. We will, however, apply Kirchhoff's current law and Ohm's law to determine various unknown quantities in the circuit.

Following our approach with the single-loop circuit, we will begin with the simplest case and then generalize our analysis. Consider the circuit shown in Fig. 2.22. Here we have an independent current source in parallel with two resistors.

Figure 2.22 Simple parallel circuit.

Since all of the circuit elements are in parallel, the voltage $v(t)$ appears across each of them.

Furthermore, an examination of the circuit indicates that the current $i(t)$ is into the upper node of the circuit and the currents $i_1(t)$ and $i_2(t)$ are out of the node. Since KCL essentially states that what goes in must come out, the question we must answer is how $i_1(t)$ and $i_2(t)$ divide the input current $i(t)$.

Applying Kirchhoff's current law to the upper node, we obtain

$$i(t) = i_1(t) + i_2(t)$$

And, employing Ohm's law, we have

$$
\begin{aligned}
i(t) &= \frac{v(t)}{R_1} + \frac{v(t)}{R_2} \\
&= \left(\frac{1}{R_1} + \frac{1}{R_2} \right) v(t) \\
&= \frac{v(t)}{R_p}
\end{aligned}
$$

where

$$\frac{1}{R_p} = \frac{1}{R_1} + \frac{1}{R_2} \qquad \textbf{2.16} \quad \text{The parallel resistance equation}$$

$$R_p = \frac{R_1 R_2}{R_1 + R_2} \qquad \textbf{2.17}$$

Therefore, the equivalent resistance of two resistors connected in parallel is equal to the product of their resistances divided by their sum. Note also that this equivalent resistance R_p is always less than either R_1 or R_2. Hence, by connecting resistors in parallel we reduce the overall resistance. In the special case when $R_1 = R_2$, the equivalent resistance is equal to half of the value of the individual resistors.

The manner in which the current $i(t)$ from the source divides between the two branches is called *current division* and can be found from the preceding expressions. For example,

$$
\begin{aligned}
v(t) &= R_p i(t) \\
&= \frac{R_1 R_2}{R_1 + R_2} i(t) \qquad \textbf{2.18}
\end{aligned}
$$

and

$$i_1(t) = \frac{v(t)}{R_1}$$

The manner in which current divides between two parallel resistors

$$i_1(t) = \frac{R_2}{R_1 + R_2} i(t) \qquad \textbf{2.19}$$

and

DRILL

D2.5 Find I_1 and I_2 in the following circuit:

ANSWER: $I_1 = 1\,\text{mA}$

$I_2 = 4\,\text{mA}$

$$i_2(t) = \frac{v(t)}{R_2}$$

$$= \frac{R_1}{R_1 + R_2} i(t) \qquad \textbf{2.20}$$

Equations (2.19) and (2.20) are mathematical statements of the current-division rule.

EXAMPLE 2.17

Given the network in Fig. 2.23a, let us find I_1, I_2 and V_o.

(a)   (b)

Figure 2.23 Circuits used in Example 2.17.

SOLUTION First, it is important to recognize that the current source feeds two parallel paths. To emphasize this point, the circuit is redrawn as shown in Fig. 2.23b. Applying current division, we obtain

$$I_1 = \left[\frac{40\text{k} + 80\text{k}}{60\text{k} + (40\text{k} + 80\text{k})}\right](0.9 \times 10^{-3})$$

$$= 0.6\,\text{mA}$$

and

$$I_2 = \left[\frac{60\text{k}}{60\text{k} + (40\text{k} + 80\text{k})}\right](0.9 \times 10^{-3})$$

$$= 0.3\,\text{mA}$$

Note that the larger current flows through the smaller resistor, and vice versa. In addition, one should note that if the resistances of the two paths are equal, the current will divide equally between them. KCL is satisfied since $I_1 + I_2 = 0.9\,\text{mA}$.

The voltage V_o can be derived using Ohm's law as

$$V_o = 80\text{k}I$$

$$= 24\text{ V} \qquad \square$$

EXAMPLE 2.18

A typical car stereo consists of a 2-W audio amplifier and two speakers represented by the diagram shown in Fig. 2.24a. The output circuit of the audio amplifier is in essence a 430-mA current source and the speakers each have a resistance of 4 Ω. Let us determine the power absorbed by the speakers.

(a)   (b)

Figure 2.24 Circuits used in Example 2.18.

SOLUTION   The audio system can be modeled as shown in Fig. 2.24b. Since the speakers are both 4-Ω devices, the current will split evenly between them and the power absorbed by each speaker is

$$P = I^2R$$

$$= (215 \times 10^{-3})^2(4)$$

$$= 184.9\text{ mW} \qquad \square$$

EXTENSION EXERCISE

E2.12   Find the currents I_1 and I_2 and the power absorbed by the 40-kΩ resistor in the network in Fig. E2.12.

Figure E2.12

ANSWER: $I_1 = 12\text{ mA}$, $I_2 = -4\text{ mA}$, and $P_{40\text{k}\Omega} = 5.76\text{ W}$.

Multiple Source/Resistor Networks

Let us now extend our analysis to include a multiplicity of current sources and resistors in parallel. For example, consider the circuit shown in Fig. 2.25a. We have assumed that the upper node is $v(t)$ volts positive with respect to the lower node. Applying Kirchhoff's current law to the upper node yields

$$i_1(t) - i_2(t) - i_3(t) + i_4(t) - i_5(t) - i_6(t) = 0$$

or

$$i_1(t) - i_3(t) + i_4(t) - i_6(t) = i_2(t) + i_5(t)$$

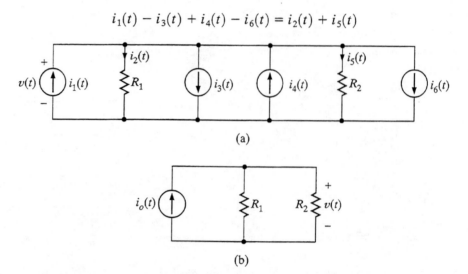

(a)

(b)

Figure 2.25 Equivalent circuits.

The terms on the left side of the equation all represent sources that can be combined algebraically into a single source; that is,

$$i_o(t) = i_1(t) - i_3(t) + i_4(t) - i_6(t)$$

which effectively reduces the circuit in Fig. 2.25a to that in Fig. 2.25b. We could, of course, generalize this analysis to a circuit with N current sources. Using Ohm's law, the currents on the right side of the equation can be expressed in terms of the voltage and individual resistances so that the KCL equation reduces to

$$i_o(t) = \left(\frac{1}{R_1} + \frac{1}{R_2} \right) v(t)$$

Now consider the circuit with N resistors in parallel, as shown in Fig. 2.26a. Applying Kirchhoff's current law to the upper node yields

(a)

(b)

Figure 2.26 Equivalent circuits.

$$i_o(t) = i_1(t) + i_2(t) + \cdots + i_N(t)$$

$$= \left(\frac{1}{R_1} + \frac{1}{R_2} + \cdots + \frac{1}{R_N} \right) v(t) \qquad \textbf{2.21}$$

or

$$i_o(t) = \frac{v(t)}{R_p} \qquad \textbf{2.22}$$

where

$$\frac{1}{R_p} = \sum_{i=1}^{N} \frac{1}{R_i} \qquad \textbf{2.23}$$

so that as far as the source is concerned, Fig. 2.26a can be reduced to an equivalent circuit, as shown in Fig. 2.26b.

The current division for any branch can be calculated using Ohm's law and the preceding equations. For example, for the jth branch in the network of Fig. 2.26a,

$$i_j(t) = \frac{v(t)}{R_j}$$

Using Eq. (2.22), we obtain

$$i_j(t) = \frac{R_p}{R_j} i_o(t) \qquad \textbf{2.24}$$

which defines the current-division rule for the general case.

EXAMPLE 2.19

Given the circuit in Fig. 2.27a, we wish to find the current in the 12-kΩ load resistor.

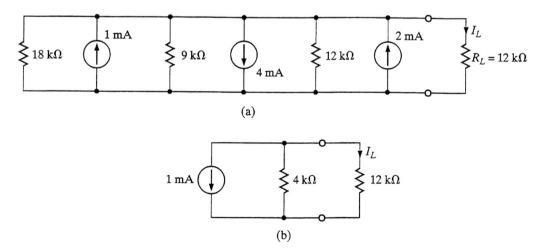

(a)

(b)

Figure 2.27 Circuits used in Example 2.19.

SOLUTION   In order to simplify the network in Fig. 2.27a, we add the current sources algebraically and combine the parallel resistors in the following manner:

$$\frac{1}{R_p} = \frac{1}{18k} + \frac{1}{9k} + \frac{1}{12k}$$

$$R_p = 4\ k\Omega$$

Using these values we can reduce the circuit in Fig. 2.27a to that in Fig. 2.27b. Now, applying current division, we obtain

$$I_L = -\left[\frac{4k}{4k + 12k}\right](1 \times 10^{-3})$$

$$= -0.25\ \text{mA}$$

EXTENSION EXERCISE

E2.13   Find the power absorbed by the 6-kΩ resistor in the network in Fig. E2.13.

Figure E2.13

ANSWER: $P = 2.67$ mW.

2.5   SERIES AND PARALLEL RESISTOR COMBINATIONS

We have shown in our earlier developments that the equivalent resistance of N resistors in series is

$$R_S = R_1 + R_2 + \cdots + R_N \qquad \textbf{2.25}$$

and the equivalent resistance of N resistors in parallel is found from

$$\frac{1}{R_p} = \frac{1}{R_1} + \frac{1}{R_2} + \cdots + \frac{1}{R_N} \qquad \textbf{2.26}$$

Let us now examine some combinations of these two cases.

EXAMPLE 2.20

We wish to determine the resistance at terminals A-B in the network in Fig. 2.28a.

Figure 2.28 Simplification of a resistance network.

SOLUTION   Starting at the opposite end of the network from the terminals and combining resistors as shown in the sequence of circuits in Fig. 2.28, we find that the equivalent resistance at the terminals is 5 kΩ.   ❑

EXTENSION EXERCISES

E2.14   Find the equivalent resistance at the terminals A-B in the network in Fig. E2.14.

ANSWER: $R_{AB} = 22$ kΩ.

E2.15   Find the equivalent resistance at the terminals A-B in the circuit in Fig. E2.15.

ANSWER: $R_{AB} = 3$ kΩ

Figure E2.14

Figure E2.15

A problem-solving strategy
for simplifying resistor combinations

When trying to determine the equivalent resistance at a pair of terminals of a network composed of an interconnection of numerous resistors, it is recommended that the analysis begin at the end of the network opposite the terminals. Two or more resistors are combined to form a single resistor, thus simplifying the network by reducing the number of components as the analysis continues in a steady progression toward the terminals. The simplification involves the following:

1. *Resistors in series.* Resistors R_1 and R_2 are in series if they are connected in tandem and carry exactly the same current. They can then be combined into a single resistor R_S, where $R_S = R_1 + R_2$.

2. *Resistors in parallel.* Resistors R_1 and R_2 are in parallel if they are connected to the same two nodes and have exactly the same voltage across their terminals. They can then be combined into a single resistor R_P, where $R_P = R_1 R_2/(R_1 + R_2)$.

These two combinations are used repeatedly, as needed, to reduce the network to a single resistor at the pair of terminals.

As a practical example of the use of series and parallel resistors, consider the following example.

EXAMPLE 2.21

A standard dc current-limiting power supply shown in Fig. 2.29a provides 0–18 V at 3 A to a load. The voltage drop, V_R, across a resistor, R, is used as a current-sensing device, fed back to the power supply and used to limit the current I. That is, if the load is adjusted so that the current tries to exceed 3 A, the power supply will act to limit the current to that value. The feedback voltage, V_R, should typically not exceed 600 mV.

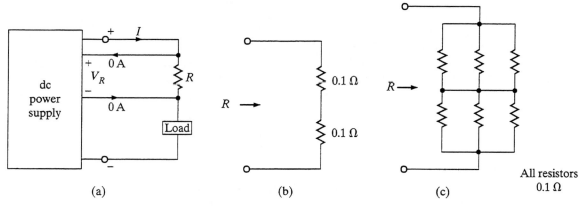

Figure 2.29 Circuits used in Example 2.21.

If we have a box of standard 0.1-Ω, 5-W resistors, let us determine the configuration of these resistors that will provide a $V_R = 600$ mV when the current is 3 A.

SOLUTION   Using Ohm's law, the value of R should be

$$R = \frac{V_R}{I}$$

$$= \frac{0.6}{3}$$

$$= 0.2\ \Omega$$

Therefore, two 0.1-Ω resistors connected in series, as shown in Fig. 2.29b, will provide the proper feedback voltage. Suppose, however, that the power supply current is to be limited to 9 A. The resistance required in this case to produce $V_R = 600$ mV is

$$R = \frac{0.6}{9}$$

$$= 0.0667\ \Omega$$

We must now determine how to interconnect the 0.1-Ω resistors to obtain $R = 0.0667\ \Omega$. Since the desired resistance is less than the components available (i.e., 0.1 Ω), we must connect the resistors in some type of parallel configuration. Since all the resistors are of equal value, note that three of them connected in parallel would provide a resistance of one-third their value, or 0.0333 Ω. Then two such combinations connected in series, as shown in Fig. 2.29c, would produce the proper resistance.

Finally, we must check to ensure that the configurations in Figs. 2.29b and c have not exceeded the power rating of the resistors. In the first case, the current $I = 3$ A is present in each of the two series resistors. Therefore, the power absorbed in each resistor is

$$P = I^2 R$$
$$= (3)^2(0.1)$$
$$= 0.9 \text{ W}$$

which is well within the 5-W rating of the resistors.

In the second case, the current $I = 9$ A. The resistor configuration for R in this case is a series combination of two sets of three parallel resistors of equal value. Using current division, we know that the current I will split equally among the three parallel paths and, hence, the current in each resistor will be 3 A. Therefore, once again, the power absorbed by each resistor is within its power rating. ❑

2.6 CIRCUITS WITH SERIES-PARALLEL COMBINATIONS OF RESISTORS

At this point we have learned many techniques that are fundamental to circuit analysis. Now we wish to apply them and show how they can be used in concert to analyze circuits. We will illustrate their application through a number of examples that will be treated in some detail.

EXAMPLE 2.22

We wish to find all the currents and voltages labeled in the ladder network shown in Fig. 2.30a.

Figure 2.30 Analysis of a ladder network.

SOLUTION   To begin our analysis of the network, we start at the right end of the circuit and combine the resistors to determine the total resistance seen by the 12-V source. This will allow us to calculate the current I_1. Then employing KVL, KCL, Ohm's law, and/or voltage and current division, we will be able to calculate all currents and voltages in the network.

At the right end of the circuit, the 9-kΩ and 3-kΩ resistors are in series and, thus, can be combined into one equivalent 12-kΩ resistor. This resistor is in parallel with the 4-kΩ resistor, and their combination yields an equivalent 3-kΩ resistor, shown at the right edge of the circuit in Fig. 2.30b. In Fig. 2.30b the two 3-kΩ resistors are in series and their combination is in parallel with the 6-kΩ resistor. Combining all three resistances yields the circuit shown in Fig. 2.30c.

Applying Kirchhoff's voltage law to the circuit in Fig. 2.30c yields

$$I_1(9k + 3k) = 12$$
$$I_1 = 1 \text{ mA}$$

V_a can be calculated from Ohm's law as

$$V_a = I_1(3k)$$
$$= 3 \text{ V}$$

or, using Kirchhoff's voltage law,

$$V_a = 12 - 9kI_1$$
$$= 12 - 9$$
$$= 3 \text{ V}$$

Knowing I_1 and V_a, we can now determine all currents and voltages in Fig. 2.30b. Since $V_a = 3$ V, the current I_2 can be found using Ohm's law as

$$I_2 = \frac{3}{6k}$$
$$= \frac{1}{2} \text{ mA}$$

Then, using Kirchhoff's current law, we have

$$I_1 = I_2 + I_3$$
$$1 \times 10^{-3} = \frac{1}{2} \times 10^{-3} + I_3$$
$$I_3 = \frac{1}{2} \text{ mA}$$

Note that the I_3 could also be calculated using Ohm's law:

$$V_a = (3k + 3k)I_3$$
$$I_3 = \frac{3}{6k}$$
$$= \frac{1}{2} \text{ mA}$$

D2.6   Find V_o in the following network:

ANSWER: $V_o = 1$ V

D2.7   Find I_o in the following circuit:

ANSWER: $I_o = 1$ A

Applying Kirchhoff's voltage law to the right-hand loop in Fig. 2.30b yields

$$V_a - V_b = 3kI_3$$

$$3 - V_b = \frac{3}{2}$$

$$V_b = \frac{3}{2} \text{ V}$$

or, since V_b is equal to the voltage drop across the 3-kΩ resistor, we could use Ohm's law as

$$V_b = 3kI_3$$

$$= \frac{3}{2} \text{ V}$$

We are now in a position to calculate the final unknown currents and voltages in Fig. 2.30a. Knowing V_b, we can calculate I_4 using Ohm's law as

$$V_b = 4kI_4$$

$$I_4 = \frac{\frac{3}{2}}{4k}$$

$$= \frac{3}{8} \text{ mA}$$

Then, from Kirchhoff's current law, we have

$$I_3 = I_4 + I_5$$

$$\frac{1}{2} \times 10^{-3} = \frac{3}{8} \times 10^{-3} + I_5$$

$$I_5 = \frac{1}{8} \text{ mA}$$

We could also have calculated I_5 using the current-division rule. For example,

$$I_5 = \frac{4k}{4k + (9k + 3k)} I_3$$

$$= \frac{1}{8} \text{ mA}$$

Finally, V_c can be computed as

$$V_c = I_5(3k)$$

$$= \frac{3}{8} \text{ V}$$

V_c can also be found using voltage division (i.e., the voltage V_b will be divided between the 9-kΩ and 3-kΩ resistors). Therefore,

$$V_c = \left[\frac{3k}{3k + 9k}\right] V_b$$

$$= \frac{3}{8} V$$

Note that Kirchhoff's current law is satisfied at every node and Kirchhoff's voltage law is satisfied around every loop, as shown in Fig. 2.30d.   ❑

The following example is, in essence, the reverse of the previous example in that we are given the current in some branch in the network and asked to find the value of the input source.

EXAMPLE 2.23

Given the circuit in Fig. 2.31 and $I_4 = \frac{1}{2}$ mA, let us find the source voltage V_o.

Figure 2.31 Example circuit for analysis.

SOLUTION   If $I_4 = \frac{1}{2}$ mA, then from Ohm's law, $V_b = 3$ V. V_b can now be used to calculate $I_3 = 1$ mA. Kirchhoff's current law applied at node y yields

$$I_2 = I_3 + I_4$$

$$= 1.5 \text{ mA}$$

Then, from Ohm's law, we have

$$V_a = (1.5 \times 10^{-3})(2k)$$

$$= 3 \text{ V}$$

Since $V_a + V_b$ is now known, I_5 can be obtained:

$$I_5 = \frac{V_a + V_b}{3k + 1k}$$

$$= 1.5 \text{ mA}$$

Applying Kirchhoff's current law at node x yields

$$I_1 = I_2 + I_5$$
$$= 3 \text{ mA}$$

Now KVL applied to any closed path containing V_o will yield the value of this input source. For example, if the path is the outer loop, KVL yields

$$-V_o + 6kI_1 + 3kI_5 + 1kI_5 + 4kI_1 = 0$$

Since $I_1 = 3$ mA and $I_5 = 1.5$ mA,

$$V_o = 36 \text{ V}$$

If we had selected the path containing the source and the points $x, y,$ and z, we would obtain

$$-V_o + 6kI_1 + V_a + V_b + 4kI_1 = 0$$

Once again, this equation yields

$$V_o = 36 \text{ V} \qquad \square$$

EXTENSION EXERCISES

E2.16 Find V_o in the network in Fig. E2.16.

Figure E2.16

ANSWER: $V_o = 2$ V.

E2.17 Find V_S in the circuit in Fig. E2.17.

Figure E2.17

ANSWER: $V_S = 9$ V.

E2.18 Find I_S in the circuit in Fig. E2.18.

Figure E2.18

ANSWER: $I_S = 0.3$ mA.

Problem-solving strategy for analyzing circuits containing a single source and a series-parallel interconnection of resistors

Step 1. Systematically reduce the resistive network so that the resistance seen by the source is represented by a single resistor.

Step 2. Determine the source current for a voltage source or the source voltage if a current source is present.

Step 3. Expand the network, retracing the simplification steps, and apply Ohm's law, KVL, KCL, voltage division, and current division to determine all currents and voltages in the network.

2.7 CIRCUITS WITH DEPENDENT SOURCES

In Chapter 1 we outlined the different kinds of dependent sources. These controlled sources are extremely important because they are used to model physical devices such as *npn* and *pnp* bipolar junction transistors (BJTs) and field-effect transistors (FETs) that are either metal-oxide-semiconductor field-effect transistors (MOSFETs) or insulated-gate field-effect transistors (IGFETs). These basic structures are, in turn, used to make analog and digital devices. A typical analog device is an operational amplifier (op-amp). Typical digital devices are random access memories (RAMs), read-only memories (ROMs), and microprocessors. We will now show how to solve simple one-loop and one-node circuits that contain these dependent sources. Although the following examples are fairly simple, they will serve to illustrate the basic concepts.

Problem-solving strategy for circuits with dependent sources

Step 1. When writing the KVL and/or KCL equations for the network, treat the dependent source as though it were an independent source.

Step 2. Write the equation that specifies the relationship of the dependent source to the controlling parameter.

Step 3. Solve the equations for the unknowns. Be sure that the number of linearly independent equations matches the number of unknowns.

The following example employs a current-controlled voltage source.

Let us determine the voltage V_o in the circuit in Fig. 2.32.

Figure 2.32 Circuit used in Example 2.24.

SOLUTION   Applying KVL, we obtain

$$-12 + 3kI_1 - V_A + 5kI_1 = 0$$

where

$$V_A = 2000I_1$$

and the units of the multiplier, 2000, are ohms. Solving these equations yields

$$I_1 = 2 \text{ mA}$$

Then

$$V_o = (5k)I_1$$

$$= 10 \text{ V}$$

Given the circuit in Fig. 2.33 containing a current-controlled current source, let us find the voltage V_o.

Figure 2.33 Circuit used in Example 2.25.

SOLUTION   Applying KCL at the top node, we obtain

$$10 \times 10^{-3} + \frac{V_S}{2k + 4k} + \frac{V_S}{3k} - 4I_o = 0$$

where

$$I_o = \frac{V_S}{3k}$$

Substituting this expression for the controlled source into the KCL equation yields

$$\frac{10}{1k} + \frac{V_S}{6k} + \frac{V_S}{3k} - \frac{4V_S}{3k} = 0$$

Solving this equation for V_S, we obtain

$$V_S = 12 \text{ V}$$

The voltage V_o can now be obtained using a simple voltage divider; that is,

$$V_o = \left[\frac{4k}{2k + 4k} \right] V_S$$

$$= 8 \text{ V} \qquad \square$$

EXAMPLE 2.26

The network in Fig. 2.34 contains a voltage-controlled voltage source. We wish to find V_o in this circuit.

Figure 2.34 Circuit used in Example 2.26.

SOLUTION   Applying KVL to this network yields

$$-12 + 3kI + 2V_o + 1kI = 0$$

where

$$V_o = 1kI$$

Hence, the KVL equation can be written as

$$-12 + 3kI + 2kI + 1kI = 0$$

or

$$I = 2 \text{ mA}$$

Therefore,

$$V_o = 1kI$$

$$= 2 \text{ V} \qquad \square$$

EXAMPLE 2.27

An equivalent circuit for a FET common-source amplifier or BJT common-emitter amplifier can be modeled by the circuit shown in Fig. 2.35a. We wish to determine an expression for the gain of the amplifier, which is the ratio of the output voltage to the input voltage.

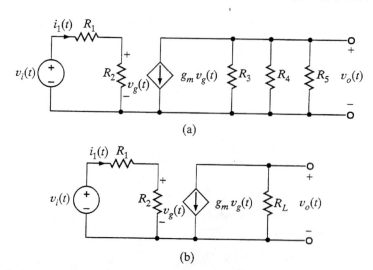

(a)

(b)

Figure 2.35 Example circuit containing a voltage-controlled current source.

SOLUTION Note that although this circuit, which contains a voltage-controlled current source, appears to be somewhat complicated, we are actually in a position now to solve it with techniques we have studied up to this point. The loop on the left, or input to the amplifier, is essentially detached from the output portion of the amplifier on the right. The voltage across R_2 is $v_g(t)$, which controls the dependent current source.

To simplify the analysis, let us replace the resistors $R_3, R_4,$ and R_5 with R_L such that

$$\frac{1}{R_L} = \frac{1}{R_3} + \frac{1}{R_4} + \frac{1}{R_5}$$

Then the circuit reduces to that shown in Fig. 2.35b. Applying Kirchhoff's voltage law to the input portion of the amplifier yields

$$v_i(t) = i_1(t)(R_1 + R_2)$$

and

$$v_g(t) = i_1(t)R_2$$

Solving these equations for $v_g(t)$ yields

$$v_g(t) = \frac{R_2}{R_1 + R_2} v_i(t)$$

From the output circuit, note that the voltage $v_o(t)$ is given by the expression

$$v_o(t) = -g_m v_g(t)R_L$$

Combining this equation with the preceding one yields

$$v_o(t) = \frac{-g_m R_L R_2}{R_1 + R_2} v_i(t)$$

Therefore, the amplifier gain, which is the ratio of the output voltage to the input voltage, is given by

$$\frac{v_o(t)}{v_i(t)} = -\frac{g_m R_L R_2}{R_1 + R_2}$$

Reasonable values for the circuit parameters in Fig. 2.35a are $R_1 = 100\ \Omega$, $R_2 = 1\ k\Omega, g_m = 0.04\ S, R_3 = 50\ k\Omega$, and $R_4 = R_5 = 10\ k\Omega$. Hence, the gain of the amplifier under these conditions is

$$\frac{v_o(t)}{v_i(t)} = \frac{-(0.04)(4.545)(10^3)(1)(10^3)}{(1.1)(10^3)}$$

$$= -165.29$$

Thus, the magnitude of the gain is 165.29.   ❑

At this point it is perhaps helpful to point out again that when analyzing circuits with dependent sources, we first treat the dependent source as though it were an independent source when we write a Kirchhoff's current or voltage law equation. Once the equation is written, we then write the controlling equation that specifies the relationship of the dependent source to the unknown variable. For instance, the first equation in Example 2.25 treats the dependent source like an independent source. The second equation in the example specifies the relationship of the dependent source to the voltage, which is the unknown in the first equation.

EXTENSION EXERCISES

E2.19   Find V_o in the circuit in Fig. E2.19.

Figure E2.19

ANSWER: $V_o = 12\ V.$

E2.20   Find V_o in the network in Fig. E2.20.

Figure E2.20

ANSWER: $V_o = 8\ V.$

2.8 DC MEASUREMENTS

We now describe the instruments that are employed to physically measure current, voltage, resistance, and power. However, before proceeding with a discussion of these devices, it is instructive to describe the meter movement used in many instruments for electrical measurements.

The D'Arsonval Meter Movement

The D'Arsonval meter movement, which is a fundamental component in many electric instruments, is shown in Fig. 2.36. This is a permanent-magnet-moving-coil (PMMC) mechanism in which a coil of fine wire is allowed to rotate between the poles of a permanent magnet. If a current is passed through the moving coil, the resulting magnetic field reacts with the magnetic field of the permanent magnet, producing a torque, which is counterbalanced by a restoring spring. The deflection of the pointer attached to the coil is proportional to the coil current, produced by the quantity being measured. A meter rated at 100 mV, 30 mA is an example of a typical standard meter.

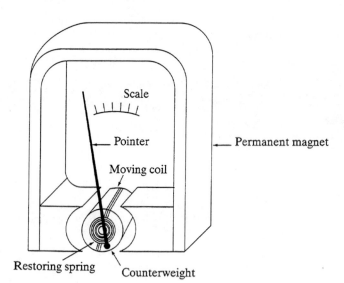

Scale

Pointer — Permanent magnet

Moving coil

Restoring spring — Counterweight

Figure 2.36 D'Arsonval meter movement.

Current Measurement. The ammeter is a device employed to measure current, and a variety of meter movements is commercially available. When connected in a network, the ammeter is inserted directly into the path of the current to be measured. DC ammeters should be connected such that the current enters the positive terminal for upscale deflection.

In the ideal case the internal resistance of an ammeter would be zero. However, practical ammeters do have some small internal resistance, R_m, which can produce a loading effect in low-voltage, low-resistance circuits; that is, the current measured by the meter is changed by inserting the meter. Therefore, it is convenient to represent the ammeter as shown in Fig. 2.37, in which all the resistance of the meter is lumped into R_m.

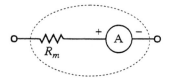

Figure 2.37 Ammeter circuit representation.

EXAMPLE 2.28

Let us determine the internal resistance of a standard ammeter that has a full-scale rating of 30 mA at 100 mV.

SOLUTION   The internal resistance for this meter is

$$R_m = \frac{100 \times 10^{-3}}{30 \times 10^{-3}}$$
$$= 3.33 \ \Omega$$

Suppose now that we wish to use the meter mechanism in Example 2.28 to measure a current, as shown in Fig. 2.38a, which is larger than the meter's current rating. We can do this by shunting a large part of the current to be measured around the meter movement, as shown in Fig. 2.38b. Note that the parallel combination of R_m and R_{SH} form a current divider. By selecting the proper value of the shunt resistor, we can use the meter movement to measure a current larger than the rating of the meter mechanism. For example, the configuration shown in Fig. 2.38c is typical of a multirange ammeter. The following example illustrates the selection of a shunt resistor to increase the current range.

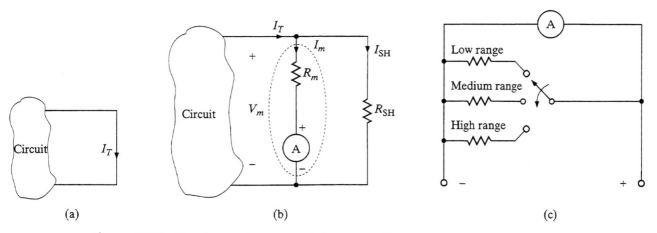

(a)   (b)   (c)

Figure 2.38   Circuits used to expand the range of a meter.

EXAMPLE 2.29

We wish to employ a 5-V, 500-mA dc adjustable current power supply to provide the power for the ringing circuit for a doorbell, as shown in Fig. 2.39a. The coil for the ringing circuit has a current limit of 250 mA. The only meter movement available to us is the 30-mA, 100-mV meter described previously. Therefore, we must select a shunt resistance, which when placed in parallel with the meter will permit us to measure 250 mA full scale.

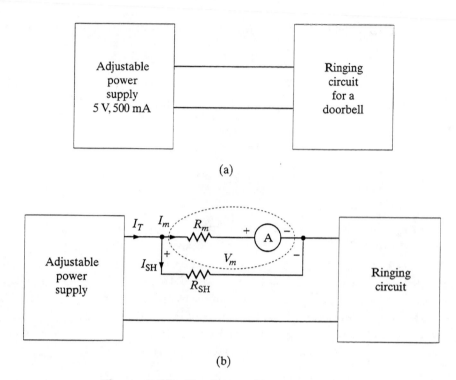

(a)

(b)

Figure 2.39 Circuits used in Example 2.29.

SOLUTION With reference to Fig. 2.39b, the total current $I_T = 250$ mA. The current in the meter movement $I_m = 30$ mA. Therefore, using KCL

$$I_{SH} = I_T - I_m$$

$$= 220 \text{ mA}$$

Since the voltage $V_m = 100$ mV is across both the meter resistance and the shunt resistance, the required value of R_{SH} is

$$R_{SH} = \frac{V_m}{I_{SH}}$$

$$= \frac{0.1}{0.22}$$

$$= 0.4545 \ \Omega$$

Note that if the meter is removed the power supply must be dialed back by 0.1 V in order to limit the current to 250 mA. This illustrates the point made earlier about the loading effect in low voltage circuits. ❏

EXTENSION EXERCISES

E2.21   A dc ammeter has a full-scale rating of 200 mV, 20 mA. Find the meter resistance.

ANSWER: $R_m = 10\ \Omega$

E2.22   Determine the amount of power that the meter in drill problem E2.21 dissipates at full-scale deflection.

ANSWER: $P = 4\ \text{mW}.$

Voltage Measurement

A voltmeter is the basic device used to measure voltage and is connected in parallel with the element whose voltage is desired. In order to obtain an upscale meter deflection, the + terminal of the voltmeter should be connected to the point of higher potential.

Once again the resistance of the meter movement is R_m and is represented as shown in Fig. 2.37. A resistance R_{se}, called a multiplier, is connected in series with the meter movement. This resistance R_{se} and the meter movement form a voltage divider and permit the meter to be used in a manner that extends the voltage range of the meter. The following example illustrates this technique.

EXAMPLE 2.30

The remote control for our television set does not work, and we are suspicious that the 9-V battery in the remote is bad. We wish to measure this voltage, but all we have available is a 100-mV, 30-mA D'Arsonval meter movement. Let us determine the value of the resistor, R_{se}, which must be placed in series with the meter movement to measure this voltage.

SOLUTION   The circuit for this application is shown in Fig. 2.40 and should produce full-scale deflection of 9 V using a 100-mV, 30-mA meter movement. For full-scale deflection, $V_m = 100$ mV and, hence, using KVL,

Figure 2.40 Circuit used in Example 2.30.

$$9 = I_m R_{se} + V_m$$

and since $V_m = 100$ mV and $I_m = 30$ mA,

$$R_{se} = \frac{9 - 0.100}{0.03}$$
$$= 296.7 \ \Omega$$

Recall from Example 2.28 that the meter resistance is 3.33 Ω. Therefore, if the battery voltage has dropped to 6 V, the current read by the meter would be

$$I_m = \frac{6}{296.7 + 3.33}$$
$$= 20 \text{ mA}$$

which is 6/9 or 2/3 of full scale.   ❏

EXTENSION EXERCISE

E2.23   A meter has a resistance of 50 Ω and a full-scale current of 1 mA. Select the multiplier resistance needed to measure 12 V full scale.
ANSWER: $R_{se} = 11.95$ kΩ.

Finally, the voltmeter, like the ammeter, has a loading effect on the circuit. The magnitude of the effect is dependent upon the resistance of the voltmeter and the resistance of the circuit elements the voltmeter is connected across. When a voltmeter is connected across an element, some current will flow in the meter. This current will be negligible if the resistance of the meter is much higher than that of the elements across which it is connected.

Voltmeter sensitivity is a quantitative measure of the loading effect. It is defined as the reciprocal of the current required for full-scale deflection; that is,

$$\text{Sensitivity} = \frac{1}{I_m} \ \Omega/\text{V}$$

Therefore, a voltmeter with a 50-V range, which has a 100-μA meter movement, has a sensitivity of

$$\text{Sensitivity} = \frac{1}{100 \times 10^{-6}}$$
$$= 10{,}000 \ \Omega/\text{V}$$

and, hence, the meter resistance is

$$R_m = (\text{sensitivity})(\text{voltage rating})$$
$$= (10{,}000)(50)$$
$$= 0.5 \ \text{M}\Omega$$

The sensitivity of a meter in ohms/volt is normally printed on the meter scale, which aids us in selecting a meter that will not load a circuit and cause measurement errors.

Resistance Measurement

An ohmmeter is an instrument used to measure resistance. An ohmmeter circuit, shown in Fig. 2.41a, typically consists of a meter movement in series with a battery and regulating resistor. The meter scale typically reads in ohms from right to left, as shown in Fig. 2.41b. If $R_x = \infty$ (that is, the ohmmeter leads or terminals are open), there is no meter deflection. If, however, $R_x = 0$ (that is, the meter leads are shorted), maximum current flows in the circuit and full-scale deflection occurs. The regulating resistor R_{reg} is a calibration device, which allows us to zero-adjust the meter movement when the leads are shorted and, thus, compensate for such things as meter resistance.

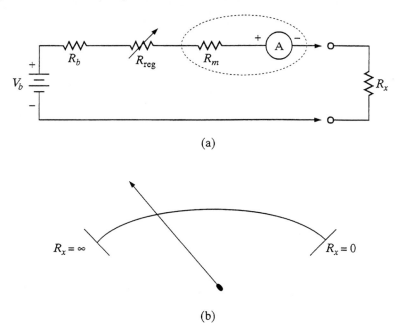

(a)

(b)

Figure 2.41 Ohmmeter circuit and scale.

The ohmmeter differs from the ammeter and voltmeter in one important respect: It must never be connected to an energized circuit. That is, the element to be measured must be disconnected from the original circuit when the measurement is made.

Although this type of ohmmeter is very useful, it is not a precision instrument. A much more accurate device for measuring resistance values over a wide range is the *Wheatstone bridge*. The Wheatstone bridge circuit is shown in Fig. 2.42. In this network, the resistors R_1, R_2, and R_3 are known. R_x is the unknown resistor. The device in the center leg of the bridge is a sensitive D'Arsonval mechanism in the microamp range called a galvanometer.

Figure 2.42 The Wheatstone bridge circuit.

The bridge is used in the following manner. The unknown resistance R_x is connected as shown in Fig. 2.42, and then R_3 is adjusted until there is no current in the galvanometer. At this point the bridge is said to be balanced. Under this balanced condition, $I_G = 0$ and, hence, KCL applied to the center nodes of the bridge yields

$$I_1 = I_3$$

and

$$I_2 = I_x$$

Furthermore, since $I_G = 0$, there is no voltage drop across the galvanometer and, therefore, KVL requires that

$$I_1 R_1 = I_2 R_2$$

and

$$I_3 R_3 = I_x R_x$$

Dividing one equation by the other and using the fact that $I_1 = I_3$ and $I_2 = I_x$ yields the relationship

$$\frac{R_1}{R_3} = \frac{R_2}{R_x}$$

so that

$$R_x = \left(\frac{R_2}{R_1}\right) R_3$$

It is interesting to note that R_x is independent of the voltage source and dependent only upon the ratio of R_2 to R_1 and not their specific values. In commercial units the ratio is made equal to a power of 10 by switching into the bridge circuit resistors with values of 1, 10, 100, and so on.

In addition to using the Wheatstone bridge to measure an unknown resistance, it is employed by engineers to measure strain in a solid material, as illustrated in the following example.

EXAMPLE 2.31

A system used to determine the weight of a truck is shown in Fig. 2.43a. The platform is supported by cylinders upon which are mounted strain gauges. The strain gauges, which measure strain when the cylinder deflects under load, are connected to a Wheatstone bridge as shown in Fig. 2.43b. The strain gauge has a resistance of 120 Ω under no-load conditions and changes value under load. The variable resistor in the bridge is a calibrated precision device.

Weight is determined in the following manner. The ΔR_3 required to balance the bridge represents the Δ strain, which when multiplied by the modulus of elasticity yields the Δ stress. The Δ stress multiplied by the cross-sectional area of the cylinder produces the Δ load, which is used to determine weight.

Let us determine the value of R_3 under no load when the bridge is balanced and its value when the resistance of the strain gauge changes to 120.24 Ω under load.

Figure 2.43 Diagrams used in Example 2.31.

SOLUTION   Using the balance equation for the bridge, the value of R_3 at no load is

$$R_3 = \left(\frac{R_1}{R_2}\right) R_x$$

$$= \left(\frac{100}{110}\right)(120)$$

$$= 109.0909 \ \Omega$$

Under load, the value of R_3 is

$$R_3 = \left(\frac{100}{110}\right)(120.24)$$

$$= 109.3091 \ \Omega$$

Therefore, the ΔR_3 is

$$\Delta R_3 = 109.3091 - 109.0909$$

$$= 0.2182 \ \Omega$$

❏

EXTENSION EXERCISE

E2.24   Given the bridge circuit in Fig. 2.42, find the value of R_x if the bridge is balanced at $R_1 = 200 \ \Omega$, $R_2 = 240 \ \Omega$, and $R_3 = 220 \ \Omega$.

ANSWER: $R_x = 264 \ \Omega$.

Power Measurements

In a dc circuit, power is the product of voltage and current. Therefore, power can be derived from current and voltage measurements. DC power can be measured directly using a wattmeter. However, since very accurate measurements can be obtained using a voltmeter in conjunction with an ammeter, the wattmeter is normally not used in dc measurements. The wattmeter is used extensively to measure power in ac circuits, and we will discuss this device in some detail in ac circuit analysis later in this text.

There are many instruments on the market that are capable of measuring voltage, current, and resistance. Two such devices are the digital multimeters shown in Fig. 2.44a. The source meter shown in Fig. 1.14 is also capable of these measurements. Instruments are also available for measurements in a specific range. For example, the high-precision instrument shown in Fig. 2.44b makes micro-ohm measurements. Thus, the modern instruments available today provide us with enormous capability in our analysis of electric circuits.

2.9 APPLICATIONS

Throughout this book we endeavor to present a wide variety of examples that demonstrate the usefulness of the material under discussion in a practical environment. In order to enhance our presentation of the practical aspects of circuit analysis and design, we have dedicated sections, such as this one, in most chapters for the specific purpose of presenting additional application-oriented examples.

EXAMPLE 2.32

The eyes (heating elements) of an electric range are frequently made of resistive nichrome strips. Operation of the eye is quite simple. A current is passed through the heating element causing it to dissipate power in the form of heat. Also, a four-position selector switch, shown in Fig. 2.45, controls the power (heat) output. In this case the eye consists of two nichrome strips modeled by the resistors R_1 and R_2, where $R_1 < R_2$.

1. How should positions $A, B, C,$ and D be labeled with regard to high, medium, low, and off settings?
2. If we desire that high and medium correspond to 2000 W and 1200 W power dissipation, respectively, what are the values of R_1 and R_2?
3. What is the power dissipation at the low setting?

SOLUTION   Position A is the off setting since no current flows to the heater elements. In position B, current flows through R_2 only, while in position C current flows through R_1 only. Since $R_1 < R_2$, more power will be dissipated when the switch is at position C. Thus, position C is the medium setting, B is the low setting, and, by elimination, position D is the high setting.

When the switch is at the medium setting, only R_1 dissipates power, and we can write R_1 as

$$R_1 = \frac{V_s^2}{P_1} = \frac{230^2}{1200}$$

(a)

(b)

Figure 2.44 (a) Two modern digital multimeters. (Courtesy of Fluke Corporation and Keithley Instruments, Inc.) (b) A micro-ohm meter. (Courtesy of Keithley Instruments, Inc.)

Figure 2.45 Simple resistive heater selector circuit.

or

$$R_1 = 44.08 \ \Omega$$

On the high setting, 2000 W of total power is delivered to R_1 and R_2. Since R_1 dissipates 1200 W, R_2 must dissipate the remaining 800 W. Therefore, R_2 is

$$R_2 = \frac{V_S^2}{P_2} = \frac{230^2}{800}$$

or

$$R_2 = 66.13 \ \Omega$$

Finally, at the low setting, only R_2 is connected to the voltage source; thus, the power dissipation at this setting is 800 W.   ❏

EXAMPLE 2.33

An equivalent circuit for a transistor amplifier used in a portable tape player is shown in Fig. 2.46. The ideal independent source, V_S, and R_S represent the magnetic head playback circuitry. The dependent source, R_o and R_{in}, model the transistor. Finally, the resistor R_L models the load, which in this case is another transistor circuit. Let us find the voltage gain of the network.

Figure 2.46 Transistor amplifier circuit model.

SOLUTION   The gain of the transistor amplifier can be derived as follows. The output voltage can be expressed as

$$V_o = -\frac{V}{100}\left(R_o//R_L\right)$$

where $\left(R_o//R_L\right)$ represents R_o in parallel with R_L.

From the input voltage divider, we can express V as a function of V_S.

$$V = V_S\left(\frac{R_{\text{in}}}{R_{\text{in}} + R_S}\right)$$

Therefore,

$$V_o = -\frac{V_S}{100}\left(\frac{R_{\text{in}}}{R_{\text{in}} + R_S}\right)\left(R_o//R_L\right)$$

Given the component values in Fig. 2.46, the voltage gain is

$$A_V = \frac{V_o}{V_S} = -160 \qquad \qquad \square$$

2.10 CIRCUIT DESIGN

The vast majority of this text is concerned with circuit analysis; that is, given a circuit in which all the components are specified, analysis involves finding such things as the voltage across some element or the current through another. Furthermore, the solution of an analysis problem is generally unique. In contrast, design involves determining the circuit configuration that will meet certain specifications. In addition, the solution is generally not unique in that there may be many ways to satisfy the circuit/performance specifications. It is also possible that there is no solution that will meet the design criteria.

In addition to meeting certain technical specifications, designs normally must also meet other criteria, such as economic, environmental, and safety constraints. For example, if a circuit design that meets the technical specifications is either too expensive or unsafe, it is not viable regardless of its technical merit.

At this point, the number of elements that we can employ in circuit design is limited primarily to the linear resistor and the active elements we have presented. However, as we progress through the text we will introduce a number of other elements (for example, the op-amp, capacitor, and inductor), which will significantly enhance our design capability.

We begin our discussion of circuit design by considering a couple of simple examples that demonstrate the selection of specific components to meet certain circuit specifications.

EXAMPLE 2.34

An electronics hobbyist who has built his own stereo amplifier wants to add a back-lit display panel to his creation for that professional look. His panel design requires seven light bulbs—two operate at 12 V/15 mA and five at 9 V/5 mA. Luckily, his stereo design already has a quality 12-V dc supply; however, there is no 9-V supply. Rather than building a new dc power supply, let us use the inexpensive circuit shown in Fig. 2.47a to design a 12-V to 9-V converter with the restriction that the

(a)

(b)

Figure 2.47 12-V to 9-V converter circuit for powering panel lighting.

variation in V_2 be no more than $\pm 5\%$. In particular, we must determine the necessary values of R_1 and R_2.

SOLUTION   First, lamps L_1 and L_2 have no effect on V_2. Second, when lamps L_3–L_7 are on, they each have an equivalent resistance of

$$R_{eq} = \frac{V_2}{I} = \frac{9}{0.005} = 1.8 \text{ k}\Omega$$

As long as V_2 remains fairly constant, the lamp resistance will also be fairly constant. Thus, the requisite model circuit for our design is shown in Fig. 2.47b. The voltage V_2 will be at its maximum value of $9 + 5\% = 9.45$ V when L_3–L_7 are all off. In this case R_1 and R_2 are in series and V_2 can be expressed by simple voltage division as

$$V_2 = 9.45 = 12\left[\frac{R_2}{R_1 + R_2}\right]$$

Rearranging the equation yields

$$\frac{R_1}{R_2} = 0.27$$

A second expression involving R_1 and R_2 can be developed by considering the case when L_3–L_7 are all on, which causes V_2 to reach its minimum value of $9 - 5\%$, or 8.55 V. Now, the effective resistance of the lamps is five 1.8-kΩ resistors in parallel, or 360 Ω. The corresponding expression for V_2 is

$$V_2 = 8.55 = 12\left[\frac{R_2//360}{R_1 + (R_2//360)}\right]$$

which can be rewritten in the form

$$\frac{\dfrac{360R_1}{R_2} + 360 + R_1}{360} = \frac{12}{8.55} = 1.4$$

Substituting the value determined for R_1/R_2 into the preceding equation yields

$$R_1 = 360[1.4 - 1 - 0.27]$$

or

$$R_1 = 48.1 \ \Omega$$

and so for R_2

$$R_2 = 178.3 \ \Omega \qquad \qquad \square$$

EXAMPLE 2.35

A circuit board within a stereo amplifier requires the use of three different voltages at the circuit nodes. The voltages needed are 4 V, 16 V, and 24 V. Criteria such as weight, cost, and size dictate that these voltages be supplied with a single voltage source. Furthermore, heat constraints dictate that the supply must dissipate less than 5 W. Assuming no loading effect from the balance of the network, design a voltage string (multiple-resistor voltage divider) that will satisfy the requirements.

SOLUTION   A 24-V source can directly provide one of the voltages, and the voltage divider in Fig. 2.48 can be used to generate the other two. The power requirement limits our choices to current less than 208.33 mA. The exact value we choose is somewhat arbitrary and, therefore, we select $I = 200$ mA. The total resistance is then

$$R_1 + R_2 + R_3 = \frac{24}{0.2} = 120 \ \Omega$$

Using voltage division, we find that

$$\frac{R_3}{R_1 + R_2 + R_3} = \frac{R_3}{120} = \frac{4}{24}$$

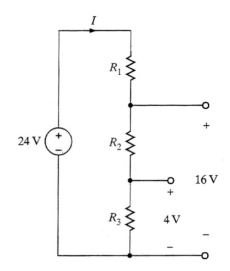

Figure 2.48 Voltage string for generating intermediate voltages.

or

$$R_3 = 20 \ \Omega$$

Similarly,

$$\frac{R_2}{120} = \frac{16 - 4}{24}$$

or

$$R_2 = 60 \ \Omega$$

Finally, since the total resistance is 120 Ω, R_1 = 40 Ω.   ❑

EXAMPLE 2.36

The network in Fig. 2.49 is an equivalent circuit for a transistor amplifier used in a stereo preamplifier. The input circuitry, consisting of a 2-mV source in series with a 500-Ω resistor, models the output of a compact disk player. The dependent source, R_{in}, and R_o model the transistor, which amplifies the signal and then sends it to the power amplifier. The 10-kΩ load resistor models the input to the power amplifier that actually drives the speakers. We must design a transistor amplifier as shown in Fig. 2.49 that will provide an overall gain of −200. In practice we do not actually vary the device parameters to achieve the desired gain; rather we select a transistor from the manufacturer's data books that will satisfy the required specification. The model parameters for three different transistors are listed as follows:

Figure 2.49 Transistor amplifier circuit model.

Manufacturer's transistor parameter values

Part Number	R_{in} (kΩ)	R_o (kΩ)	g_m (mA/V)
1	1.0	50	50
2	2.0	75	30
3	8.0	80	20

Design the amplifier by choosing the transistor that produces the most accurate gain. What is the percent error of your choice?

SOLUTION   The output voltage can be written

$$V_o = -g_m V (R_o // R_L)$$

Using voltage division at the input to find V,

$$V = V_S \left(\frac{R_{in}}{R_{in} + R_S} \right)$$

Combining these two expressions, we can solve for the gain:

$$A_V = \frac{V_o}{V_S} = -g_m \left(\frac{R_{in}}{R_{in} + R_S} \right) (R_o // R_L)$$

Using the parameter values for the three transistors, we find that

Part Number	Gain
1	−277.8
2	−211.8
3	−167.3

Obviously, the best alternative is transistor number 2, which has a gain error of

$$\text{Percent error} = \left(\frac{211.8 - 200}{200} \right) \times 100\% = 5.9\%$$

2.11 SUMMARY

Ohm's law $V = IR$

The passive sign convention with Ohm's law　The current enters the resistor terminal with the positive voltage reference.

Kirchhoff's current law (KCL)　The algebraic sum of the currents leaving (entering) a node is zero.

Kirchhoff's voltage law (KVL)　The algebraic sum of the voltages around any closed path is zero.

Solving a single-loop circuit　Determine the loop current by applying KVL and Ohm's law.

Solving a single-node-pair circuit　Determine the voltage between the pair of nodes by applying KCL and Ohm's law.

The voltage division rule　The voltage is divided between two series resistors in direct proportion to their resistance.

The current division rule　The current is divided between two parallel resistors in reverse proportion to their resistance.

The equivalent resistance of a network of resistors　Combine resistors in series by adding their resistances. Combine resistors in parallel by adding their conductances.

Short circuit　Zero resistance, zero voltage; the current in the short is determined by the rest of the circuit.

Open circuit　Zero conductance, zero current; the voltage across the open terminals is determined by the rest of the circuit.

PROBLEMS

Section 2.1

2.1　Determine the current and power dissipated in the resistor in Fig. P2.1.

Ohm's law is applicable.

Figure P2.1

2.2　Determine the current and power dissipated in the resistors in Fig. P2.2.

Once the total resistance is known the problem is similar to P2.1.

Figure P2.2

2.3   Determine the voltage across the resistor in Fig. P2.3 and the power dissipated.

Figure P2.3

Similar to P2.1.

2.4   Given the circuit in Fig. P2.4, find the voltage across each resistor and the power dissipated in each.

Figure P2.4

Similar to P2.2.

2.5   A model for a standard two D-cell flashlight is shown in Fig. P2.5. Find the power dissipated in the lamp.

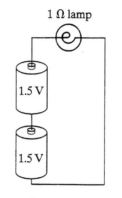

Figure P2.5

Find the total voltage across the lamp and use this voltage and the resistance to determine the power.

2.6   Determine the power dissipated in the resistor in Fig. P2.6.

Figure P2.6

Similar to P2.5.

2.7   An automobile uses two halogen headlights connected as shown in Fig. P2.7. Determine the power supplied by the battery if each headlight draws 3 A of current.

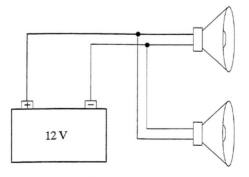

Figure P2.7

The current and voltage will yield the power.

86   Chapter 2   Resistive Circuits

2.8 Find the current I and the power supplied by the source in the network in Fig. P2.8.

Ohm's law is applicable.

Figure P2.8

2.9 In the circuit in Fig. P2.9, find the voltage across the current source and the power absorbed by the resistor.

Similar to P2.8.

Figure P2.9

2.10 If the 5-kΩ resistor in the network in Fig. P2.10 absorbs 200 mW, find V_S.

The power expression is applicable.

Figure P2.10

2.11 In the network in Fig. P2.11, the power absorbed by R_x is 20 mW. Find R_x.

Similar to P2.10.

Figure P2.11

2.12 Many years ago a string of Christmas tree lights was manufactured in the form shown in Fig. P2.12a. Today the lights are manufactured as shown in Fig. P2.12b. Is there a good reason for this change?

What happens if a bulb malfunctions?

(a)

(b)

Figure P2.12

Section 2.2

2.13 Find I_1 in the network in Fig. P2.13.

Figure P2.13

KCL is applicable.

2.14 Find I_1 in the network in Fig. P2.14.

Figure P2.14

Similar to P2.13.

2.15 Find I_1 in the network in Fig. P2.15.

Figure P2.15

Similar to P2.13.

2.16 Find I_1 and I_2 in the circuit in Fig. P2.16.

Figure P2.16

KCL is applicable.

2.17 Find I_1 and I_2 in the circuit in Fig. P2.17.

Figure P2.17

Similar to P2.16.

2.18 Find I_1 in the circuit in Fig. P2.18.

Similar to P2.16.

Figure P2.18

2.19 Find I_o in the circuit in Fig. P2.19.

Similar to P2.16.

Figure P2.19

2.20 Find I_o and I_1 in the circuit in Fig. P2.20.

Similar to P2.16.

Figure P2.20

2.21 Find I_x in the circuit in Fig. P2.21.

KCL yields a solution.

Figure P2.21

2.22 Find I_x in the circuit in Fig. P2.22.

Similar to P2.21.

Figure P2.22

2.23 Find I_x in the circuit in Fig. P2.23.

Figure P2.23

Similar to P2.21.

2.24 Find I_x in the circuit in Fig. P2.24.

Figure P2.24

Similar to P2.21.

2.25 Find V_{bd} in the circuit in Fig. P2.25.

Figure P2.25

KVL is applicable in this case.

2.26 Find V_{ae} and V_{cf} in the circuit in Fig. P2.26.

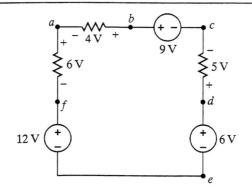

Figure P2.26

Similar to P2.25.

2.27 Find V_x in the circuit in Fig. P2.27.

Figure P2.27

Similar to P2.25.

2.28 Find V_{ac} in the circuit in Fig. P2.28.

Similar to P2.25.

Figure P2.28

2.29 Find V_{ad} and V_{ce} in the circuit in Fig. P2.29.

Similar to P2.25.

Figure P2.29

Section 2.3

2.30 Find V_{ab} in the network in Fig. P2.30.

The voltage division rule yields an immediate solution.

Figure P2.30

2.31 Find V_{ac} in the network in Fig. P2.31.

Similar to P2.30.

Figure P2.31

2.32 Find V_{ab} in the network in Fig. P2.32.

Combine sources and use the voltage division rule.

Figure P2.32

2.33 Find both I and V_{bd} in the circuit in Fig. P2.33.

Figure P2.33

Similar to P2.32.

2.34 Find the power supplied by the 12-V source and the power absorbed by the 40-kΩ resistor in the network in Fig. P2.34

Figure P2.34

Similar to P2.32.

2.35 Find V_x in the network in Fig. P2.35.

Figure P2.35

Use KVL to find the current in the loop that will, in turn, yield V_x.

2.36 Find V_1 in the network in Fig. P2.36.

Figure P2.36

Similar to P2.25.

2.37 Find the power absorbed by the 30-kΩ resistor in the circuit in Fig. P2.37.

Figure P2.37

Similar to P2.35.

Section 2.4

2.38 Find I_o in the circuit in Fig. P2.38.

The current-division rule is
immediately applicable.

Figure P2.38

2.39 Determined I_L in the circuit in Fig. P2.39.

Combine sources and use the
current-division rule.

Figure P2.39

2.40 Determined I_L in the circuit in Fig. P2.40.

Assume the voltage across the
parallel elements is V_L and
use KCL.

Figure P2.40

2.41 Determine I_L in the circuit in Fig. P2.41.

Similar to P2.40.

Figure P2.41

Section 2.5

2.42 If the conductance of each resistor in the network in Fig. P2.42 is 250 μS, find the
resistance at the terminals A-B.

Use the problem-solving
strategy for resistor
combinations.

Figure P2.42

2.43 Find R_{AB} in the circuit in Fig. P2.43.

Figure P2.43

Similar to P2.42.

2.44 Find R_{AB} in the circuit in Fig. P2.44.

Figure P2.44

Similar to P2.42.

2.45 Find R_{AB} in the network in Fig. P2.45.

Figure P2.45

Similar to P2.42.

2.46 Determine R_{AB} in the circuit in Fig. P2.46.

Figure P2.46

Similar to P2.42.

2.47 Find R_{AB} in the network in Fig. P2.47.

Similar to P2.42.

Figure P2.47

Section 2.6

2.48 Find I_1 and V_o in the circuit in Fig. P2.48.

Use the problem-solving strategy for circuits with series-parallel combinations of resistors.

Figure P2.48

2.49 Determine V_o in the network in Fig. P2.49.

Similar to P2.48.

Figure P2.49

2.50 Find I_1 and V_o in the circuit in Fig. P2.50.

Similar to P2.48.

Figure P2.50

2.51 Find I_1 in the circuit in the Fig. P2.51.

Figure P2.51

Similar to P2.48.

2.52 Determine I_o in the circuit in Fig. P2.52.

Figure P2.52

Similar to P2.48.

2.53 Determine V_o in the network in Fig. P2.53.

Figure P2.53

Similar to P2.48.

2.54 Determine V_o in the circuit in Fig. P2.54.

Figure P2.54

Similar to P2.48.

2.55 If $V_o = 3$ V in the circuit in Fig. P2.55, find V_S.

Reverse the voltage division rule or use Ohm's law and KVL to determine V_s.

Figure P2.55

2.56 If $I_o = 5$ mA in the circuit in Fig. P2.56, find I_S.

Reverse the current-division rule or use Ohm's law and KCL to find I_S.

Figure P2.56

2.57 If $V_o = 2$ V in the circuit in Fig. P2.57, find V_S.

Similar to P2.55.

Figure P2.57

2.58 If the power absorbed by the 4-kΩ resistor in the circuit in Fig. P2.58 is 36 mW find I_o.

The current in the 4-kΩ resistor and Ohm's law yield a solution.

Figure P2.58

2.59 If $V_o = 6$ V in the circuit in Fig. P2.59, find I_S.

The current in the 3-kΩ resistor, Ohm's law, and KCL yield a solution.

Figure P2.59

2.60 If $I_o = 2$ mA in the circuit in Fig. P2.60, find V_S.

Figure P2.60

Similar to P2.59.

Section 2.7

2.61 Find V_o in the circuit in Fig. P2.61.

Figure P2.61

Use the problem-solving strategy for circuits with dependent sources.

2.62 Find V_o in the network in Fig. P2.62.

Figure P2.62

Similar to P2.61.

2.63 Find V_o in the network in Fig. P2.63.

Figure P2.63

Similar to P2.61.

2.64 Find V_o in the circuit in Fig. P2.64.

Similar to P2.61.

Figure P2.64

2.65 A single-stage transistor amplifier is modeled as shown in Fig. P2.65. Find the current in the load R_L.

Similar to P2.61.

Figure P2.65

Section 2.8

Ohm's law yields the resistance.

2.66 Given a meter movement rated at 50 mV, 20 mA full-scale, find the meter resistance and power dissipated at full-scale deflection.

Similar to P2.66.

2.67 A 50-mV, 20-mA meter movement is used to measure currents between 0 and 150 mA. Find the shunt resistance required for the measurements.

Similar to P2.66.

2.68 A 30-mA, 100-mV meter movement is used in conjunction with a shunt resistance of 0.4 Ω. Find the magnitude of the current that can be measured by the meter at full scale.

Similar to P2.66.

2.69 A 100-mV, 30-mA meter movement is to be used to measure 100 mA full scale. Find the shunt resistance required for this application.

2.70 A 50-mV, 20-mA meter movement is to be used to check the voltage on 1.5-V flashlight batteries, as shown in Fig P2.70. Find the value of R_{se} required for this meter to read 1.5 V full scale.

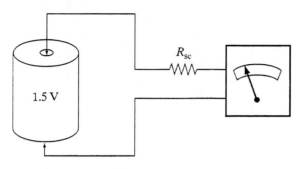

Figure P2.70

Similar to P2.66.

2.71 A resistance of 50 Ω is connected in series with a 50-mV, 20-mA meter movement. What full-scale voltage will this combination measure?

Similar to P2.66.

2.72 The ideal ammeter in Fig. P2.72 has no internal resistance and a maximum range of 10 mA. In conjunction with the series resistance R_{se}, it functions as a voltmeter to measure the voltage V_x. Choose R_{se} such that the ammeter reads its maximum value when V_x is 20 V. If the ammeter reads 2.7 mA, what is the voltage at V_x?

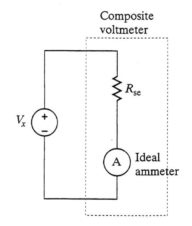

Figure P2.72

Similar to P2.66.

Advanced Problems

2.73 Find R_{AB} in the circuit in Fig. P2.73.

Figure P2.73

2.74 Find R_{AB} in the circuit in Fig. P2.74.

Figure P2.74

2.75 Find I_o in the circuit in Fig. P2.75.

Figure P2.75

2.76 Find V_o In the circuit in Fig. P2.76.

Figure P2.76

2.77 Find V_o in the circuit in Fig. P2.77.

Figure P2.77

2.78 Find V_o in the circuit in Fig. P2.78.

Figure P2.78

2.79 Find V_o in the circuit in Fig. P2.79.

Figure P2.79

2.80 Find I_o in the circuit in Fig. P2.80.

Figure P2.80

2.81 Find I_o in the circuit in Fig. P2.81.

Figure P2.81

2.82 Find I_o in the circuit in Fig. P2.82.

Figure P2.82

2.83 Find V_o in the circuit in Fig. P2.83.

Figure P2.83

2.84 Find I_o in the circuit in Fig. P2.84.

Figure P2.84

2.85 Determine the value of V_o in the circuit in Fig. P2.85

Figure P2.85

2.86 Find V_o in the circuit in Fig. P2.86.

Figure P2.86

2.87 If $V_1 = 5$ V in the circuit in Fig. P2.87, find I_S.

Figure P2.87

2.88 If $I_o = 2$ mA in the circuit in Fig. P2.88, find V_S.

Figure P2.88

2.89 If $I_o = 4$ mA in the circuit in Fig. P2.89, find I_S.

Figure P2.89

2.90 If $V_o = 6$ V in the circuit in Fig. P2.90, find V_S.

Figure P2.90

2.91 If $V_o = 2$ V in the network in Fig. P2.91, find V_S.

Figure P2.91

2.92 If $V_o = 6$ V in the circuit in Fig. P2.92, find I_S.

Figure P2.92

2.93 Given that $V_o = 4$ V in the network in Fig. P2.93, find V_S.

Figure P2.93

2.94 Find I_o in the circuit in Fig. P2.94.

Figure P2.94

2.95 Find I_o in the circuit in Fig. P2.95.

Figure P2.95

2.96 Find V_o in the circuit in Fig. P2.96.

Figure P2.96

2.97 Find the power supplied by the 24-V source in the circuit in Fig. P2.97.

Figure P2.97

2.98 If all the resistors in the network in Fig. P2.98 are 12 kΩ, find V_o.

Figure P2.98

2.99 A typical transistor amplifier is shown in Fig. P2.99. Find the amplifier gain G (i.e., the ratio of the output voltage to the input voltage).

Figure P2.99

2.100 In many amplifier applications we are concerned not only with voltage gain, but also with power gain.

Power gain $= A_p = $ (power delivered to the load)/(power delivered by the input)

Find the power gain for the circuit in Fig. P2.100, where $R_L = 50$ kΩ.

Figure P2.100

2.101 For the network in Fig. P2.101, choose the values of R_{in} and R_o such that V_o is maximized. What is the resulting ratio, V_o/V_S?

Figure P2.101

2.102 Find the power absorbed by the 12-kΩ resistor in the circuit in Fig. P2.102.

Figure P2.102

2.103 Find the power absorbed by the 12-kΩ resistor in the circuit in Fig. P2.103.

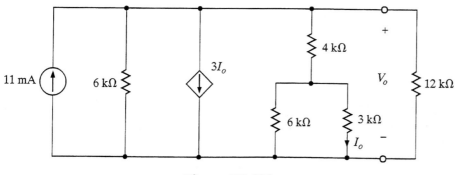

Figure P2.103

2.104 Find the power supplied by the 21-V source in the circuit in Fig. 2.104.

Figure P2.104

2.105 The Wheatstone bridge in Fig. 2.42 is used to measure an unknown resistor R_x. The bridge is balanced with $R_1 = 1$ kΩ, $R_2 = 1.2$ kΩ, and $R_3 = 1.8$ kΩ. Find the value of R_x.

2.106 If the Wheatstone bridge in Fig. 2.42 balanced with $R_x = 140$ Ω, $R_1 = 100$ Ω, and $R_2 = 200$ Ω, what is the value of the variable resistor R_3?

2.107 Redesign the voltage divider of Example 2.35 if the node voltage values are 18 V, 15 V, and 4 V with 60 mW of power dissipation.

Nodal and Loop Analysis Techniques

3

In Chapter 2 we analyzed the simplest possible circuits, those containing only a single-node pair or a single loop. We found that these circuits can be completely analyzed via a single algebraic equation. In the case of the single-node-pair circuit (i.e., one containing two nodes, one of which is a reference node), once the node voltage is known, we can calculate all the currents. In a single-loop circuit, once the loop current is known, we can calculate all the voltages.

In this chapter we extend our capabilities in a systematic manner so that we can calculate all currents and voltages in circuits that contain multiple nodes and loops. Our analyses are based primarily on two laws with which we are already familiar: Kirchhoff's current law (KCL) and Kirchhoff's voltage law (KVL). In a nodal analysis we employ KCL to determine the node voltages, and in a loop analysis we use KVL to determine the loop currents.

A very important commercially available circuit known as the operational amplifier, or op-amp, is presented and discussed. Op-amps are used in literally thousands of applications, including such things as compact disk (CD) players, random access memories (RAMs), analog-to-digital (A/D) and digital-to-analog (D/A) converters, headphone amplifiers, and electronic instrumentation of all types. Finally, we will discuss the terminal characteristics of this circuit and demonstrate its use in practical applications as well as circuit design.

3.1 NODAL ANALYSIS

In a nodal analysis the variables in the circuit are selected to be the node voltages. The node voltages are defined with respect to a common point in the circuit. One node is selected as the reference node, and all other node voltages are defined with respect to that node. Quite often this node is the one to which the largest number of branches are connected. It is commonly called *ground* because it is said to be at ground-zero potential, and it sometimes represents the chassis or ground line in a practical circuit.

We will select our variables as being positive with respect to the reference node. If one or more of the node voltages are actually negative with respect to the reference node, the analysis will indicate it.

In order to understand the value of knowing all the node voltages in a network, we consider once again the network in Fig. 2.30, which is redrawn in Fig. 3.1. The voltages, V_S, V_a, V_b, and V_c, are all measured with respect to the bottom node, which is selected as the reference and labeled with the ground symbol $\perp$. Therefore, the voltage at node 1 is $V_S = 12$ V with respect to the reference node 5; the voltage at node 2 is $V_a = 3$ V with respect to the reference node 5, and so on. Now note carefully that once these node voltages are known, we can immediately calculate any branch current or the power supplied or absorbed by any element, since we know the voltage across every element in the network. For example, the voltage V_1 across the left-most 9-kΩ resistor is the difference in potential between the two ends of the resistor; that is,

$$V_1 = V_S - V_a$$
$$= 12 - 3$$
$$= 9 \text{ V}$$

This equation is really nothing more than an application of KVL around the left-most loop; that is,

$$-V_S + V_1 + V_a = 0$$

In a similar manner, we find that

$$V_3 = V_a - V_b$$

and

$$V_5 = V_b - V_c$$

Then the currents in the resistors are

Figure 3.1 Circuit with known node voltages.

$$I_1 = \frac{V_1}{9k} = \frac{V_S - V_a}{9k}$$

$$I_3 = \frac{V_3}{3k} = \frac{V_a - V_b}{3k}$$

$$I_5 = \frac{V_5}{9k} = \frac{V_b - V_c}{9k}$$

In addition,

$$I_2 = \frac{V_a - 0}{6k}$$

$$I_4 = \frac{V_b - 0}{4k}$$

since the reference node 5 is at zero potential.

Thus, as a general rule, if we know the node voltages in a circuit, we can calculate the current through any resistive element using Ohm's law; that is,

$$i = \frac{v_m - v_N}{R} \qquad \textbf{3.1}$$

as illustrated in Fig. 3.2.

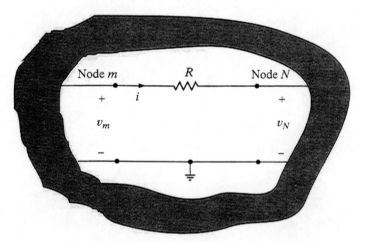

Figure 3.2 Circuit used to illustrate Ohm's law in a multiple-node network.

In Example 2.5 we illustrated that the number of linearly independent KCL equations for an N-node network was $N - 1$. Furthermore, we found that in a two-node circuit, in which one node was the reference node, only one equation was required to solve for the unknown node voltage. What is illustrated in this simple case is true in general; that is, in an N-node circuit one linearly independent KCL equation is written for each of the $N - 1$ nonreference nodes, and this set of $N - 1$ linearly independent simultaneous equations, when solved, will yield the $N - 1$ unknown node voltages.

It is instructive to treat nodal analysis by examining several different types of circuits and illustrating the salient features of each. We begin with the simplest case. However, as a prelude to our discussion of the details of nodal analysis, experience indicates that it is worthwhile to diverge for a moment to ensure that the concept of node voltage is clearly understood.

At the outset it is important to specify a reference. For example, to state that the voltage at node A is 12 V means nothing unless we provide the reference point; that is, the voltage at node A is 12 V with respect to what. The circuit in Fig. 3.3 illustrates a portion of a network containing three nodes, one of which is the reference node.

Figure 3.3 An illustration of node voltages.

The voltage $V_1 = 4$ V is the voltage at node 1 with respect to the reference node 3. Similarly, the voltage $V_2 = -2$ V is the voltage at node 2 with respect to node 3. In addition, however, the voltage at node 1 with respect to node 2 is $+6$ V and the voltage at node 2 with respect to node 1 is -6 V. Furthermore, since the current will flow from the node of higher potential to the node of lower potential, the current in R_1 is from top to bottom, the current in R_2 is from left to right, and the current in R_3 is from bottom to top.

These concepts have important ramifications in our daily lives. If a man were hanging in midair with one hand on one line and one hand on another and the dc line voltage of each line was exactly the same, the voltage across his heart would be zero and he would be safe. If, however, he let go of one line and let his feet touch the ground, the dc line voltage would then exist from his hand to his foot with his heart in the middle. He would probably be dead the instant his foot hit the ground.

In the town where we live, a young man tried to retrieve his parakeet that had escaped its cage and was outside sitting on a power line. He stood on a metal ladder and with a metal pole reached for the parakeet; when the metal pole touched the power line, the man was killed instantly. Electric power is vital to our standard of living, but it is also very dangerous. The material in this book *does not* qualify you to handle it safely. Therefore, always be extremely careful around electric circuits.

Now as we begin our discussion of nodal analysis, our approach will be to begin with simple cases and proceed in a systematic manner to those that are more challenging. Numerous examples will be the vehicle used to demonstrate each facet of this approach. Finally, at the end of this section, we will outline a strategy for attacking any circuit using nodal analysis.

Circuits Containing Only Independent Current Sources

Consider the network shown in Fig. 3.4. There are three nodes, and the bottom node is selected as the reference node. The branch currents are assumed to flow in the directions indicated in the figures. If one or more of the branch currents are actually flowing in a direction opposite to that assumed, the analysis will simply produce a branch current that is negative.

Figure 3.4 A three-node circuit.

Applying KCL at node 1 yields

$$-i_A + i_1 + i_2 = 0$$

Employing the passive sign convention.

Using Ohm's law $(i = Gv)$ and noting that the reference node is at zero potential, we obtain

$$-i_A + G_1(v_1 - 0) + G_2(v_1 - v_2) = 0$$

or

$$(G_1 + G_2)v_1 - G_2 v_2 = i_A$$

KCL at node 2 yields

$$-i_2 + i_B + i_3 = 0$$

or

$$-G_2(v_1 - v_2) + i_B + G_3(v_2 - 0) = 0$$

which can be expressed as

$$-G_2 v_1 + (G_2 + G_3)v_2 = -i_B$$

Therefore, the two equations for the two unknown node voltages v_1 and v_2 are

$$(G_1 + G_2)v_1 - G_2 v_2 = i_A$$
$$-G_2 v_1 + (G_2 + G_3)v_2 = -i_B \tag{3.2}$$

Note that the analysis has produced two simultaneous equations in the unknowns v_1 and v_2. They can be solved using any convenient technique, and modern calculators and personal computers are very efficient tools for their application. Two such techniques, Gaussian elimination and matrix analysis, will be demonstrated here. For example, Eq. (3.2) may be rewritten in matrix form as

$$\mathbf{A}\mathbf{v} = \mathbf{i} \tag{3.3}$$

where

$$\mathbf{A} = \begin{bmatrix} G_1 + G_2 & -G_2 \\ -G_2 & G_2 + G_3 \end{bmatrix}, \quad \mathbf{v} = \begin{bmatrix} v_1 \\ v_2 \end{bmatrix}, \quad \mathbf{i} = \begin{bmatrix} i_A \\ -i_B \end{bmatrix}$$

Thus,

$$\begin{bmatrix} G_1 + G_2 & -G_2 \\ -G_2 & G_2 + G_3 \end{bmatrix} \begin{bmatrix} v_1 \\ v_2 \end{bmatrix} = \begin{bmatrix} i_A \\ -i_B \end{bmatrix} \tag{3.4}$$

In general, the solution to Eq. (3.3) is

D3.1   For the following network, write the KCL equations for nodes 1 and 2.

ANSWER: $i_1 + \dfrac{v_1}{R_2} + \dfrac{v_1 - v_2}{R_3} + \dfrac{v_1 - v_2}{R_4} = 0$

$i_2 + \dfrac{v_1 - v_2}{R_3} + \dfrac{v_1 - v_2}{R_4} = 0$

$$\mathbf{v} = \mathbf{A}^{-1}\mathbf{i} \qquad\qquad 3.5$$

where $\mathbf{A}^{-1}$ is the inverse of the matrix $\mathbf{A}$. This calculation is both easy and straightforward for modern hand-held calculators and personal computers.

The KCL equations at nodes 1 and 2 produced two linearly independent simultaneous equations:

$$-i_A + i_1 + i_2 = 0$$
$$-i_2 + i_B + i_3 = 0$$

The KCL equation for the third node (reference) is

$$+i_A - i_1 - i_B - i_3 = 0$$

Note that if we add the first two equations, we obtain the third. Furthermore, any two of the equations can be used to derive the remaining equation. Therefore, in this $N = 3$ node circuit, only $N - 1 = 2$ of the equations are linearly independent and required to determine the $N - 1 = 2$ unknown node voltages.

Note that a nodal analysis employs KCL in conjunction with Ohm's law. Once the direction of the branch currents has been *assumed*, then Ohm's law, as illustrated by Fig. 3.2 and expressed by Eq. (3.1), is used to express the branch currents in terms of the unknown node voltages. We can assume the currents to be in any direction. However, once we assume a particular direction, we must be very careful to write the currents correctly in terms of the node voltages using Ohm's law.

EXAMPLE 3.1

Suppose that the network in Fig. 3.4 has the following parameters: $I_A = 1$ mA, $R_1 = 12$ kΩ, $R_2 = 6$ kΩ, $I_B = 4$ mA, and $R_3 = 6$ kΩ. Let us determine all node voltages and branch currents.

SOLUTION   Using the parameter values Eq. (3.2) becomes

$$V_1\left[\frac{1}{12k} + \frac{1}{6k}\right] - V_2\left[\frac{1}{6k}\right] = 1 \times 10^{-3}$$

$$-V_1\left[\frac{1}{6k}\right] + V_2\left[\frac{1}{6k} + \frac{1}{6k}\right] = -4 \times 10^{-3}$$

where we employ capital letters because the voltages are constant. The equations can be written as

$$\frac{V_1}{4k} - \frac{V_2}{6k} = 1 \times 10^{-3}$$

$$-\frac{V_1}{6k} + \frac{V_2}{3k} = -4 \times 10^{-3}$$

Using Gaussian elimination, we solve the first equation for V_1 in terms of V_2:

$$V_1 = V_2\left(\frac{2}{3}\right) + 4$$

This value is then substituted into the second equation to yield

$$\frac{-1}{6k}\left(\frac{2}{3}V_2 + 4\right) + \frac{V_2}{3k} = -4 \times 10^{-3}$$

or

$$V_2 = -15\,\text{V}$$

This value for V_2 is now substituted back into the equation for V_1 in terms of V_2, which yields

$$V_1 = \frac{2}{3}V_2 + 4$$

$$= -6\,\text{V}$$

The circuit equations can also be solved using matrix analysis. In matrix form the equations are

$$\begin{bmatrix} \dfrac{1}{4k} & \dfrac{-1}{6k} \\ \dfrac{-1}{6k} & \dfrac{1}{3k} \end{bmatrix} \begin{bmatrix} V_1 \\ V_2 \end{bmatrix} = \begin{bmatrix} 1 \times 10^{-3} \\ -4 \times 10^{-3} \end{bmatrix}$$

and therefore,

$$\begin{bmatrix} V_1 \\ V_2 \end{bmatrix} = \begin{bmatrix} \dfrac{1}{4k} & \dfrac{-1}{6k} \\ \dfrac{-1}{6k} & \dfrac{1}{3k} \end{bmatrix}^{-1} \begin{bmatrix} 1 \times 10^{-3} \\ -4 \times 10^{-3} \end{bmatrix}$$

To calculate the inverse of **A**, we need the adjoint and the determinant. The adjoint is

$$\text{Adj } \mathbf{A} = \begin{bmatrix} \dfrac{1}{3k} & \dfrac{1}{6k} \\ \dfrac{1}{6k} & \dfrac{1}{4k} \end{bmatrix}$$

and the determinant is

$$|\mathbf{A}| = \left(\frac{1}{3k}\right)\left(\frac{1}{4k}\right) - \left(\frac{-1}{6k}\right)\left(\frac{-1}{6k}\right)$$

$$= \frac{1}{18k^2}$$

Therefore,

$$\begin{bmatrix} V_1 \\ V_2 \end{bmatrix} = 18k^2 \begin{bmatrix} \dfrac{1}{3k} & \dfrac{1}{6k} \\[2mm] \dfrac{1}{6k} & \dfrac{1}{4k} \end{bmatrix} \begin{bmatrix} 1 \times 10^{-3} \\ -4 \times 10^{-3} \end{bmatrix}$$

$$= 18k^2 \begin{bmatrix} \dfrac{1}{3k^2} - \dfrac{4}{6k^2} \\[2mm] \dfrac{1}{6k^2} - \dfrac{1}{k^2} \end{bmatrix}$$

$$= \begin{bmatrix} -6 \\ -15 \end{bmatrix}$$

Knowing the node voltages, we can determine all the currents using Ohm's law:

$$I_1 = \frac{V_1}{R_1} = \frac{-6}{12k} = -\frac{1}{2}\,\text{mA}$$

$$I_2 = \frac{V_1 - V_2}{6k} = \frac{-6 - (-15)}{6k} = \frac{3}{2}\,\text{mA}$$

and

$$I_3 = \frac{V_2}{6k} = \frac{-15}{6k} = -\frac{5}{2}\,\text{mA}$$

Figure 3.5 illustrates the results of all the calculations. Note that KCL is satisfied at every node.

Figure 3.5 Circuit used in Example 3.1.

Let us now examine the circuit in Fig. 3.6. The current directions are assumed as shown in the figure.

At node 1 KCL yields

Figure 3.6 A four-node circuit.

$$i_1 - i_A + i_2 - i_3 = 0$$

or

$$\frac{v_1}{R_1} - i_A + \frac{v_1 - v_2}{R_2} - \frac{v_3 - v_1}{R_3} = 0$$

$$v_1\left(\frac{1}{R_1} + \frac{1}{R_2} + \frac{1}{R_3}\right) - v_2\frac{1}{R_2} - v_3\frac{1}{R_3} = i_A$$

At node 2 KCL yields

$$-i_2 + i_4 - i_5 = 0$$

or

$$-\frac{v_1 - v_2}{R_2} + \frac{v_2}{R_4} - \frac{v_3 - v_2}{R_5} = 0$$

$$-v_1\frac{1}{R_2} + v_2\left(\frac{1}{R_2} + \frac{1}{R_4} + \frac{1}{R_5}\right) - v_3\frac{1}{R_5} = 0$$

At node 3 the equation is

$$i_3 + i_5 + i_B = 0$$

or

$$\frac{v_3 - v_1}{R_3} + \frac{v_3 - v_2}{R_5} + i_B = 0$$

$$-v_1\frac{1}{R_3} - v_2\frac{1}{R_5} + v_3\left(\frac{1}{R_3} + \frac{1}{R_5}\right) = -i_B$$

Grouping the node equations together, we obtain

$$v_1\left(\frac{1}{R_1} + \frac{1}{R_2} + \frac{1}{R_3}\right) - v_2\frac{1}{R_2} - v_3\frac{1}{R_3} = i_A$$

$$-v_1\frac{1}{R_2} + v_2\left(\frac{1}{R_2} + \frac{1}{R_4} + \frac{1}{R_5}\right) - v_3\frac{1}{R_5} = 0 \qquad \textbf{3.6}$$

$$-v_1\frac{1}{R_3} - v_2\frac{1}{R_5} + v_3\left(\frac{1}{R_3} + \frac{1}{R_5}\right) = -i_B$$

Note that our analysis has produced three simultaneous equations in the three unknown node voltages $v_1, v_2,$ and v_3. The equations can also be written in matrix form as

$$
\begin{bmatrix}
\dfrac{1}{R_1} + \dfrac{1}{R_2} + \dfrac{1}{R_3} & -\dfrac{1}{R_2} & -\dfrac{1}{R_3} \\[2mm]
-\dfrac{1}{R_2} & \dfrac{1}{R_2} + \dfrac{1}{R_4} + \dfrac{1}{R_5} & -\dfrac{1}{R_5} \\[2mm]
-\dfrac{1}{R_3} & -\dfrac{1}{R_5} & \dfrac{1}{R_3} + \dfrac{1}{R_5}
\end{bmatrix}
\begin{bmatrix} v_1 \\ v_2 \\ v_3 \end{bmatrix}
=
\begin{bmatrix} i_A \\ 0 \\ -i_B \end{bmatrix}
\qquad \textbf{3.7}
$$

At this point it is important that we note the symmetrical form of the equations that describe the two previous networks. Equations (3.2) and (3.4) and Eqs. (3.6) and (3.7) exhibit the same type of symmetrical form. The **A** matrix for each network (3.4) and (3.7) is a symmetrical matrix. This symmetry is not accidental. The node equations for networks containing only resistors and independent current sources can always be written in this symmetrical form. We can take advantage of this fact and learn to write the equations by inspection. Note in the first equation of (3.2) that the coefficient of v_1 is the sum of all the conductances connected to node 1 and the coefficient of v_2 is the negative of the conductances connected between node 1 and node 2. The right-hand side of the equation is the sum of the currents entering node 1 through current sources. This equation is KCL at node 1. In the second equation in (3.2), the coefficient of v_2 is the sum of all the conductances connected to node 2, the coefficient of v_1 is the negative of the conductance connected between node 2 and node 1, and the right-hand side of the equation is the sum of the currents entering node 2 through current sources. This equation is KCL at node 2. Similarly, in the first equation in (3.6) the coefficient of v_1 is the sum of the conductances connected to node 1, the coefficient of v_2 is the negative of the conductance connected between node 1 and node 2, the coefficient of v_3 is the negative of the conductance connected between node 1 and node 3, and the right-hand side of the equation is the sum of the currents entering node 1 through current sources. The other two equations in (3.6) are obtained in a similar manner. In general, if KCL is applied to node j with node voltage v_j, the coefficient of v_j is the sum of all the conductances connected to node j and the coefficients of the other node voltages (e.g., v_{j-1}, v_{j+1}) are the negative of the sum of the conductances connected directly between these nodes and node j. The right-hand side of the equation is equal to the sum of the currents entering the node via current sources. Therefore, the left-hand side of the equation represents the sum of the currents leaving node j and the right-hand side of the equation represents the currents entering node j.

EXAMPLE 3.2

Let us apply what we have just learned to write the equations for the network in Fig. 3.7 by inspection.

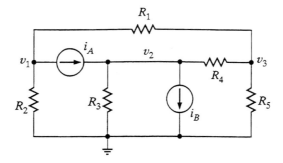

Figure 3.7 Circuit used in Example 3.2.

SOLUTION The equations are

$$v_1\left(\frac{1}{R_1} + \frac{1}{R_2}\right) - v_2(0) - v_3\left(\frac{1}{R_1}\right) = -i_A$$

$$-v_1(0) + v_2\left(\frac{1}{R_3} + \frac{1}{R_4}\right) - v_3\left(\frac{1}{R_4}\right) = i_A - i_B$$

$$-v_1\left(\frac{1}{R_1}\right) - v_2\left(\frac{1}{R_4}\right) + v_3\left(\frac{1}{R_1} + \frac{1}{R_4} + \frac{1}{R_5}\right) = 0$$

which can also be written directly in matrix form as

$$\begin{bmatrix} \dfrac{1}{R_1} + \dfrac{1}{R_2} & 0 & -\dfrac{1}{R_1} \\ 0 & \dfrac{1}{R_3} + \dfrac{1}{R_4} & -\dfrac{1}{R_4} \\ -\dfrac{1}{R_1} & -\dfrac{1}{R_4} & \dfrac{1}{R_1} + \dfrac{1}{R_4} + \dfrac{1}{R_5} \end{bmatrix} \begin{bmatrix} v_1 \\ v_2 \\ v_3 \end{bmatrix} = \begin{bmatrix} -i_A \\ i_A - i_B \\ 0 \end{bmatrix}$$

Both the equations and the **A** matrix exhibit the symmetry that will always be present in circuits that contain only resistors and current sources. ❑

EXTENSION EXERCISES

E3.1 Write the node equations for the circuit in Fig. E3.1.

Figure E3.1

ANSWER: $\dfrac{1}{4k} V_1 - \dfrac{1}{12k} V_2 = 4 \times 10^{-3}$, $\dfrac{-1}{12k} V_1 + \dfrac{1}{4k} V_2$

$$= -2 \times 10^{-3}$$

E3.2 Find all the branch currents in the network in Fig. E3.2.

Figure E3.2

ANSWER: Note that KCL and current division yield the following currents immediately: $I_{2k\Omega} = 8\,\text{mA}\uparrow$, $I_{6k\Omega} = 2\,\text{mA}\downarrow$, and $I_{3k\Omega} = 4\,\text{mA}\downarrow$.

Circuits Containing Dependent Current Sources

The presence of a dependent source may destroy the symmetrical form of the nodal equations that define the circuit. Consider the circuit shown in Fig. 3.8, which contains a current-controlled current source. The KCL equations for the nonreference nodes are

Figure 3.8 Circuit with a dependent source.

$$\beta i_o + \frac{v_1}{R_1} + \frac{v_1 - v_2}{R_2} = 0$$

and

$$\frac{v_2 - v_1}{R_2} + i_o - i_A = 0$$

where $i_o = v_2/R_3$. Simplifying the equations, we obtain

$$(G_1 + G_2)v_1 - (G_2 - \beta G_3)v_2 = 0$$
$$-G_2 v_1 + (G_2 + G_3)v_2 = i_A$$

or in matrix form

$$\begin{bmatrix} (G_1 + G_2) & -(G_2 - \beta G_3) \\ -G_2 & (G_2 + G_3) \end{bmatrix} \begin{bmatrix} v_1 \\ v_2 \end{bmatrix} = \begin{bmatrix} 0 \\ i_A \end{bmatrix}$$

Note that the presence of the dependent source has destroyed the symmetrical nature of the node equation.

EXAMPLE 3.3

Let us determine the node voltages for the network in Fig. 3.8 given the following parameters:

$$\beta = 2 \qquad R_2 = 6\,\text{k}\Omega \qquad i_A = 2\,\text{mA}$$
$$R_1 = 12\,\text{k}\Omega \qquad R_3 = 3\,\text{k}\Omega$$

SOLUTION   Using these values with the equations for the network yields

$$\frac{1}{4k} V_1 + \frac{1}{2k} V_2 = 0$$

$$-\frac{1}{6k} V_1 + \frac{1}{2k} V_2 = 2 \times 10^{-3}$$

Solving these equations using any convenient method yields $V_1 = -24/5$ V and $V_2 = 24/5$ V. We can check these answers by determining the branch currents in the network and then using that information to test KCL at the nodes. For example, the current from top to bottom through R_3 is

$$I_o = \frac{V_2}{R_3} = \frac{12/5}{3k} = \frac{4}{5k} \text{ A}$$

Similarly, the current from right to left through R_2 is

$$I_2 = \frac{V_2 - V_1}{R_2} = \frac{12/5 - (-24/5)}{6k} = \frac{6}{5k} \text{ A}$$

All the results are shown in Fig. 3.9. Note that KCL is satisfied at every node.

Figure 3.9 Circuit used in Example 3.3.

Consider now the network in Fig. 3.10. Applying KCL at each of the nonreference nodes yields the equations

$$G_3 v_1 + G_1(v_1 - v_2) - i_A = 0$$
$$i_A + G_1(v_2 - v_1) + \alpha v_x + G_2(v_2 - v_3) = 0$$
$$G_2(v_3 - v_2) + G_4 v_3 - i_B = 0$$

where $v_x = v_2 - v_3$. Simplifying these equations, we obtain

Figure 3.10 Circuit containing a
voltage-controlled current source.

$$(G_1 + G_3)v_1 - G_1 v_2 = i_A$$
$$-G_1 v_1 + (G_1 + \alpha + G_2)v_2 - (\alpha + G_2)v_3 = -i_A$$
$$-G_2 v_2 + (G_2 + G_4)v_3 = i_B$$

or in matrix form

$$\begin{bmatrix} (G_1 + G_3) & -G_1 & 0 \\ -G_1 & (G_1 + \alpha + G_2) & -(\alpha + G_2) \\ 0 & -G_2 & (G_2 + G_4) \end{bmatrix} \begin{bmatrix} v_1 \\ v_2 \\ v_3 \end{bmatrix} = \begin{bmatrix} i_A \\ -i_A \\ i_B \end{bmatrix}$$

Once again, we find that the symmetry of the equation is destroyed by the dependent source.

EXTENSION EXERCISES

E3.3 Find the node voltages in the circuit in Fig. E3.3.

Figure E3.3

ANSWER: $V_1 = 16$ V, $V_2 = -8$ V.

E3.4 Find the voltages V_o in the network in Fig. E3.4.

Figure E3.4

ANSWER: $V_o = 4$ V.

Circuits Containing Independent Voltage Sources

As is our practice, in our discussion of this topic we will proceed from the simplest case to those that are more complicated. The simplest case is that in which an independent voltage source is connected to the reference node. The following example illustrates this case.

EXAMPLE 3.4

Consider the circuit shown in Fig. 3.11a. Let us determine all node voltages and branch currents.

Figure 3.11 Circuit used in Example 3.4.

SOLUTION   This network has three nonreference nodes with labeled node voltages V_1, V_2, and V_3. Based upon our previous discussions, we would assume that in order to find all the node voltages we would need to write a KCL equation at each of the nonreference nodes. The resulting three linearly independent simultaneous equations would produce the unknown node voltages. However, note that V_1 and V_3 are known quantities because an independent voltage source is connected directly between the nonreference node and each of these nodes. Therefore, $V_1 = 12$ V and $V_3 = -6$ V. Furthermore, note that the current through the 9-kΩ resistor is $[12 - (-6)]/9k = 2$ mA from left to right. We do not know V_2 or the current in the remaining resistors. However, since only one node voltage is unknown, a single-node equation will produce it. Applying KCL to this center node yields

$$\frac{V_2 - V_1}{12k} + \frac{V_2 - 0}{6k} + \frac{V_2 - V_3}{12k} = 0$$

or

$$\frac{V_2 - 12}{12k} + \frac{V_2}{6k} + \frac{V_2 - (-6)}{12k} = 0$$

from which we obtain

$$V_2 = \frac{3}{2}\,\text{V}$$

Anytime an independent voltage source is connected between the reference node and a nonreference node, the nonreference node voltage is known.

Once all the node voltages are known, Ohm's law can be used to find the branch currents shown in Fig. 3.11b. The diagram illustrates that KCL is satisfied at every node.

Note that the presence of the voltage sources in this example has simplified the analysis, since two of the three linear independent equations are $V_1 = 12$ V and $V_3 = -6$ V. We will find that as a general rule, anytime voltage sources are present between nodes, the node voltage equations that describe the network will be simpler.

EXTENSION EXERCISES

E3.5 Find the current I_o in the network in Fig. E3.5.

ANSWER: $I_o = \dfrac{3}{4}$ mA.

Figure E3.5

Next let us consider the case in which an independent voltage source is connected between two nonreference nodes. Once again, we will use an example to illustrate the approach.

EXAMPLE 3.5

We wish to find the currents in the two resistors in the circuit in Fig. 3.12a.

(a)

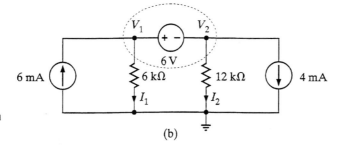

(b)

Figure 3.12 Circuits used in Example 3.5.

SOLUTION If we try to attack this problem in a brute force manner, we immediately encounter a problem. Thus far, branch currents were either known source values or could be expressed as the branch voltage divided by the branch resistance.

However, the branch current through the 6-V source is certainly not known and cannot be directly expressed using Ohm's law. We can, of course, give this current a name and write the KCL equations at the two nonreference nodes in terms of this current. However, this approach is no panacea because this technique will result in *two* linearly independent simultaneous equations in terms of *three* unknowns; that is, the two node voltages and the current in the voltage source.

To solve this dilemma, we recall that $N - 1$ linearly independent equations are required to determine the $N - 1$ nonreference node voltages in an N-node circuit. Since our network has three nodes, we need two linearly independent equations. Now note that if somehow one of the node voltages is known, we immediately know the other; that is, if V_1 is known, then $V_2 = V_1 - 6$. And if V_2 is known, then $V_1 = V_2 + 6$. Therefore, the difference in potential between the two nodes is *constrained* by the voltage source and, hence,

$$V_1 - V_2 = 6$$

This constraint equation is one of the two linearly independent equations needed to determine the node voltages.

Next consider the network in Fig. 3.12b, in which the 6-V source is completely enclosed within the dashed surface. The constraint equation governs this dashed portion of the network. The remaining equation is obtained by applying KCL to this dashed surface, which is commonly called a *supernode*. Recall that in Chapter 2 we demonstrated that KCL must hold for a surface, and this technique eliminates the problem of dealing with a current through a voltage source. KCL for the supernode is

$$-6 \times 10^{-3} + \frac{V_1}{6k} + \frac{V_2}{12k} + 4 \times 10^{-3} = 0$$

Solving these equations yields $V_1 = 10$ V and $V_2 = 4$ V and, hence, $I_1 = 5/3$ mA and $I_2 = 1/3$ mA. A quick check indicates that KCL is satisfied at every node. ❑

The supernode technique

- Used when a branch between two nonreference nodes contains a voltage source.
- First encircle the voltage source and the two connecting nodes to form the supernode.
- Write the equation that defines the voltage relationship between the two nonreference nodes as a result of the presence of the voltage source.
- Write the KCL equation for the supernode.
- If the voltage source is dependent, then the controlling equation for the dependent source is also needed.

EXAMPLE 3.6

Let us write the node equations for the circuit in Fig. 3.13.

Figure 3.13 Circuit used in Example 3.6.

SOLUTION The equation for the node labeled v_1 is

$$(v_1 - v_3)G_1 + (v_1 - v_2)G_2 - i_A = 0$$

The constraint equation for the supernode is

$$v_2 - v_3 = v_A$$

And the KCL equation for the supernode is

$$(v_2 - v_1)G_2 + v_2 G_3 + (v_3 - v_1)G_1 + v_3 G_4 = 0$$

The three equations will yield the node voltages.   ❏

EXAMPLE 3.7

Let us determine the current I_o in the network in Fig. 3.14a.

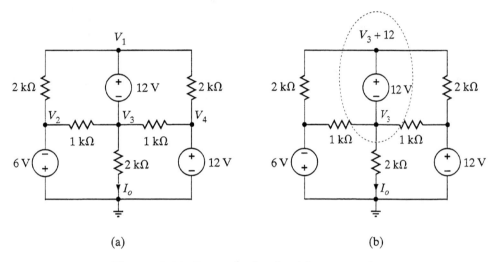

(a)   (b)

Figure 3.14 Example circuit with supernodes.

SOLUTION   Examining the network, we note that node voltages V_2 and V_4 are known and the node voltages V_1 and V_3 are constrained by the equation

$$V_1 - V_3 = 12$$

The network is redrawn in Fig. 3.14b.

Since we want to find the current I_o, V_1 (in the supernode containing V_1 and V_3) is written as $V_3 + 12$. The KCL equation at the supernode is then

$$\frac{V_3 + 12 - (-6)}{2k} + \frac{V_3 + 12 - 12}{2k} + \frac{V_3 - (-6)}{1k} + \frac{V_3 - 12}{1k} + \frac{V_3}{2k} = 0$$

Solving the equation for V_3 yields

$$V_3 = -\frac{6}{7} \, \text{V}$$

I_o can then be computed immediately as

$$I_o = \frac{-\dfrac{6}{7}}{2k} = -\frac{3}{7} \, \text{mA}$$ ❏

EXTENSION EXERCISES

E3.6 Use nodal analysis to find I_o in the network in Fig. E3.6.

Figure E3.6

ANSWER: $I_o = 3.8 \text{ mA}$.

Circuits Containing Dependent Voltage Sources

As the following examples will indicate, networks containing dependent (controlled) sources are treated in the same manner as described earlier.

EXAMPLE 3.8

We wish to find I_o in the network in Fig. 3.15.

Figure 3.15 Circuit used in Example 3.8.

SOLUTION As a general rule, when writing the circuit equations we first treat the dependent source as though it were an independent source and then write the controlling equation. Since the independent voltage source is connected between the nonreference node labeled V_1 and the reference, one of the two linearly independent equations is

$$V_1 = 3 \text{ V}$$

KCL applied to the second nonreference node yields

$$\frac{V_2 - 3}{3k} + \frac{V_2}{6k} = 2I_x$$

where the controlling equation is

$$I_x = \frac{V_2}{6k}$$

solving these equations yields $V_2 = 6\,\text{V}$ and, hence,

$$I_o = \frac{3-6}{3k} = -1\,\text{mA}$$ ❑

EXAMPLE 3.9

Let us find the current I_o in the network in Fig. 3.16.

Figure 3.16 Circuit used in Example 3.9.

SOLUTION   This circuit contains both an independent voltage source and a voltage-controlled voltage source. Note that $V_3 = 6\,\text{V}$, $V_2 = V_x$, and a supernode exists between the nodes labeled V_1 and V_2.

Applying KCL to the supernode, we obtain

$$\frac{V_1 - V_3}{6k} + \frac{V_1}{12k} + \frac{V_2}{6k} + \frac{V_2 - V_3}{12k} = 0$$

where the constraint equation for the supernode is

$$V_1 - V_2 = 2V_x$$

The final equation is

$$V_3 = 6$$

Solving these equations, we find that

$$V_1 = \frac{9}{2}\,\text{V}$$

and, hence,

$$I_o = \frac{V_1}{12k} = \frac{3}{8}\,\text{mA}$$ ❑

EXTENSION EXERCISES

E3.7   Find I_o in the circuit in Fig. E3.7.

Figure E3.7

ANSWER: $I_o = \dfrac{4}{3}$ mA.

Problem-Solving Strategies Nodal Analysis

* Select one node in the N-node circuit as the reference node. Assume that the node voltage is zero and measure all node voltages with respect to this node.

* If only independent current sources are present in the network, write the KCL equations at the $N - 1$ nonreference nodes. If dependent current sources are present, write the KCL equations as is done for networks with only independent current sources; then write the controlling equations for the dependent sources.

* If voltage sources are present in the network, they may be connected (1) between the reference node and a nonreference node or (2) between two nonreference nodes. In the former case, if the voltage source is an independent source, then the voltage at one of the nonreference nodes is known. If the source is dependent, it is treated as an independent source when writing the KCL equation, but an additional constraint equation is necessary, as described previously.

  In the latter case, if the source is independent, the voltage between the two nodes is constrained by the value of the voltage source, and an equation describing this constraint represents one of the $N - 1$ linearly independent equations required to determine the N-node voltages. The surface of the network described by the constraint equation (ie, the source and two connecting nodes) is called a supernode. One of the remaining $N - 1$ linearly independent equations is obtained by applying KCL at this supernode. If the voltage source is dependent, it is treated as an independent source when writing the KCL equations, but an additional constraint equation is necessary, as described previously.

3.2 LOOP ANALYSIS

In a nodal analysis the unknown parameters are the node voltages, and KCL is employed to determine them. In contrast to this approach, a loop analysis uses KVL to determine currents in the circuit. Once the currents are known, Ohm's law can be used to calculate the voltages. Recall that, in Chapter 2, we found that a single equation was sufficient to determine the current in a circuit containing a single loop. If the circuit contains N independent loops, we will show that N independent simultaneous equations will be required to describe the network. For now we will assume that the circuits are planar, which simply means that we can draw the circuit on a sheet of paper in a way such that no conductor crosses another conductor.

Our approach to loop analysis will mirror that used in nodal analysis (i.e., we will begin with simple cases and systematically proceed to those that are more difficult). Then at the end of this section we will outline a general strategy for employing loop analysis.

Circuits Containing Only Independent Voltage Sources

To begin our analysis, consider the circuit shown in Fig. 3.17. Let us also identify two loops, A-B-E-F-A and B-C-D-E-B. We now define a new set of current variables called *loop currents*, which can be used to find the physical currents in the circuit. Let us assume that current i_1 flows in the first loop and that current i_2 flows in the second loop. Then the branch current flowing from B to E through R_3 is $i_1 - i_2$. The directions of the currents have been assumed. As was the case in the nodal analysis, if the actual currents are not in the direction indicated, the values calculated will be negative.

Figure 3.17 A two-loop circuit.

Applying KVL to the first loop yields

$$+v_1 + v_3 + v_2 - v_{S1} = 0$$

KVL applied to loop 2 yields

$$+v_{S2} + v_4 + v_5 - v_3 = 0$$

where $v_1 = i_1 R_1$, $v_2 = i_1 R_2$, $v_3 = (i_1 - i_2)R_3$, $v_4 = i_2 R_4$, and $v_5 = i_2 R_5$.

The equations employ the passive sign convention.

DRILL

D3.2 Write the mesh equations for the following circuit.

ANSWER:
$$-v_{S1} + i_1 R_1 - v_{S2} + (i_1 - i_2)R_2 = 0$$
$$i_2 R_3 + (i_2 - i_1)R_2 + i_{2R_5} + i_2 R_4 + V_{S3} = 0$$

Substituting these values into the two KVL equations produces the two simultaneous equations required to determine the two loop currents; that is,

$$i_1(R_1 + R_2 + R_3) - i_2(R_3) = v_{S1}$$
$$-i_1(R_3) + i_2(R_3 + R_4 + R_5) = -v_{S2}$$

or in matrix form

$$\begin{bmatrix} R_1 + R_2 + R_3 & -R_3 \\ -R_3 & R_3 + R_4 + R_5 \end{bmatrix} \begin{bmatrix} i_1 \\ i_2 \end{bmatrix} = \begin{bmatrix} v_{S1} \\ -v_{S2} \end{bmatrix}$$

At this point it is important to define what is called a *mesh*. A mesh is a special kind of loop that does not contain any loops within it. Therefore, as we traverse the path of a mesh, we do not encircle any circuit elements. For example, the network in Fig. 3.17 contains two meshes defined by the paths *A-B-E-F-A* and *B-C-D-E-B*. The path *A-B-C-D-E-F-A* is a loop but it is not a mesh. Since the majority of our analysis in this section will involve writing KVL equations for meshes, we will refer to the currents as mesh currents and the analysis as a *mesh analysis*.

EXAMPLE 3.10

Consider the network in Fig. 3.18a. We wish to find the current I_o.

Figure 3.18 Circuits used in Example 3.10.

SOLUTION We will begin the analysis by writing mesh equations. Note that there are no + and − signs on the resistors. However, they are not needed, since we will apply Ohm's law to each resistive element as we write the KVL equations. The equation for the first mesh is

$$-12 + 6kI_1 + 6k(I_1 - I_2) = 0$$

The KVL equation for the second mesh is

$$6k(I_2 - I_1) + 3kI_2 + 3 = 0$$

where $I_o = I_1 - I_2$.

Solving the two simultaneous equations yields $I_1 = 5/4$ mA and $I_2 = 1/2$ mA. Therefore, $I_o = 3/4$ mA. All the voltages and currents in the network are shown in Fig. 3.18b. Recall from nodal analysis that once the node voltages were determined, we could check our analysis using KCL at the nodes. In this case we know the branch currents and can use KVL around any closed path to check our results. For example, applying KVL to the outer loop yields

$$-12 + \frac{15}{2} + \frac{3}{2} + 3 = 0$$

$$0 = 0$$

Since we want to calculate the current I_o, we could use loop analysis, as shown in Fig. 3.18c. Note that the loop current I_1 passes through the center leg of the network and, therefore, $I_1 = I_o$. The two loop equations in this case are

$$-12 + 6k(I_1 + I_2) + 6kI_1 = 0$$

and

$$-12 + 6k(I_1 + I_2) + 3kI_2 + 3 = 0$$

Solving these equations yields $I_1 = 3/4$ mA and $I_2 = 1/2$ mA. Since the current in the 12-V source is $I_1 + I_2 = 5/4$ mA, these results agree with the mesh analysis.

Finally, for purposes of comparison, let us find I_o using nodal analysis. The presence of the two voltage sources would indicate that this is a viable approach. Applying KCL at the top center node, we obtain

$$\frac{V_o - 12}{6k} + \frac{V_o}{6k} + \frac{V_o - 3}{3k} = 0$$

and hence,

$$V_o = \frac{9}{2} \text{ V}$$

and then

$$I_o = \frac{V_o}{6k} = \frac{3}{4} \text{ mA}$$

Note that in this case we had to solve only one equation instead of two. ❏

Once again we are compelled to note the symmetrical form of the mesh equations that describe the circuit in Fig. 3.17. Note that the **A** matrix for this circuit is symmetrical.

Since this symmetry is generally exhibited by networks containing resistors and independent voltage sources, we can learn to write the mesh equations by inspection. In the first equation the coefficient of i_1 is the sum of the resistances through which mesh current 1 flows, and the coefficient of i_2 is the negative of the sum of the resistances common to mesh current 1 and mesh current 2. The right-hand side of the equation is the algebraic sum of the voltage sources in mesh 1. The sign of the voltage source is positive if it aids the assumed direction of the current flow and negative if it opposes the assumed flow. The first equation is KVL for mesh 1. In the second equation, the coefficient of i_2 is the sum of all the resistances in mesh 2, the coefficient of i_1 is the negative of the sum of the resistances common to mesh 1 and mesh 2, and the right-hand side of the equation is the algebraic sum of the voltage sources in mesh 2. In general, if we assume all of the mesh currents to be in the same direction (clockwise or counterclockwise), then if KVL is applied to mesh j with mesh current i_j, the coefficient of i_j is the sum of the resistances in mesh j and the coefficients of the other mesh currents $(e.g., i_{j-1}, i_{j+1})$ are the negatives of the resistances common to these meshes and mesh j. The right-hand side of the equation is equal to the algebraic sum of the voltage sources in mesh j. These voltage sources have a positive sign if they aid the current flow i_j and a negative sign if they oppose it.

EXAMPLE 3.11

Let us write the mesh equations by inspection for the network in Fig. 3.19.

Figure 3.19 Circuit used in Example 3.11.

SOLUTION The three linearly independent simultaneous equations are

$$(4k + 6k)I_1 - (0)I_2 - (6k)I_3 = -6$$
$$-(0)I_1 + (9k + 3k)I_2 - (3k)I_3 = 6$$
$$-(6k)I_1 - (3k)I_2 + (3k + 6k + 12k)I_3 = 0$$

or in matrix form

$$\begin{bmatrix} 10k & 0 & -6k \\ 0 & 12k & -3k \\ -6k & -3k & 21k \end{bmatrix} \begin{bmatrix} I_1 \\ I_2 \\ I_3 \end{bmatrix} = \begin{bmatrix} -6 \\ 6 \\ 0 \end{bmatrix}$$

Note the symmetrical form of the equations. ❏

Circuits Containing Independent Current Sources

Just as the presence of a voltage source in a network simplified the nodal analysis, the presence of a current source simplifies a loop analysis. The following examples illustrate the point.

EXTENSION EXERCISES

E3.8 Use mesh equations to find V_o in the circuit in Fig. E3.8.

ANSWER: $V_o = \dfrac{33}{5}$ V.

Figure E3.8

EXAMPLE 3.12

Let us find both V_o and V_1 in the circuit in Fig. 3.20.

Figure 3.20 Circuit used in Example 3.12.

SOLUTION   Although it appears that there are two unknown mesh currents, the current I_1 goes directly through the current source and, therefore, I_1 is constrained to be 2 mA. Hence, only the current I_2 is unknown. KVL for the right-most mesh is

$$2k(I_2 - I_1) - 2 + 6kI_2 = 0$$

And, of course,

$$I_1 = 2 \times 10^{-3}$$

These equations yield

$$I_2 = \frac{3}{4} \, \text{mA}$$

And hence,

$$V_o = 6kI_2 = \frac{9}{2} \, \text{V}$$

To obtain V_1 we apply KVL around any closed path. If we use the outer loop, the KVL equation is

$$-V_1 + 4kI_1 - 2 + 6kI_2 = 0$$

And therefore,

$$V_1 = \frac{21}{2} \text{ V}$$

Note that since the current I_1 is known, the 4-kΩ resistor did not enter the equation in finding V_o. However, it appears in every loop containing the current source and, thus, is used in finding V_1. ❏

EXAMPLE 3.13

We wish to find V_o in the network in Fig. 3.21.

Figure 3.21 Circuit used in Example 3.13.

SOLUTION Since the currents I_1 and I_2 pass directly through a current source, two of the three required equations are

$$I_1 = 4 \times 10^{-3}$$
$$I_2 = -2 \times 10^{-3}$$

The third equation is KVL for the mesh containing the voltage source; that is,

$$4k(I_3 - I_2) + 2k(I_3 - I_1) + 6kI_3 - 3 = 0$$

These equations yield

$$I_3 = \frac{1}{4} \text{ mA}$$

and hence,

$$V_o = 6kI_3 - 3 = \frac{-3}{2} \text{ V}$$ ❏

What we have demonstrated in the previous examples is the general approach for dealing with independent current sources when writing KVL equations; that is, use one loop through each current source. The number of "window panes" in the network tells us how many equations we need. Additional KVL equations are written to cover the remaining circuit elements in the network. The following example illustrates this approach.

EXAMPLE 3.14

Let us find I_o in the network in Fig. 3.22a.

In this case the 4-mA current source is located on the boundary between two meshes. Thus we will demonstrate two techniques for dealing with this type of situation. One is a special loop technique and the other is known as the supermesh approach.

Figure 3.22 Circuits used in Example 3.14.

SOLUTION   First, we select two loop currents I_1 and I_2 such that I_1 passes directly through the 2-mA source, and I_2 passes directly through the 4-mA source, as shown in Fig. 3.22b. Therefore, two of our three linearly independent equations are

$$I_1 = 2 \times 10^{-3}$$
$$I_2 = 4 \times 10^{-3}$$

The remaining loop current I_3 must pass through the circuit elements not covered by the two previous equations and cannot, of course, pass through the current sources. The path for this remaining loop current can be obtained by open-circuiting the current sources, as shown in Fig. 3.22c. When all current are labeled on the original circuit, the KVL equation for this last loop, as shown in Fig. 3.22d, is

$$-6 + 1kI_3 + 2k(I_2 + I_3) + 2k(I_3 + I_2 - I_1) + 1k(I_3 - I_1) = 0$$

Solving the equations yields

$$I_3 = \frac{-2}{3}\,mA$$

And therefore,

$$I_o = I_1 - I_2 - I_3 = \frac{-4}{3}\,mA$$

Next consider the supermesh technique. In this case the three mesh currents are specified as shown in Fig. 3.22e, and since the voltage across the 4-mA current source is unknown, it is assumed to be V_x. The mesh currents constrained by the current sources are

$$I_1 = 2 \times 10^{-3}$$
$$I_2 - I_3 = 4 \times 10^{-3}$$

The KVL equations for meshes 2 and 3, respectively, are

$$2kI_2 + 2k(I_2 - I_1) - V_x = 0$$
$$-6 + 1kI_3 + V_x + 1k(I_3 - I_1) = 0$$

Adding the last two equations yields

$$-6 + 1kI_3 + 2kI_2 + 2k(I_2 - I_1) + 1k(I_3 - I_1) = 0$$

Note that the unknown voltage V_x has been eliminated. The two constraint equations, together with this latter equation, yield the desired result.

The purpose of the supermesh approach is to avoid introducing the unknown voltage V_x. The supermesh is created by mentally removing the 4-mA current source, as shown in Fig. 3.22f. Then writing the KVL equation around the dotted path, which defines the supermesh, using the original mesh currents as shown in Fig. 3.22e, yields

$$-6 + 1kI_3 + 2kI_2 + 2k(I_2 - I_1) + 1k(I_3 - I_1) = 0$$

Note that this supermesh equation is the same as that obtained earlier by introducing the voltage V_x.   ❑

EXTENSION EXERCISES

E3.9   Find V_o in the network in Fig. E3.9.

Figure E3.9

ANSWER: $V_o = \dfrac{33}{5}\,\text{V}.$

E3.10   Find V_o in the network in Fig. E3.10.

Figure E3.10

ANSWER: $V_o = 8\,\text{V}.$

Circuits Containing Dependent Sources

We deal with circuits containing dependent sources just as we have in the past. First, we treat the dependent source as though it were an independent source when writing the KVL equations. Then we write the controlling equation for the dependent source. The following examples illustrate the point.

EXAMPLE 3.15

Let us find V_o in the circuit in Fig. 3.23, which contains a voltage-controlled voltage source.

Figure 3.23 Circuit used in Example 3.15.

SOLUTION   The equations for the loop currents shown in the figure are

$$-2V_x + 2\text{k}(I_1 + I_2) + 4\text{k}I_1 = 0$$
$$-2V_x + 2\text{k}(I_1 + I_2) - 3 + 6\text{k}I_2 = 0$$

where

$$V_x = 4\text{k}I_1$$

Solving these equations yields

$$I_2 = \frac{3}{2}\,\text{mA}$$

and therefore,

$$V_o = 6kI_2 = 9\,\text{V}$$

For comparison, we will also solve the problem using nodal analysis. The presence of the voltage sources indicates that this method could be simpler. Treating the 3-V source and its connecting nodes as a supernode and writing the KCL equation for this supernode yields

$$\frac{V_x - 2V_x}{2k} + \frac{V_x}{4k} + \frac{V_x + 3}{6k} = 0$$

where

$$V_o = V_x + 3$$

These equations also yield $V_o = 9\,\text{V}$.   ❏

EXAMPLE 3.16

Let us find V_o in the circuit in Fig. 3.24, which contains a voltage-controlled current source.

Figure 3.24 Circuit used in Example 3.16.

SOLUTION   The currents I_1 and I_2 are drawn through the current sources. Therefore, two of the equations needed are

$$I_1 = \frac{V_x}{2000}$$

$$I_2 = 2 \times 10^{-3}$$

The KVL equation for the third mesh is

$$-3 + 2k(I_3 - I_1) + 6kI_3 = 0$$

where

$$V_x = 4k(I_1 - I_2)$$

Solving these equations yields

$$I_3 = \frac{11}{8}\,\text{mA}$$

and hence,

$$V_o = \frac{33}{4} \text{ V}$$ ❏

As a final point, it is very important to examine the circuit carefully before selecting an analysis approach. One method could be much simpler than another, and a little time invested up front may save a lot of time in the long run.

EXTENSION EXERCISES

E3.11   Use mesh analysis to find V_o in the circuit in Fig. E3.11.

Figure E3.11

ANSWER: $V_o = 12$ V.

Problem-Solving Strategies for Loop Analysis

- One loop current is assigned to each independent loop in a circuit that contains N independent loops.

- If only independent voltage sources are present in the network, write the N linearly independent KVL equations, one for each loop. If dependent voltage sources are present, write the KVL equation as is done for circuits with only independent voltage sources; then write the controlling equations for the dependent sources.

- If current sources are present in the network, either of two techniques can be used. In the first case, one loop current is selected to pass through one of the current sources. This is done for each current source in the network. The remaining loop currents (N − the number of current sources) are determined by open-circuiting the current sources in the network and using this modified network to select them. Once all these currents are defined in the original network, the N loop equations can be written. The second approach is similar to the first with the exception that if two mesh currents pass through a particular current source, a supermesh is formed around this source. The two required equations for the meshes containing this source are the constraint equations for the two mesh currents that pass through the source and the supermesh equation. As indicated earlier, if dependent current sources are present, the controlling equations for these sources are also necessary.

EXTENSION EXERCISES

E3.12 Use loop analysis to solve the network in Example 3.4 and compare the time and effort involved in the two solution techniques.

E3.13 Use nodal analysis to solve the circuit in Example 3.13 and compare the time and effort involved in the two solution strategies.

Finally, it is important to note that the techniques for nodal and loop analysis are firmly rooted in topology, and the manner in which the network equations are derived via network topology is illustrated in Appendix A.

3.3 CIRCUITS WITH OPERATIONAL AMPLIFIERS

The operational amplifier, or op-amp as it is commonly known, is an extremely important circuit. It is a versatile interconnection of devices that vastly expands our capabilities in circuit design. It is employed in everything from engine control systems to microwave ovens. The op-amp was first introduced in the 1940s for use in analog computers and now finds wide application in circuit design as a result of the advances made in integrated circuit technology. Although we will model the op-amp as a fairly simple device, its actual construction involves the use of numerous components, including resistors, capacitors, and transistors. By adding resistors and capacitors to the op-amp terminals, we can use this simple op-amp model to design networks to perform many functions, such as addition, subtraction, differentiation, and integration, as well as voltage scaling and current-to-voltage conversion. These "building blocks" can then be employed to satisfy more complex applications.

The purpose of our model is to describe the terminal characteristics (the voltage–current relationships) of the op-amp. Therefore, let us begin by considering the terminals of the commercially available LM101 op-amp. Figure 3.25a shows an outline of the LM101 DIP (dual in-line package), which has eight pins, or terminals. Pins 1, 5, and 8 are used for fine adjustments of the device's characteristics; however, a discussion of them at this point would only serve to complicate the presentation unnecessarily. Pins 4 and 7 are voltage terminals for dc power supply (typically 10 to 15 V); and, in addition, they provide a return path to ground for ac current. The op-amp has two inputs—pins 2 and 3. A positive voltage on pin 3 produces a positive output. Accordingly, pin 3 is termed the *noninverting input*. A positive voltage at pin 2 yields a negative output; hence, it is called the *inverting input*. Based on the preceding description, the op-amp, unlike the elements we have introduced thus far, is essentially a five-terminal device. The circuit symbol for the op-amp with its five terminal connections is shown in Fig. 3.25b.

Functionally, the op-amp operates like the equivalent circuit shown in Fig. 3.25c. This equivalent circuit implies a unilateral device; that is, the output is determined by the difference in input voltages, but the input is unaffected by voltages at the output. The resistance R_i is very large, the gain A is very high, and the resistance R_o is very low; typical values for these parameters are 10^5 to 10^{12} Ω, 10^5 to 10^7 V/V, and 1 to 50 Ω, respectively.

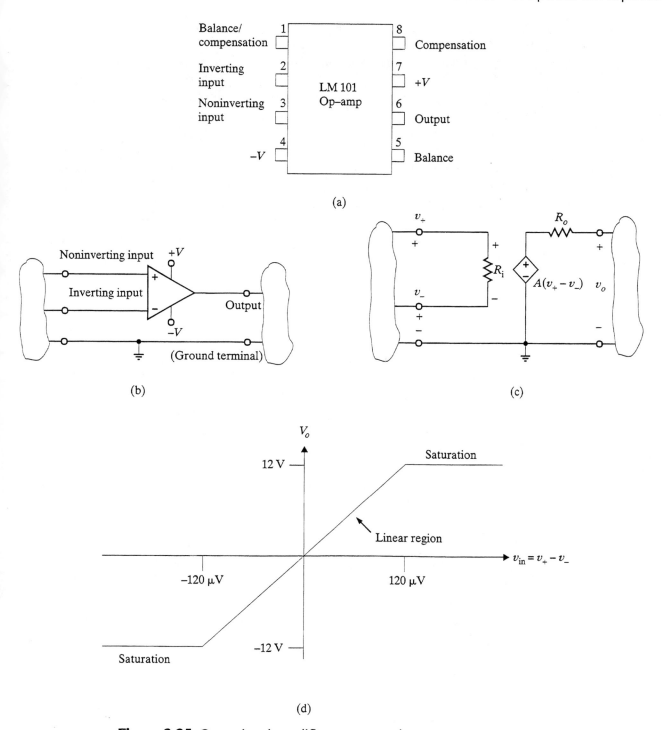

Figure 3.25 Operational amplifier representations.

The dc transfer curve (a plot of the output versus the input) for an op-amp with $A = 10^5$, $V_+ = 12\,\text{V}$ and $V_- = -12\,\text{V}$ is shown in Fig. 3.25d. The input voltage, V_{in}, is the noninverting input minus the inverting input voltages. Note first that the output voltage cannot exceed the power supply values. This is due to circuitry inside the op-amp.

Once the output reaches either power supply value, further increases in V_{in} have no effect on the output and the op-amp is said to be saturated. The portion of the curve of interest to us is the nonsaturated linear region. Here the ratio of output to input is the gain, $A = 10^5$.

The values listed previously for R_i, R_o, and A suggest an even simpler model known as the *ideal op-amp model*. R_i is very large and, therefore, can be assumed to be an open circuit. R_o is very low with respect to the recommended output load connection and, therefore, can be assumed to be zero. Finally, the gain A is a large amplification factor and can also be assumed to be infinite. At this point in our education, however, the infinite gain causes some concern. Does this mean that if the input is 1 μV the output is unlimited? Our common sense leads us to question such a condition. The reality of the situation is that the output voltages will always be between the dc power supply voltages, as shown in Fig. 3.25d. Therefore, if the op-amp is to remain in the linear region of operation, the input voltage (the difference between the noninverting and inverting input voltages) must remain within the limits.

$$\frac{V_-}{A} \leq V_{in} \leq \frac{V_+}{A}$$

For very large values of A, V_{in} approaches zero. Thus, an infinite gain yields a finite output voltage from a zero-volt input. We call this condition at the input terminals a *virtual short*.

Another important aspect of the op-amp configuration shown in Fig. 3.25c is that if the gain is assumed to be infinite, the output voltage is impossible to control. For the myriad of applications in which we need amplifiers, control of the output voltage is vital. The manner in which we accomplish this control is through feedback; that is, a portion of the output signal is fed back to the input. Later, we will show that through this mechanism the performance of the entire op-amp circuit can be controlled by external elements such as resistors. To accomplish this, the op-amp is connected to external components with a signal path from the output to the inverting input. With the op-amp connected in this manner, we will see that the high gain serves to maintain the input–output voltage relationships despite changes in load characteristics. As long as the amplifier gain A is very large, the circuit's functionality will be set by the externally connected components.

Under the assumption just stated, the ideal model for the op-amp is reduced to that shown in Fig. 3.26. The important characteristics of the model are as follows: (1) Since R_i is extremely large, the input currents to the op-amp are approximately zero (i.e., $i_+ \approx i_- \approx 0$); and (2) if the output voltage is to remain bounded, then as the gain becomes very large and approaches infinity, the voltage across the input terminals must simultaneously become infinitesimally small so that as $A \to \infty$, $v_+ - v_- \to 0$ (i.e., $v_+ - v_- = 0$ or $v_+ = v_-$). The difference between these input voltages is often called the *error signal* for the op-amp (i.e., $v_+ - v_- = v_e$).

The ground terminal $\perp$ shown on the op-amp is necessary for signal current return, and it guarantees that Kirchhoff's current law is satisfied at both the op-amp and the ground node in the circuit.

In summary, then, our ideal model for the op-amp is simply stated by the following conditions:

$$i_+ = i_- = 0 \qquad \qquad \textbf{3.8}$$
$$v_+ = v_-$$

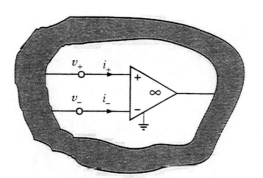

Figure 3.26 Ideal model for an operational amplifier. Model parameters: $i_+ = i_- = 0$, $v_+ = v_-$.

These simple conditions are extremely important because they form the basis of our analysis of op-amp circuits.

The following examples will serve to illustrate the validity of our assumptions when analyzing op-amp circuits.

EXAMPLE 3.17

Let us determine the gain of the basic inverting op-amp configuration shown in Fig. 3.27a using both the nonideal and ideal op-amp models.

SOLUTION   Our model for the op-amp is shown generically in Fig. 3.27b and specifically in terms of the parameters R_i, A, and R_o in Fig. 3.27c. If the model is inserted in the network in Fig. 3.27a, we obtain the circuit shown in Fig. 3.27d, which can be redrawn as shown in Fig. 3.27e.

The node equations for the network are

$$\frac{v_1 - v_S}{R_1} + \frac{v_1}{R_i} + \frac{v_1 - v_o}{R_2} = 0$$

$$\frac{v_o - v_1}{R_2} + \frac{v_o - Av_e}{R_o} = 0$$

where $v_e = -v_1$. The equations can be written in matrix form as

$$\begin{bmatrix} \dfrac{1}{R_1} + \dfrac{1}{R_i} + \dfrac{1}{R_2} & -\left(\dfrac{1}{R_2}\right) \\[2mm] -\left(\dfrac{1}{R_2} - \dfrac{A}{R_o}\right) & \dfrac{1}{R_2} + \dfrac{1}{R_o} \end{bmatrix} \begin{bmatrix} v_1 \\ v_0 \end{bmatrix} = \begin{bmatrix} \dfrac{v_S}{R_1} \\ 0 \end{bmatrix}$$

Solving for the node voltages, we obtain

$$\begin{bmatrix} v_1 \\ v_0 \end{bmatrix} = \frac{1}{\Delta} \begin{bmatrix} \dfrac{1}{R_2} + \dfrac{1}{R_o} & \dfrac{1}{R_2} \\[2mm] \dfrac{1}{R_2} - \dfrac{A}{R_o} & \dfrac{1}{R_1} + \dfrac{1}{R_i} + \dfrac{1}{R_o} \end{bmatrix} \begin{bmatrix} \dfrac{v_S}{R_1} \\ 0 \end{bmatrix}$$

where

$$\Delta = \left(\frac{1}{R_1} + \frac{1}{R_i} + \frac{1}{R_2}\right)\left(\frac{1}{R_2} + \frac{1}{R_o}\right) - \left(\frac{1}{R_2}\right)\left(\frac{1}{R_2} - \frac{A}{R_o}\right)$$

Figure 3.27 Op-amp circuit.

Hence,

$$v_o = \frac{\left(\dfrac{1}{R_2} - \dfrac{A}{R_o}\right)\left(\dfrac{v_S}{R_1}\right)}{\left(\dfrac{1}{R_1} + \dfrac{1}{R_i} + \dfrac{1}{R_2}\right)\left(\dfrac{1}{R_2} + \dfrac{1}{R_o}\right) - \left(\dfrac{1}{R_2}\right)\left(\dfrac{1}{R_2} - \dfrac{A}{R_o}\right)}$$

which can be written as

$$\frac{v_o}{v_S} = \frac{-(R_2/R_1)}{1 - \left[\left(\frac{1}{R_1} + \frac{1}{R_i} + \frac{1}{R_2}\right)\left(\frac{1}{R_2} + \frac{1}{R_o}\right)\Big/\left(\frac{1}{R_2}\right)\left(\frac{1}{R_2} - \frac{A}{R_o}\right)\right]}$$

If we now employ typical values for the circuit parameters (e.g., $A = 10^5, R_i = 10^8$ Ω, $R_o = 10$ Ω, $R_1 = 1$ kΩ, and $R_2 = 5$ kΩ) the voltage gain of the network is

$$\frac{v_o}{v_S} = -4.9996994 \approx -5.000$$

However, the ideal op-amp has infinite gain and, therefore, if we take the limit of the gain equation as $A \rightarrow \infty$, we obtain

$$\lim_{A \rightarrow \infty}\left(\frac{v_o}{v_S}\right) = -\frac{R_2}{R_1} = -5.000$$

Note that the ideal op-amp yielded a result accurate to within four significant digits of that obtained from an exact solution of a typical op-amp model. These results are easily repeated for the vast array of useful op-amp circuits.

We now analyze the network in Fig. 3.27a using the ideal op-amp model. In this model

$$i_+ = i_- = 0$$

$$v_+ = v_-$$

As shown in Fig. 3.27a, $v_+ = 0$ and, therefore, $v_- = 0$. If we now write a node equation at the negative terminal of the op-amp, we obtain

$$\frac{v_S - 0}{R_1} + \frac{v_o - 0}{R_2} = 0$$

or

$$\frac{v_o}{v_S} = -\frac{R_2}{R_1}$$

and we have immediately obtained the results derived previously.

Notice that the gain is a simple resistor ratio. This fact makes the amplifier very versatile in that we can control the gain accurately and alter its value by changing only one resistor. Also, the gain is essentially independent of op-amp parameters. Since the precise values of A, R_i, and R_o are sensitive to such factors as temperature, radiation, and age, their elimination results in a gain that is stable regardless of the immediate environment. Since it is much easier to employ the ideal op-amp model rather than the nonideal model, unless otherwise stated we will use the ideal op-amp assumptions to analyze circuits that contain operational amplifiers.   ❑

EXAMPLE 3.18

Consider the op-amp circuit shown in Fig. 3.28. Let us determine an expression for the output voltage.

Figure 3.28 Differential amplifier operational amplifier circuit.

SOLUTION The node equation at the inverting terminal is

$$\frac{v_1 - v_-}{R_1} + \frac{v_o - v_-}{R_2} = i_-$$

At the noninverting terminal KCL yields

$$\frac{v_2 - v_+}{R_3} = \frac{v_+}{R_4} + i_+$$

However, $i_+ = i_- = 0$ and $v_+ = v_-$. Substituting these values into the two preceding equations yields

$$\frac{v_1 - v_-}{R_1} + \frac{v_o - v_-}{R_2} = 0$$

and

$$\frac{v_2 - v_-}{R_3} = \frac{v_-}{R_4}$$

Solving these two equations for v_o results in the expression

$$v_o = \frac{R_2}{R_1}\left(1 + \frac{R_1}{R_2}\right)\frac{R_4}{R_3 + R_4}v_2 - \frac{R_2}{R_1}v_1$$

Note that if $R_4 = R_2$ and $R_3 = R_1$, the expression reduces to

$$v_o = \frac{R_2}{R_1}(v_2 - v_1)$$

Therefore, this op-amp can be employed to subtract two input voltages. ❏

EXAMPLE 3.19

The circuit shown in Fig. 3.29a is a precision differential voltage-gain device. It is used to provide a single-ended input for an analog-to-digital converter. We wish to derive an expression for the output of the circuit in terms of the two inputs.

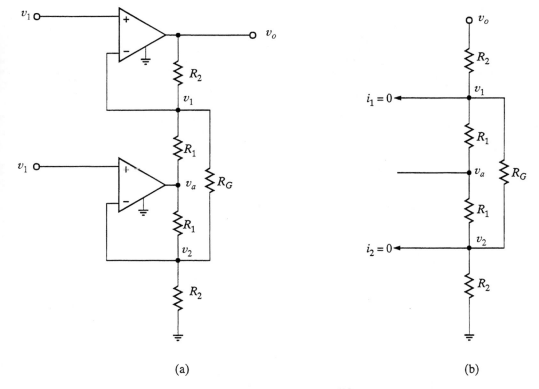

(a)   (b)

Figure 3.29 Instrumentation amplifier circuit.

SOLUTION   To accomplish this, we draw the equivalent circuit shown in Fig. 3.29b. Recall that the voltage across the input terminals of the op-amp is approximately zero and the currents into the op-amp input terminals are approximately zero. Note that we can write node equations for node voltages v_1 and v_2 in terms of v_o and v_a. Since we are interested in an expression for v_o in terms of the voltages v_1 and v_2, we simply eliminate the v_a terms from the two node equations. The node equations are

$$\frac{v_1 - v_o}{R_2} + \frac{v_1 - v_a}{R_1} + \frac{v_1 - v_2}{R_G} = 0$$

$$\frac{v_2 - v_a}{R_1} + \frac{v_2 - v_1}{R_G} + \frac{v_2}{R_2} = 0$$

Combining the two equations to eliminate v_a, and then writing v_o in terms of v_1 and v_2, yields

$$v_o = (v_1 - v_2)\left(1 + \frac{R_2}{R_1} + \frac{2R_2}{R_G}\right) \qquad \square$$

EXAMPLE 3.20

We wish to calculate the input resistance of the op-amp circuit shown in Fig. 3.30 without any assumptions about the sizes of the parameters A, R_i, and R_o.

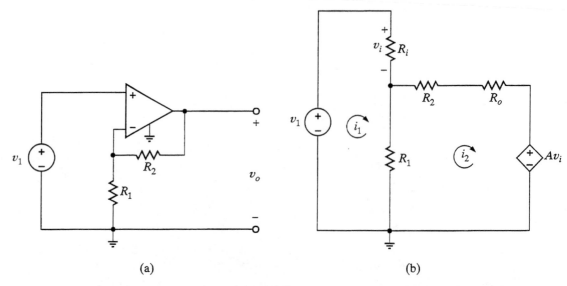

Figure 3.30 Circuit and model for a noninverting operational amplifier.

SOLUTION The input resistance is the resistance seen by the input voltage source v_1. Using the op-amp model shown in Fig. 3.25c, we obtain the circuit shown in Fig. 3.30b. The mesh equations for this circuit are

$$v_1 = i_1(R_i + R_1) - i_2 R_1$$
$$-Av_i = -i_1 R_1 + i_2(R_1 + R_2 + R_o)$$
$$v_i = R_i i_1$$

Solving these equations for i_1 yields

$$i_1 = \frac{v_1(R_1 + R_2 + R_o)}{(R_i + R_1)(R_1 + R_2 + R_o) + R_1(AR_i - R_1)}$$

The input resistance is

$$R_{in} = \frac{v_1}{i_1} = \frac{(R_i + R_1)(R_1 + R_2 + R_o) + R_1(AR_i - R_1)}{R_1 + R_2 + R_o}$$
$$= R_i + \frac{R_1(R_2 + R_o + AR_i)}{R_1 + R_2 + R_o}$$

Since A and R_i are normally very large and R_o is very small, the expression for R_{in} can be approximated by the equation

$$R_{in} \simeq \frac{AR_i}{1 + R_2/R_1} \qquad \square$$

EXTENSION EXERCISES

E3.14 Find I_o in the network in Fig. E3.14.

ANSWER: $I_o = 8.4$ mA.

Figure E3.14

E3.15 Determine the gain of the op-amp circuit in Fig. E3.15.

Figure E3.15

ANSWER: $\dfrac{V_o}{V_S} = 1 + \dfrac{R_2}{R_1}$.

E3.16 Determine both the gain and the output voltage of the op-amp configuration shown in Fig. E3.16.

Figure E3.16

ANSWER: $V_o = 0.101$ V, gain $= 101$.

An important special case of the noninverting amplifier is the configuration shown in Fig. 3.31. Note carefully that this circuit is equivalent to that shown in Fig. E3.15 with $R_1 = \infty$ (i.e., open circuited) and $R_2 = 0$ (i.e., short circuited). Under these conditions we note that the gain of the circuit is $1 + R_2/R_1 = 1$ (i.e., $v_o = v_S$). Since v_o follows v_S, the circuit is called a *voltage follower*.

Figure 3.31 Voltage-follower operational amplifier configuration.

An obvious question at this point is this: If $v_o = v_S$, why not just connect v_S to v_o via two parallel connection wires; why do we need to place an op-amp between them? The answer to this question is fundamental and provides us with some insight that will aid us in circuit analysis and design.

Consider the circuit shown in Fig. 3.32a. In this case v_o is not equal to v_S because of the voltage drop across R_S:

$$v_o = v_S - iR_S$$

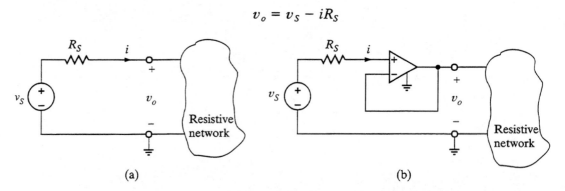

(a)   (b)

Figure 3.32 Illustration of the isolation capability of a voltage follower.

However, in Fig. 3.32b, the input current to the op-amp is zero and, therefore, v_S appears at the op-amp input. Since the gain of the op-amp configuration is 1, $v_o = v_S$. In Fig. 3.32a the resistive network's interaction with the source caused the voltage v_o to be less than v_S. In other words, the resistive network loads the source voltage. However, in Fig. 3.32b the op-amp isolates the source from the resistive network, and therefore the voltage follower is referred to as a *buffer amplifier* because it can be used to isolate one circuit from another. The energy supplied to the resistive network in the first case must come from the source v_S, whereas in the second case it comes from the power supplies that supply the amplifier, and little or no energy is drawn from v_S.

EXAMPLE 3.21

Often we present amplifier gain in graphical format. Plots of the output signal versus the input are called *transfer plots* or *transfer curves*. Let us find an expression for v_o in Fig. 3.33 and plot the transfer curve.

Figure 3.33 Circuit used in Example 3.21.

SOLUTION   Note that we have deviated from our normal model for the op-amp and specifically shown the dc supply voltages in the circuit. We have made an exception in this case because of their relevance to the transfer curves.

The nodal equation at the inverting input is

$$\frac{v_1 - 0.5}{1k} + \frac{v_1 - v_o}{4k} = 0$$

The output voltage is then given by the equation

$$v_o = 5v_1 - 2$$

The transfer curve for this circuit, shown in Fig. 3.34, contains several key features. First, the slope of the line,

$$\frac{dv_o}{dv_1} = 5 \text{ V/V}$$

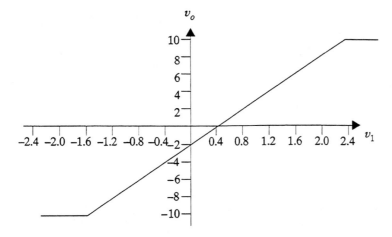

Figure 3.34 Transfer curve for an op-amp configuration.

is the gain of the circuit. Second, the output voltage is shifted by the 0.5-V source. Since v_1 is 0.4 V when v_o is 0 V, we say that the amplifier is *offset* by +0.40 V. Third, the output voltage cannot exceed the dc power supply voltage value. The output voltage obeys the transfer equation until v_o reaches either +10 V or −10 V. Beyond that point, v_o saturates at the power supply voltages. Since the ideal op-amp model does not include any information about how large the output voltage can be, it is left to the circuit designers to be conscious of the limitations of the model we are using. This is a common theme in the modeling of physical devices with mathematical expressions. Models tend to be valid only under certain conditions. It is paramount that the engineer understand the model's limitations and its implications. ❑

3.4 APPLICATIONS

At this point, we have a new element, the op-amp, which we can effectively employ in both applications and circuit design. This device is an extremely useful element that vastly expands our capability in these areas. Because of its ubiquitous nature, the addition of the op-amp to our repertoire of circuit elements permits us to deal with a wide spectrum of practical circuits. Thus, we will not only employ it here, but we will use it throughout this text.

EXAMPLE 3.22

In a light meter, a sensor produces a current proportional to the intensity of the incident radiation. We wish to obtain a voltage proportional to the light's intensity using the circuit in Fig. 3.35. Thus, we select a value of R that will produce an output voltage of 1 V for each 10μA of sensor current. Assume the sensor has zero resistance.

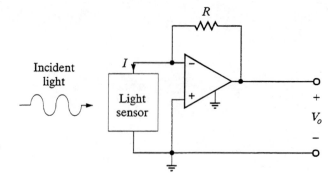

Figure 3.35 Light intensity to voltage converter.

SOLUTION Applying KCL at the op-amp input,

$$I = V_o/R$$

Since V_o/I is 10^5,

$$R = 100 \text{ k}\Omega$$

❑

EXAMPLE 3.23

The circuit in Fig. 3.36 is an electronic ammeter. It operates as follows: The unknown current, I through R_I produces a voltage, V_I. V_I is amplified by the op-amp to produce a voltage, V_o, which is proportional to I. The output voltage is measured with a simple voltmeter. We want to find the value of R_2 such that 10 V appears at V_o for each milliamp of unknown current.

SOLUTION   Since the current into the op-amp $+$ terminal is zero, the relationship between V_I and I is

$$V_I = IR_I$$

The relationship between the input and output voltages is

$$V_o = V_I\left(1 + \frac{R_2}{R_1}\right)$$

or, solving the equation for V_o/I, we obtain

$$\frac{V_o}{I} = R_I\left(1 + \frac{R_2}{R_1}\right)$$

Using the required ratio V_o/I of 10^4 and resistor values from Fig. 3.36, we can find that

$$R_2 = 9\ k\Omega$$

Figure 3.36 Electronic ammeter.   ❏

3.5 CIRCUIT DESIGN

EXAMPLE 3.24

A typical stereo system is shown in block form in Fig. 3.37a. The phonograph output signal is only about 2 µV. The standard input voltage for stereo power amplifiers is about 2 mV. Therefore, the phonograph signal must be amplified by a factor of 1000 before it reaches the poweramp. To accomplish this, a special-purpose amp, called the phono preamp, is used. Stereo manufacturers place them within the pre-amplifier cabinet, as shown in Fig. 3.37a.

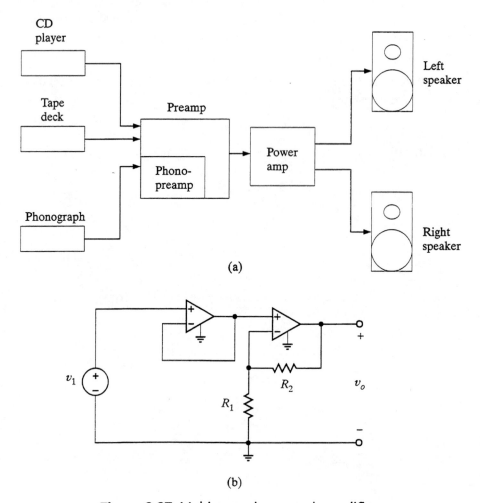

Figure 3.37 Multistage phonograph amplifier.

Let us design a phono preamp using an ideal op-amp that has an input resistance of at least 1 MΩ and a voltage gain of 1000. All resistors must be less than 10 MΩ in order to limit noise.

SOLUTION One possible network is shown in Fig. 3.37b. The input resistance requirement can be easily met with a voltage follower as the first stage of the amplifier. The second stage, or gain stage, can be a noninverting op-amp configuration. We will show in later chapters that the overall voltage gain is the product of the gains of the two stages,

$$A_V = A_{V1}A_{V2} = (1)(1 + (R_2/R_1))$$

To achieve a gain of 1000, we select $R_1 = 1$ kΩ and $R_2 = 999$ kΩ. ❑

EXAMPLE 3.25

A thermocouple is a two-wire element that produces a small voltage proportional to temperature. A particular thermocouple generates a voltage of 6 mV for every 1°C rise in temperature. This signal is to be used to activate a voltage-controlled switch that requires an input signal of 5 V to operate. Fig. 3.38a shows a block diagram of the system. Let us design an op-amp circuit that will trigger the switch at the boiling point of water (100°C) if the thermocouple generates 0 V at 0°C.

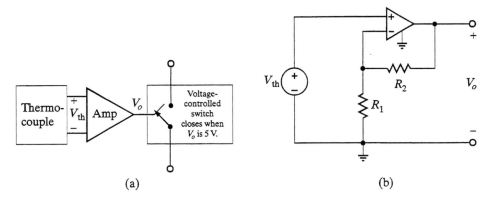

(a)   (b)

Figure 3.38 Amplifier for an electronic thermometer switch driver.

SOLUTION   The single-stage noninverting op-amp configuration shown in Fig. 3.38b should satisfy the requirement. At 100°C, the thermocouple voltage, V_{th} is

$$V_{th} = (100°)(6 \text{ mV}/°) = 600 \text{ mV}$$

This 600-mV signal must be amplified to 5 V to trigger the switch.
Therefore, the op-amp gain is

$$A_V = \frac{5}{0.6} = 8.33$$

Selecting $R_1 = 3 \text{ k}\Omega$ and $R_2 = 22 \text{ k}\Omega$ in the op-amp circuit in Fig. 3.38b satisfies the gain requirement.   ❏

EXAMPLE 3.26

Nerve fibers carry small electrical impulses from the brain to the muscles. These impulses cause muscle contraction and body movement. Physicians can monitor voltages in a muscle as a diagnostic process to look for proper brain–nerve–muscle interaction. A simplified version of the medical instrument used for this purpose is shown in Fig. 3.39. Conductive needles are inserted into a muscle and the voltage difference is measured. The signals on these needles are modeled in Fig. 3.39 by the voltage sources V_1 and V_2. The two voltage followers are used to buffer the patient from the instrument. Let us design the amplifier such that the output voltage is $100(V_1 - V_2)$.

SOLUTION The input–output relationship for the circuit in Fig. 3.39 is

$$V_o = (R_2/R_1)(V_1 - V_2) = 100(V_1 - V_2)$$

Figure 3.39 Neuromuscular voltage amplifier.

We have two unknowns and only one equation. Thus, if we select $R_1 = 1$ kΩ, then $R_2 = 100$ kΩ. The configuration with these values satisfies the requirements. ❏

EXAMPLE 3.27

The amplitude of the signal received by the antenna of a portable AM-radio receiver is 1 μV. The signal is then amplified to supply 8 V to a speaker. What is the required voltage gain of this system? Let us design a four-stage amplifier using identical inverting op-amp configurations to meet the gain requirement.

SOLUTION The required voltage gain is

$$A_V = 8/10^{-6} = 8 \times 10^6$$

As mentioned earlier, a four-stage amplifier has an overall gain equal to the product of the gains of the individual stages; that is,

$$A_V = A_{V1} A_{V2} A_{V3} A_{V4}$$

Since in this case the four stages are identical, we can write

$$A_{V1} = A_{V2} = A_{V3} = A_{V4} = A$$

where A represents the gain of a single stage. The inverting op-amp has a gain of

$$A = -\frac{R_2}{R_1}$$

The gain of a single stage is then

$$-\frac{R_2}{R_1} = -\sqrt[4]{(8 \times 10^6)} = -53.2$$

Therefore, the four-stage op-amp configuration shown in Fig. 3.40 will satisfy the design requirements.

Figure 3.40 Four-stage AM-radio amplifier.   ❏

3.6 SUMMARY

Nodal analysis for an *N*-node circuit

- Select one node in the *N*-node circuit as the reference node. Assume that the node voltage is zero and measure all node voltages with respect to this node.

- If only independent current sources are present in the network, write the KCL equations as the *N* − 1 nonreference nodes. If dependent current sources are present, write the KCL equations as is done for networks with only independent current sources; then write the controlling equations for the dependent sources.

- If voltage sources are present in the network, they may be connected (1) between the reference node and a nonreference node or (2) between two nonreference nodes. In the former case, if the voltage source is an independent source, then the voltage at one of the nonreference nodes is known. If the source is dependent, it is treated as an independent source when writing the KCL equations, but an additional constraint equation is necessary.

  In the latter case, if the source is independent, the voltage between the two nodes is constrained by the value of the voltage source and an equation describing this constraint represents one of the *N* − 1 linearly independent equations required to determine the *N*-node voltages. The surface of the network described by the constraint equation (i.e., the source and two connecting nodes) is called a supernode. One of the remaining *N* − 1 linearly independent equations is obtained by applying KCL at this supernode. If the voltage source is dependent, it is treated as an independent source when writing the KCL equations, but an additional constraint equation is necessary.

Loop analysis for a *N*-loop circuit

- One loop current is assigned to each independent loop in a circuit that contains N independent loops.

- If only independent voltage sources are present in the network, write the N linearly independent KVL equations, one for each loop. If dependent voltage sources are present, write the KVL equations as is done for circuits with only independent voltage sources; then write the controlling equations for the dependent sources.

- If current sources are present in the network, either of two techniques can be used. In the first case, one loop current is selected to pass through one of the current sources. This is done for each current source in the network. The remaining loop currents (N − the number of current sources) are determined by open-circuiting the current sources in the network and using this modified network to select them. Once all these currents are defined in the original network, the N loop equations can be written. The second approach is similar to the first with the exception that if two mesh currents pass through a particular current source, a supermesh is formed around this source. The two required equations for the meshes containing this source are the constraint equations for the two mesh currents that pass through the source and the supermesh equation. If dependent current sources are present, the controlling equations for these sources are also necessary.

Ideal Op-Amp Model   For an ideal op-amp, $i_x + i_- = 0$ and $v_x + v_-$. Both nodal and loop analysis are useful in solving circuits containing operational amplifiers.

PROBLEMS

Section 3.1

3.1   Find I_o in the circuit in Fig. P3.1 using nodal analysis.

Write the KCL equations for nodes 1 and 2 in terms of the node voltages and solve the equations for I_o.

Figure P3.1

3.2 Use nodal analysis to find V_1 in the circuit in Fig. P3.2.

Figure P3.2

Similar to P3.1.

3.3 Find V_2 in the circuit in Fig. P3.3 using nodal analysis.

Figure P3.3

Similar to P3.1.

3.4 Use nodal analysis to find V_o in the circuit in Fig. P3.4.

Figure P3.4

Similar to P3.1. Note that V_o, is the difference between two node voltages.

3.5 Find V_1 and V_2 in the circuit in Fig. P3.5 using nodal analysis.

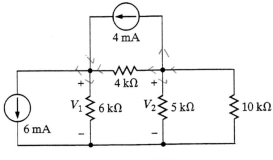

Figure P3.5

Similar to P3.1.

3.6 Use nodal analysis to find V_2 in the circuit in Fig. P3.6.

Similar to P3.1.

Figure P3.6

3.7 Find I_o in the circuit in Fig. P3.7 using nodal analysis.

Similar to P3.1.

Figure P3.7

3.8 Use nodal analysis to find V_o in the circuit in Fig. P3.8.

Similar to P3.1.

Figure P3.8

3.9 Find V_o in the circuit in Fig. P3.9 using nodal analysis.

Similar to P3.1.

Figure P3.9

3.10 Use nodal analysis to find V_o in the circuit in Fig. P3.10.

Similar to P3.1.

Figure P3.10

3.11 Find V_o in the circuit in Fig. P3.11 using nodal analysis.

Figure P3.11

Similar to P3.1. Note that voltage division is useful in this case.

3.12 Use nodal analysis to find V_o in the circuit in Fig. P3.12.

Figure P3.12

Similar to P3.4.

3.13 Find V_o in the circuit in Fig. P3.13 using nodal analysis.

Figure P3.13

Similar to P3.1. The 12-kΩ and 4-kΩ resistors can be combined.

3.14 Use nodal analysis to find V_o in the circuit in Fig. P3.14.

Figure P3.14

Similar to P3.13.

3.15 Find V_o in the network in Fig. P3.15 using nodal analysis.

V_o is the difference between two node voltages.

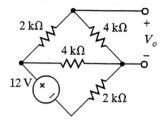

Figure P3.15

3.16 Use nodal analysis to find V_o in the circuit in Fig. P3.16.

Similar to P3.15.

Figure P3.16

3.17 Find V_o in the network in Fig. P3.17 using nodal analysis.

Similar to P3.15.

Figure P3.17

3.18 Find I_o in the circuit in Fig. P3.18 using nodal analysis.

Similar to P3.1 if the locations of the 3V source and 6KΩ resistor are exchanged.

Figure P3.18

3.19 Use nodal analysis to find I_o in the circuit in Fig. P3.19.

Similar to P3.1.

Figure P3.19

3.20 Find V_o in the circuit in Fig. P3.20 using nodal analysis.

Figure P3.20

Similar to P3.1.

3.21 Use nodal analysis to find I_o and I_S in the circuit in Fig. P3.21.

Figure P3.21

Similar to P3.1.

3.22 Find V_o in the network in Fig. P3.22 using nodal analysis.

Figure P3.22

Similar to P3.1.

3.23 Find I_o in the circuit in Fig. P3.23 using nodal analysis.

Figure P3.23

Similar to P3.1.

3.24 Use nodal analysis to find V_o in the network in Fig. P3.24.

Figure P3.24

Similar to P3.1.

3.25 Use nodal analysis to find V_o in the circuit in Fig. P3.25.

Similar to P3.15.

Figure P3.25

3.26 Find I_o in the circuit in Fig. P3.26 using nodal analysis.

Use the supernode technique.

Figure P3.26

3.27 Use nodal analysis to find I_o in the circuit in Fig. P3.27.

Similar to P3.26.

Figure P3.27

3.28 Using nodal analysis find V_o in the network in Fig. P3.28.

Similar to P3.26.

Figure P3.28

3.29 Find V_o in the network in Fig. P3.29 using nodal analysis.

Figure P3.29

Similar to P3.26.

3.30 Use nodal analysis to find V_o in Fig. P3.30.

Figure P3.30

Treat the dependent source as an independent source and then include the controlling parameter.

3.31 Find I_o in the circuit in Fig. P3.31 using nodal analysis.

Figure P3.31

Similar to P3.30.

3.32 Find V_o in the network in Fig. P3.32 using nodal analysis.

Figure P3.32

Similar to P3.30.

3.33 Use nodal analysis to find V_o in the circuit in Fig. P3.33. In addition, find all branch currents and check your answers using KCL at every node.

Similar to P3.30. Use the supernode technique.

Figure P3.33

3.34 Find V_o in the network in Fig. P3.34. In addition, determine all branch currents and check KCL at every node.

Similar to P3.33.

Figure P3.34

3.35 Find V_o in the circuit in Fig. P3.35.

Similar to P3.33.

Figure P3.35

3.36 Find V_o in the circuit in Fig. P3.36.

Similar to P3.30.

Figure P3.36

3.37 Find V_o in the circuit in Fig. P3.37.

Figure P3.37

Similar to P3.30.

Section 3.2

3.38 Use mesh analysis to find V_o in the circuit in Fig. P3.38.

Figure P3.38

Write the mesh current equations and solve them to find V_o.

3.39 Use mesh analysis to find V_o in the network in Fig. P3.39.

Figure P3.39

Similar to P3.38.

3.40 Use loop analysis to find V_o in the circuit in Fig. P3.40.

Figure P3.40

Use one loop each through the current source and the outer loop, or apply the supermesh technique.

3.41 Find I_o in the network in Fig. P3.41 using mesh analysis.

Similar to P3.39.

Figure P3.41

3.42 Find I_o in the network in Fig. P3.42.

Similar to P3.40.

Figure P3.42

3.43 Use loop analysis to find I_o in the circuit in Fig. P3.43.

Similar to P3.40.

Figure P3.43

3.44 Use both nodal analysis and mesh analysis to find I_o in the circuit in Fig. P3.44.

The supernode is useful for nodal analysis, and the problem is similar to P3.39 for mesh analysis.

Figure P3.44

3.45 Find I_o in the network in Fig. P3.45 using mesh analysis.

Figure P3.45

The supermesh technique is applicable here.

3.46 Find V_o in the network in Fig. P3.46 using both mesh and nodal analysis.

Figure P3.46

Similar to P3.44.

3.47 Use mesh analysis to find V_o in the circuit in Fig. P3.47.

Figure P3.47

Treat the dependent source as an independent source and then include the controlling parameter.

3.48 Use loop analysis to find V_o in the circuit in Fig. P3.48.

Figure P3.48

Similar to P3.40.

3.49 Find V_o in the circuit in Fig. P3.49 using mesh analysis.

Similar to P3.47.

Figure P3.49

3.50 Using mesh analysis find V_o in the circuit in Fig. P3.50.

Similar to P3.44.

Figure P3.50

3.51 Use both nodal analysis and mesh analysis to find V_o in the circuit in Fig. P3.51.

Use a supernode for nodal analysis and a supermesh for mesh analysis.

Figure P3.51

Section 3.3

3.52 Find V_o in the circuit in Fig. P3.52 assuming the op-amp is ideal.

Use $v_- = v_+ = 0$, $i_- = i_+ = 0$, and nodal analysis to find v_o.

Figure P3.52

3.53 The network in Fig. P3.53 is a current-to-voltage converter or transconductance amplifier. Find v_o/i_S for this network.

Figure P3.53

Similar to P3.52.

3.54 Find v_o in the network in Fig. P3.54 and explain what effect R_1 has on the output.

Figure P3.54

Similar to P3.52. The use of $i_+ = 0$ indicates R_1's effect.

3.55 Find V_o in the network in Fig. P3.55.

Figure P3.55

Similar to P3.52 with $v_- = v_+ = 4$.

3.56 Find the voltage gain of the op-amp circuit shown in Fig. P3.56.

Find v_+ using $i_+ = 0$. Find v_- using $i_- = 0$, and use $v_+ = v_-$ to find v_o.

Figure P3.56

3.57 For the circuit in Fig. P3.57 find the value of R_1 that produces a voltage gain of 10.

Similar to P3.55.

Figure P3.57

The circuit in P3.52 can generate $v_o = -2v_1$. Then $-4v_2$ can be obtained from the V_- terminal.

3.58 Given a box of 10-kΩ resistors and an op-amp, design a circuit that will have an output voltage of

$$v_o = -2v_1 - 4v_2$$

3.59 Find the expression for the output voltage in the op-amp circuit shown in Fig. P3.59.

Use $v_- = v_+ = v_1$, $i_- = i_+ = 0$, and nodal analysis to obtain v_o. Obtain v_{out} in a similar manner.

Figure P3.59

Advanced Problems

3.60 Find V_o in the circuit in Fig. P3.60 using both nodal and loop analysis.

Figure P3.60

3.61 Find V_o in the circuit in Fig. P3.61 using both nodal and loop analysis.

Figure P3.61

3.62 Use both nodal and loop analysis to find V_o in the circuit in Fig. P3.62.

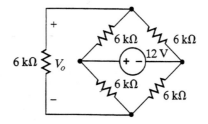

Figure P3.62

3.63 Use loop analysis to find V_o in the circuit in Fig. P3.63.

Figure P3.63

3.64 Find V_o in the circuit in Fig. P3.64.

Figure P3.64

3.65 Find all branch currents and node voltages in the circuit in Fig. P3.65.

Figure P3.65

3.66 Find V_o in the circuit in Fig. P3.66 using both nodal and loop analysis.

Figure P3.66

3.67 Use loop analysis to find V_o in the circuit in Fig. P3.67.

Figure P3.67

3.68 Find V_o in the network in Fig. P3.68.

Figure P3.68

3.69 Use loop analysis to find V_o in the network in Fig. P3.69.

Figure P3.69

3.70 Find V_o in the circuit in Fig. P3.70.

Figure P3.70

3.71 Find I_o in the circuit in Fig. P3.71.

Figure P3.71

3.72 Find V_o in the network in Fig. P3.72.

Figure P3.72

3.73 Draw a network defined by the following node equations.

$$V_1\left(\frac{1}{1\,k\Omega} + \frac{1}{2\,k\Omega} + \frac{1}{2\,k\Omega}\right) - V_2\left(\frac{1}{2\,k\Omega}\right)$$

$$= \frac{4\,V}{1\,k\Omega} + \frac{2\,V}{1\,k\Omega}; \qquad -V_1\left(\frac{1}{2\,k\Omega}\right) + V_2\left(\frac{1}{2\,k\Omega} + \frac{1}{4\,k\Omega}\right)$$

$$= -\frac{1\,V}{1\,k\Omega}$$

3.74 Draw a network defined by the following node equations.

$$V_1\left(\frac{1}{8\,k\Omega} + \frac{1}{1\,k\Omega}\right) - V_2\left(\frac{1}{1\,k\Omega}\right) = -\frac{2\,V}{1\,k\Omega} - \frac{1\,V}{1\,k\Omega}$$

$$-V_1\left(\frac{1}{1\,k\Omega}\right) + V_2\left(\frac{1}{1\,k\Omega} + \frac{1}{2\,k\Omega}\right) = \frac{2\,V}{1\,k\Omega} + \frac{4\,V}{1\,k\Omega}$$

3.75 Draw a network defined by the following mesh equations.

$$(12\,k\Omega)I_1 - (4\,k\Omega)I_3 = -28\,V$$

$$(3\,k\Omega)I_2 - (2\,k\Omega)I_3 = 16\,V$$

$$I_3 = -\frac{4\,V}{1\,k\Omega}$$

3.76 Draw a circuit defined by the following mesh equations.

$$I_1 = -\frac{6\,V}{1\,k\Omega}$$

$$-(4\,k\Omega)I_1 + (12\,k\Omega)I_2 - (2\,k\Omega)I_4 = 0$$

$$(4\,k\Omega)I_3 - (1\,k\Omega)I_4 = -12\,V$$

$$I_4 = -\frac{4\,V}{1\,k\Omega}$$

3.77 Find I_o in the circuit in Fig. P3.77.

Figure P3.77

3.78 Find V_o in the circuit in Fig. P3.78.

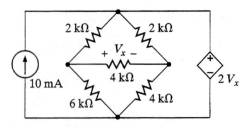

Figure P3.78

3.79 Find I_o in the circuit in Fig. P3.79.

Figure P3.79

3.80 Find V_o in the circuit in Fig. P3.80.

Figure P3.80

3.81 Find V_o in the circuit in Fig. P3.81.

Figure P3.81

3.82 Find V_o in the circuit in Fig. P3.82.

Figure P3.82

3.83 The electronic ammeter in Example 3.23 has been modified and is shown in Fig. P3.83. The selector switch allows the user to change the range of the meter. Using values for R_1 and R_2 from Example 3.23, find the values of R_A and R_B that will yield a 10-V output when the current being measured is 100 mA and 10 mA, respectively.

Figure P3.83

3.84 Design an op-amp circuit that has a gain of −50 using resistors no smaller than 1 kΩ.

3.85 Design a two-stage op-amp network that has a gain of −50,000 while drawing no current into its input terminal. Use no resistors smaller than 1 kΩ.

3.86 A 170°C maximum temperature digester is used in a paper mill to process wood chips that will eventually become paper. As shown in Fig. P3.86a, three electronic thermometers are placed along its length. Each thermometer outputs 0 V at 0°C and the voltage changes 25 mV/°C. We will use the average of the three thermometer voltages to find an aggregate digester temperature. Furthermore, 1 volt should appear at V_o for every 10°C of average temperature. Design such an averaging circuit using the op-amp configuration shown in Fig. 3.86 if the final output voltage must be positive.

Figure P3.86

Additional Analysis Techniques

4

A t this point, we have mastered the ability to solve networks containing both independent and dependent sources using either nodal or loop analysis. In this chapter we introduce several new analysis techniques that bolster our arsenal of circuit analysis tools. We will find that in some situations these techniques lead to a quick solution and in other cases they do not. However, these new techniques in many cases do provide an insight into the circuit's operation that cannot be gained from a nodal or loop analysis.

In many practical situations we are interested in the analysis of some portion of a much larger network. If we can model the remainder of the network with a simple equivalent circuit, then our task will be much simpler. For example, consider the problem of analyzing some simple electronic device that is connected to the ac wall plug in our house. In this case the complete circuit includes not only the electronic device but the utility's power grid, which is connected to the device through the circuit breakers in the home. However, if we can accurately model everything outside the device with a simple equivalent circuit, then our analysis will be tractable. Two of the theorems that we will present in this chapter will permit us to do just that.

4.1 INTRODUCTION

Before introducing additional analysis techniques, let us review some of the topics we have used either explicitly or implicitly in our analyses thus far.

Equivalence

Table 4.1 is a short compendium of some of the equivalent circuits that have been employed in our analyses. This listing serves as a quick review as we begin to look at other techniques that can be used to find a specific voltage or current somewhere in a network and provide additional insight into the network's operation. In addition to the forms listed in the table, it is important to note that a series connection of current sources or a parallel connection of voltage sources is forbidden unless the sources are pointing in the same direction and have exactly the same values.

Linearity

All the circuits we have analyzed thus far and all the circuits we will treat in this book are linear circuits. We have shown earlier that the resistor is a linear element because its current–voltage relationship is a linear characteristic; that is,

$$v(t) = Ri(t)$$

Linearity requires both additivity and homogeneity (scaling). In the case of a resistive element, if $i_1(t)$ is applied, the voltage across the resistor is

$$v_1(t) = Ri_1(t)$$

Similarly, if $i_2(t)$ is applied, then

$$v_2(t) = Ri_2(t)$$

However, if $i_1(t) + i_2(t)$ is applied, the voltage across the resistor is

$$v(t) = R[i_1(t) + i_2(t)] = Ri_1(t) + Ri_2(t) = v_1(t) + v_2(t)$$

This demonstrates the additive property. In addition, if the current is scaled by a constant K_1, the voltage is also scaled by the constant K_1, since

$$RK_1i(t) = K_1Ri(t) = K_1v(t)$$

This demonstrates homogeneity.

We have shown in the preceding chapters that a circuit that contains only independent sources, linear dependent sources, and resistors is described by equations of the form

$$a_1v_1(t) + a_2v_2(t) + \cdots + a_nv_n(t) = i(t)$$

or

$$b_1i_1(t) + b_2i_2(t) + \cdots + b_ni_n(t) = v(t)$$

Note that if the independent sources are multiplied by a constant, the node voltages or loop currents are also multiplied by the same constant. Thus, we define a linear circuit as one composed of only independent sources, linear dependent sources, and linear elements. Capacitors and inductors, which we will examine in Chapter 5, are also circuit elements that have a linear input–output relationship provided that their initial storage energy is zero.

Table 4.1 Equivalent circuit forms

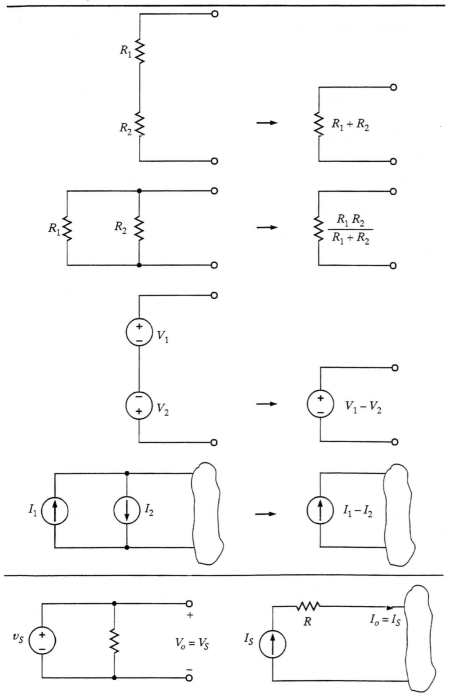

EXAMPLE 4.1

For the circuit shown in Fig. 4.1, we wish to determine the output voltage V_{out}. However, rather than approach the problem in a straightforward manner and calculate I_o, then I_1, then I_2, and so on, we will use linearity and simply assume that the output voltage is $V_{out} = 1$ V. This assumption will yield a value for the source voltage. We will then use the actual value of the source voltage and linearity to compute the actual value of V_{out}.

Figure 4.1 Circuit used in Example 4.1.

SOLUTION If we assume that $V_{out} = V_2 = 1$ V, then

$$I_2 = \frac{V_2}{2k} = 0.5 \text{ mA}$$

V_1 can then be calculated as

$$V_1 = 4kI_2 + V_2$$
$$= 3 \text{ V}$$

Hence,

$$I_1 = \frac{V_1}{3k} = 1 \text{ mA}$$

Now, applying KCL,

$$I_o = I_1 + I_2 = 1.5 \text{ mA}$$

Then

$$V_o = 2kI_o + V_1$$
$$= 6 \text{ V}$$

Therefore, the assumption that $V_{out} = 1$ V produced a source voltage of 6 V. However, since the actual source voltage is 12 V, the actual output voltage is 1 V(12/6) = 2 V. ❏

EXTENSION EXERCISE

E4.1 Use linearity and the assumption that $I_o = 1$ mA to compute the correct current I_o in the circuit in Fig. E4.1 if $I = 6$ mA.

Figure E4.1

ANSWER: $I_o = 3$ mA.

4.2 SUPERPOSITION

In order to provide motivation for this subject, let us examine a simple circuit in which two sources contribute to the current in the network.

EXAMPLE 4.2

Consider the circuit in Fig. 4.2a in which the actual values of the voltage sources are left unspecified. Let us use this network to examine the concept of superposition.

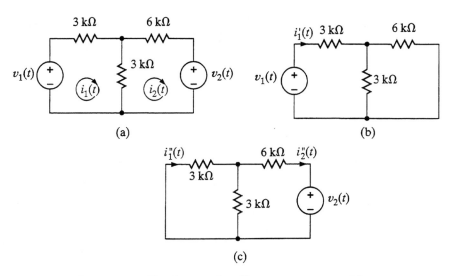

Figure 4.2 Circuits used to illustrate superposition.

SOLUTION   The mesh equations for this network are

$$6ki_1(t) - 3ki_2(t) = v_1(t)$$
$$-3ki_1(t) + 9ki_2(t) = -v_2(t)$$

Solving these equations for $i_1(t)$ yields

$$i_1(t) = \frac{v_1(t)}{5k} - \frac{v_2(t)}{15k}$$

In other words, the current $i_1(t)$ has a component due to $v_1(t)$ and a component due to $v_2(t)$. In view of the fact that $i_1(t)$ has two components, one due to each independent source, it would be interesting to examine what each source acting alone would contribute to $i_1(t)$. For $v_1(t)$ to act alone, $v_2(t)$ must be zero. As we pointed out in Chapter 2, $v_2(t) = 0$ means that the source $v_2(t)$ is replaced with a short circuit. Therefore, to determine the value of $i_1(t)$ due to $v_1(t)$ only, we employ the circuit in Fig. 4.2b and refer to this value of $i_1(t)$ as $i_1'(t)$.

$$i_1'(t) = \frac{v_1(t)}{3k + \dfrac{(3k)(6k)}{3k + 6k}}$$

$$= \frac{v_1(t)}{5k}$$

Let us now determine the value of $i_1(t)$ due to $v_2(t)$ acting alone and refer to this value as $i_1''(t)$. Using the network in Fig. 4.2c,

$$i_2''(t) = -\frac{v_2(t)}{6k + \dfrac{(3k)(3k)}{3k + 3k}} = \frac{-2v_2(t)}{15k}$$

Then, using current division, we obtain

$$i_1''(t) = \frac{-2v_2(t)}{15k}\left(\frac{3k}{3k + 3k}\right)$$

$$= \frac{-v_2(t)}{15k}$$

Now, if we add the values of $i_1'(t)$ and $i_1''(t)$, we obtain the value computed directly; that is,

$$i_1(t) = i_1'(t) + i_1''(t)$$

$$= \frac{v_1(t)}{5k} - \frac{v_2(t)}{15k}$$

Note that we have *superposed* the value of $i_1'(t)$ on $i_1''(t)$, or vice versa, to determine the unknown current. ❑

What we have demonstrated in Example 4.2 is true in general for linear circuits and is a direct result of the property of linearity. *The principle of superposition*, which provides us with this ability to reduce a complicated problem to several easier problems—each containing only a single independent source—states that

In any linear circuit containing multiple independent sources, the current or voltage at any point in the network may be calculated as the algebraic sum of the individual contributions of each source acting alone.

When determining the contribution due to an independent source, any remaining voltage sources are made zero by replacing them with short circuits, and any remaining current sources are made zero by replacing them with open circuits.

Although superposition can be used in linear networks containing dependent sources, it is not useful in this case since the dependent source is never made zero.

It is interesting to note that, as the previous example indicates, superposition provides some insight in determining the contribution of each source to the variable under investigation.

We will now demonstrate superposition with two examples and then provide a problem-solving strategy for the use of this technique. For purposes of comparison, we will also solve the networks using both node and loop analyses. Furthermore, we will employ these same networks when demonstrating subsequent techniques, if applicable.

EXAMPLE 4.3

Let us use superposition to find V_o in the circuit in Fig. 4.3a.

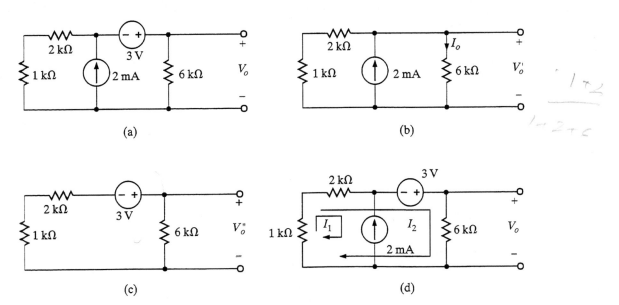

Figure 4.3 Circuits used in Example 4.3.

SOLUTION   The contribution of the 2-mA source to the output voltage is found from the network in Fig. 4.3b, using current division

$$I_o = (2 \times 10^{-3})\left(\frac{1k + 2k}{1k + 2k + 6k}\right)$$

$$= \frac{2}{3} \text{ mA}$$

and

$$V'_o = I_o(6k) = 4V$$

The contribution of the 3-V source to the output voltage is found from the circuit in Fig. 4.3c. Using voltage division,

$$V''_o = 3\left(\frac{6k}{1k + 2k + 6k}\right)$$

$$= 2 V$$

Therefore,

$$V_o = V'_o + V''_o = 6 \text{ V}$$

Although we used two separate circuits to solve the problem, both were very simple. If we use nodal analysis and Fig. 4.3a to find V_o and recognize that the 3-V source and its connecting nodes form a supernode, V_o can be found from the node equation

$$\frac{V_o - 3}{1\text{k} + 2\text{k}} - 2 \times 10^{-3} + \frac{V_o}{6\text{k}} = 0$$

which yields $V_o = 6$ V. In addition, loop analysis applied as shown in Fig. 4.3d produces the equations

$$I_1 = -2 \times 10^{-3}$$

and

$$3\text{k}(I_1 + I_2) - 3 + 6\,\text{k}I_2 = 0$$

which yield $I_2 = 1$ mA and hence $V_o = 6$ V.   ❏

DRILL

D4.1 Use superposition to find V_o in the following network.

ANSWER: $V_o = V'_o + V''_o = (9 - 3) = 6$ V

DRILL

D4.2 Use superposition to find I_o in the following network.

ANSWER: $I_o = I'_o + I''_o = (8 - 2) = 6$ mA

EXAMPLE 4.4

Consider now the network in Fig. 4.4a. Let us use superposition to find V_o.

SOLUTION The contribution of the 6-V source to V_o is found from the network in Fig. 4.4b, which is redrawn in Fig. 4.4c. The 2 kΩ + 6 kΩ = 8-kΩ resistor and 4-kΩ resistor are in parallel, and their combination is an 8/3-kΩ resistor. Then, using voltage division,

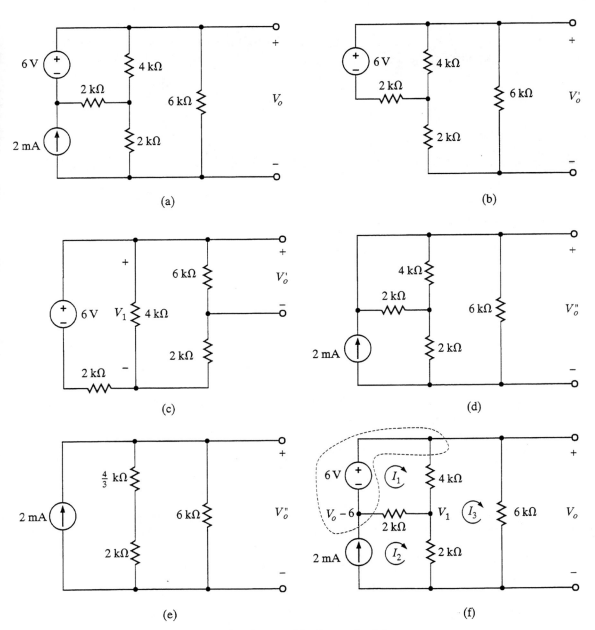

Figure 4.4 Circuits used in Example 4.4.

$$V_1 = 6\left(\frac{\dfrac{8}{3}k}{\dfrac{8}{3}k + 2k}\right) = \frac{24}{7}\,\text{V}$$

Applying voltage division again,

$$V'_o = V_1\left(\frac{6k}{6k + 2k}\right) = \frac{18}{7}\,\text{V}$$

The contribution of the 2-mA source is found from Fig. 4.4d, which is redrawn in Fig. 4.4e. V_o'' is simply equal to the product of the current source and the parallel combination of the resistors; that is,

$$V_o'' = (2 \times 10^{-3})\left(\frac{10}{3}\,\text{k}//6\text{k}\right) = \frac{30}{7}\,\text{V}$$

Then

$$V_o = V_o' + V_o'' = \frac{48}{7}\,\text{V}$$

A nodal analysis of the network can be performed using Fig. 4.4f. The equation for the supernode is

$$-2 \times 10^{-3} + \frac{(V_o - 6) - V_1}{2\text{k}} + \frac{V_o - V_1}{4\text{k}} + \frac{V_o}{6\text{k}} = 0$$

The equation for the node labeled V_1 is

$$\frac{V_1 - V_o}{4\text{k}} + \frac{V_1 - (V_o - 6)}{2\text{k}} + \frac{V_1}{2\text{k}} = 0$$

Solving these two equations, which already contain the constraint equation for the supernode, yields $V_o = 48/7$ V.

Once again, referring to the network in Fig. 4.4f, the mesh equations for the network are

$$-6 + 4\text{k}(I_1 - I_3) + 2\text{k}(I_1 - I_2) = 0$$

$$I_2 = 2 \times 10^{-3}$$

$$2\text{k}(I_3 - I_2) + 4\text{k}(I_3 - I_1) + 6\text{k}I_3 = 0$$

Solving these equations, we obtain $I_3 = 8/7$ mA and, hence, $V_o = 48/7$ V. ❏

Problem-Solving Strategy for Applying Superposition

- In a network containing multiple independent sources, each source can be applied independently with the remaining sources turned off.

- To turn off a voltage source, replace it with a short circuit, and to turn off a current source, replace it with an open circuit.

- When the individual sources are applied to the circuit, all the circuit laws and techniques we have learned, or will soon learn, can be applied to obtain a solution.

- The results obtained by applying each source independently are then added together algebraically to obtain a solution.

Superposition can be applied to a circuit with any number of dependent and independent sources. In fact, superposition can be applied to such a network in a variety of ways. For example, a circuit with three independent sources can be solved using each source acting alone, as we have just demonstrated, or we could use two at a time and sum the result with that obtained from the third acting alone. In addition, the independent sources do not have to assume their actual value or zero. However, it is mandatory that the sum of the different values chosen add to the total value of the source.

Superposition is a fundamental property of linear equations and, therefore, can be applied to any effect that is linearly related to its cause. In this regard it is important to point out that although superposition applies to the current and voltage in a linear circuit, it cannot be used to determine power because power is a nonlinear function.

EXTENSION EXERCISE

E4.2   Compute V_o in the circuit in Fig. E4.2 using superposition.

Figure E4.2

ANSWER: $V_o = \dfrac{4}{3}$ V.

4.3 SOURCE TRANSFORMATION

Before we begin discussing source transformation, it is necessary that we point out that real sources differ from the ideal models we have presented thus far. In general, a practical voltage source does not produce a constant voltage regardless of the load resistance or the current it delivers, and a practical current source does not deliver a constant current regardless of the load resistance or the voltage across its terminals.

Practical sources contain internal resistance, and this source resistance is not accessible to the user. In general, we want to minimize the resistance because it consumes power. Some typical values for internal resistance are shown in Table 4.2.

Table 4.2 Approximate Internal Resistance of Selected Sources (New Batteries) at 21°C

Battery Type	Internal Resistance
1.5-V D-cell standard carbon zinc	0.5 Ω
1.5-V D-cell heavy-duty carbon zinc	0.3 Ω
1.5-V D-cell alkaline	0.2 Ω
1.5-V D-cell nickel-cadmium	0.008 Ω
9-V standard carbon zinc	35 Ω
9-V heavy-duty carbon zinc	35 Ω
9-V alkaline	2 Ω
12-V automobile battery	0.006 Ω

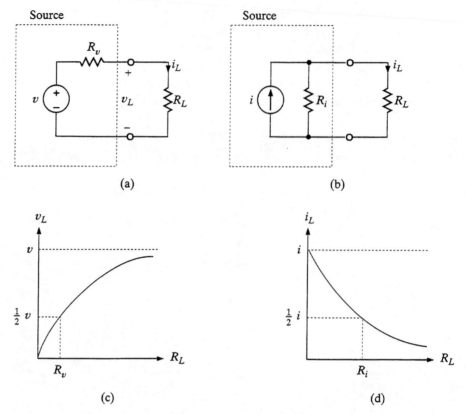

Figure 4.5 Effect of internal resistance on actual voltage and current sources.

Since practical sources contain internal resistance, the models shown in Fig. 4.5a and b more closely represent actual sources. Note that the power delivered by the practical voltage source is given by the expression

$$P_L = i_L^2 R_L$$

$$= \left(\frac{v}{R_v + R_L}\right)^2 R_L$$

which can be written as

$$P_L = \frac{v^2}{R_L}\left(\frac{1}{1 + R_v/R_L}\right)^2$$

and therefore if $R_L \gg R_v$, then

$$P_L = \frac{v^2}{R_L}$$

which is the power delivered by the ideal voltage source. Similarly, the power delivered by the practical current source is computed using current division as

$$P_L = i_L^2 R_L$$

$$= \left(\frac{i R_i}{R_i + R_L}\right)^2 R_L$$

which can be written as

$$P_L = i^2 R_L \left(\frac{1}{1 + R_L/R_i} \right)^2$$

and hence, if $R_i \gg R_L$, then

$$P_L = i^2 R_L$$

which is the power delivered by the ideal current source.

The graphs shown in Figs. 4.5c and d illustrate the effect of the internal resistance on the voltage source and current source, respectively. The graphs indicate that the output voltage approaches the ideal source voltage only for large values of load resistance and, therefore, small values of current. Similarly, the load current is approximately equal to the ideal source current only for values of the load resistance R_L that are small in comparison to the internal resistance R_i.

With this material on practical sources as background information, we now ask if it is possible to exchange one source model for another; that is, exchange a voltage source model for a current source model, or vice versa. We could exchange one source for another provided that they are equivalent; that is, each source produces exactly the same voltage and current for any load that is connected across its terminals.

Let us examine the two circuits shown in Fig. 4.6. In order to determine the conditions required for the two sources to be equivalent, let us examine the terminal conditions of each. For the network in Fig. 4.6a,

$$i = i_L + \frac{v_L}{R_i}$$

or

$$iR_i = R_i i_L + v_L$$

For the network in Fig. 4.6b,

$$v = i_L R_v + v_L$$

For the two networks to be equivalent, their terminal characteristics must be identical; that is,

$$v = iR_i \quad \text{and} \quad R_i = R_v \qquad \qquad \textbf{4.1}$$

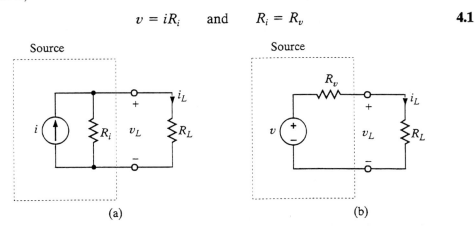

Figure 4.6 Circuits used to determine conditions for a source exchange.

The relationships specified in Eq. (4.1) and Fig. 4.6 are extremely important and the reader should not fail to grasp their significance. What these relationships tell us is that if we have embedded within a network a current source i in parallel with a resistor R, we can replace this combination with a voltage source of value $v = iR$ in series with the resistor R. The reverse is also true; that is, a voltage source v in series with a resistor R can be replaced with a current source of value $i = v/R$ in parallel with the resistor R. Parameters within the circuit (e.g., an output voltage) are unchanged under these transformations.

We must emphasize that the two equivalent circuits in Fig. 4.6, under the conditions stated in Eq. (4.1), *are equivalent only at the two external nodes*. For example, if we disconnect the load resistor, R_L, in the circuits in Fig. 4.6, the equivalent circuit in Fig. 4.6a dissipates power, but the one in Fig. 4.6b does not.

At this point let us demonstrate the utility of a source exchange using two examples.

EXAMPLE 4.5

Let us repeat Example 4.3 using source exchange.

SOLUTION   For convenience, the network is redrawn in Fig. 4.7a. Note that on the left side of the network, there is a current source in parallel with a 3-kΩ(1 kΩ + 2 kΩ) resistor, which can be transformed into a 6-V = (2 mA)(3 kΩ) source in series with the same 3-kΩ resistor. The resulting network is shown in Fig. 4.7b. The output voltage is then a simple division of the 9 volts (6 V + 3 V) between the 3-kΩ and 6-kΩ resistors. Therefore,

$$V_o = 9\left(\frac{6k}{3k + 6k}\right) = 6 \text{ V}$$

which agrees with our previous analyses. Finally, the reader is cautioned to keep the polarity of the voltage source and the direction of the current source in agreement, as shown in Fig. 4.7.

(a)   (b)

Figure 4.7 Circuits used in Example 4.5.

EXAMPLE 4.6

We will now demonstrate how to find V_o in the circuit in Fig. 4.8a using the repeated application of source transformation.

SOLUTION   If we begin at the left end of the network in Fig. 4.8a, the series combination of the 12-V source and 3-kΩ resistor is converted to a 4-mA current source in parallel with the 3-kΩ resistor. If we combine this 3-kΩ resistor with the 6-kΩ resistor, we obtain the circuit in Fig. 4.8b. Note that at this point we have eliminated one circuit element. Continuing the reduction, we convert the 4-mA source and 2-kΩ resistor into an 8-V source in series with this same 2-kΩ resistor. The two 2-kΩ resistors that are in series are now combined to produce the network in Fig. 4.8c. If we now convert the combination of the 8-V source and 4-kΩ resistor into a 2-mA source in parallel with the 4-kΩ resistor and combine the resulting current source with the other 2-mA source, we arrive at the circuit shown in Fig. 4.8d. At this point, we can simply apply current division to the two parallel resistance paths and obtain

$$I_o = (4 \times 10^{-3})\left(\frac{4k}{4k + 4k + 8k}\right) = 1 \text{ mA}$$

and hence,

$$V_o = (1 \times 10^{-3})(8k) = 8 \text{ V}$$

The reader is encouraged to consider the ramifications of working this problem using any of the other techniques we have presented.

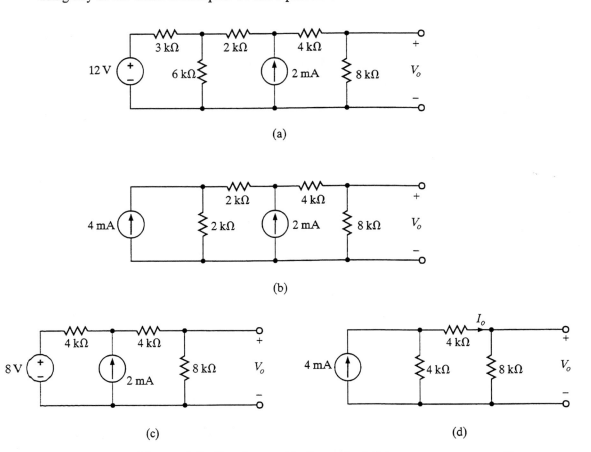

Figure 4.8 Circuits used in Example 4.6.

DRILL

D4.3 Convert the following network to a single node circuit with respect to the 12 kΩ load.

ANSWER:

DRILL

D4.4 Convert the following network to a single-loop circuit with respect to the 4-kΩ load.

ANSWER:

Note that this systematic, sometimes tedious, transformation allows us to reduce the network methodically to a simpler equivalent form with respect to some other circuit element. However, we should also realize that this technique is worthless for circuits of the form shown in Fig. 4.4. Furthermore, although applicable to networks containing dependent sources, it is not as useful as other techniques, and care must be taken not to transform the part of the circuit that contains the control variable.

Problem-Solving Strategy
for Applying Source Transformation

- When applicable (and, as indicated previously, this technique is not applicable in many cases), source transformation can be used to reduce the number of components in a network with respect to some particular load.

- The reduction in the number of components is achieved by systematically transforming (i.e., exchanging) a voltage source V in series with a resistor R for a current source $I = V/R$ in parallel with the resistor R, and vice versa.

- As this technique is applied, resistors in series or parallel can be combined as well as voltage sources in series and current sources in parallel. As this process proceeds, the number of elements in the network is continuously being reduced.

- When this technique is viable, it will simplify the network to at least some extent and in some cases will reduce it to a single-loop or single-node circuit containing the particular load that was present in the original circuit.

EXTENSION EXERCISE

E4.3   Compute V_o in the circuit in Fig. E4.2 using source transformation.

ANSWER: $V_o = \dfrac{4}{3}\,\text{V}.$

4.4 THÉVENIN'S AND NORTON'S THEOREMS

Thus far we have presented a number of techniques for circuit analysis. At this point we will add two theorems to our collection of tools that will prove to be extremely useful. The theorems are named after their authors, M. L. Thévenin, a French engineer, and E. L. Norton, a scientist formerly with Bell Telephone Laboratories.

Suppose that we are given a circuit and that we wish to find the current, voltage, or power that is delivered to some resistor of the network, which we will call the load. *Thévenin's theorem* tells us that we can replace the entire network, exclusive of the load, by an equivalent circuit that contains only an independent voltage source in series with a resistor in such a way that the current–voltage relationship at the load is unchanged. *Norton's theorem* is identical to the preceding statement except that the equivalent circuit is an independent current source in parallel with a resistor.

Note that this is a very important result. It tells us that if we examine any network from a pair of terminals, we know that with respect to those terminals, the entire network is equivalent to a simple circuit consisting of an independent voltage source in series with a resistor or an independent current source in parallel with a resistor.

In developing the theorems, we will assume that the circuit shown in Fig. 4.9a can be split into two parts, as shown in Fig. 4.9b. In general, circuit B is the load and may be linear or nonlinear. Circuit A is the balance of the original network exclusive of the load and must be linear. As such, circuit A may contain independent sources, dependent sources and resistors, or any other linear element. We require, however, that a dependent source and its control variable appear in the same circuit.

Circuit A delivers a current i to circuit B and produces a voltage v_o across the input terminals of circuit B. From the standpoint of the terminal relations of circuit A, we can replace circuit B by a voltage source of v_o volts (with the proper polarity), as shown in

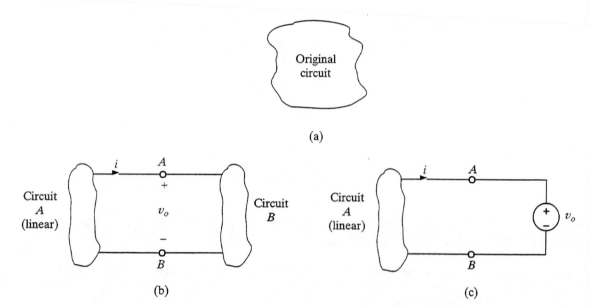

Figure 4.9 Concepts used to develop Thévenin's theorem.

Fig. 4.9c. Since the terminal voltage is unchanged and circuit A is unchanged, the terminal current i is unchanged.

Now, applying the principle of superposition to the network shown in Fig. 4.9c, the total current i shown in the figure is the sum of the currents caused by all the sources in circuit A and the source v_o that we have just added. Therefore, via superposition the current i can be written

$$i = i_o + i_{sc} \qquad \textbf{4.2}$$

where i_o is the current due to v_o with all independent sources in circuit A made zero (i.e., voltage sources replaced by short circuits and current sources replaced by open circuits), and i_{sc} is the short-circuit current due to all sources in circuit A with v_o replaced by a short circuit.

The terms i_o and v_o are related by the equation

$$i_o = \frac{-v_o}{R_{Th}} \qquad \textbf{4.3}$$

where R_{Th} is the equivalent resistance looking back into circuit A from terminals A-B with all independent sources in circuit A made zero.

Substituting Eq. (4.3) into Eq. (4.2) yields

$$i = -\frac{v_o}{R_{Th}} + i_{sc} \qquad \textbf{4.4}$$

This is a general relationship and, therefore, must hold for any specific condition at terminals A-B. As a specific case, suppose that the terminals are open circuited. For this condition, $i = 0$ and v_o is equal to the open-circuit voltage v_{oc}. Thus, Eq. (4.4) becomes

$$i = 0 = \frac{-v_{oc}}{R_{Th}} + i_{sc} \qquad \textbf{4.5}$$

Hence,

$$v_{oc} = R_{Th} i_{sc} \qquad\qquad \textbf{4.6}$$

This equation states that the open-circuit voltage is equal to the short-circuit current times the equivalent resistance looking back into circuit A with all independent sources made zero. We refer to R_{Th} as the Thévenin equivalent resistance.

Substituting Eq. (4.6) into Eq. (4.4) yields

$$i = \frac{-v_o}{R_{Th}} + \frac{v_{oc}}{R_{Th}}$$

or

$$v_o = v_{oc} - R_{Th} i \qquad\qquad \textbf{4.7}$$

Let us now examine the circuits that are described by these equations. The circuit represented by Eq. (4.7) is shown in Fig. 4.10a. The fact that this circuit is equivalent at terminals A-B to circuit A in Fig. 4.9 is a statement of *Thévenin's theorem*. The circuit represented by Eq. (4.4) is shown in Fig. 4.10b. The fact that this circuit is equivalent at terminals A-B to circuit A in Fig. 4.9 is a statement of *Norton's theorem*.

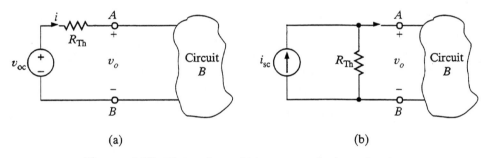

(a)   (b)

Figure 4.10   Thévenin and Norton equivalent circuits.

Note carefully that the circuits in Fig. 4.10 together with the relationship in Eq. (4.6) represent a source transformation.

The manner in which these theorems are applied depends on the structure of the original network under investigation. For example, if only independent sources are present, we can calculate the open-circuit voltage or short-circuit current and the Thévenin equivalent resistance. However, if dependent sources are also present, the Thévenin equivalent will be determined by calculating v_{oc} and i_{sc}, since this is normally the best approach for determining R_{Th} in a network containing dependent sources. Finally, if circuit A contains no *independent* sources, then both v_{oc} and i_{sc} will necessarily be zero. (Why?) Thus, we cannot determine R_{Th} by v_{oc}/i_{sc}, since the ratio is indeterminate. We must look for another approach. Notice that if $v_{oc} = 0$, then the equivalent circuit is merely the unknown resistance R_{Th}. If we apply an external source to circuit A—a test source v_t—and determine the current, i_t, which flows into circuit A from v_t, then R_{Th} can be determined from $R_{Th} = v_t/i_t$. Although the numerical value of v_t need not be specified, we could let $v_t = 1\text{ V}$ and then $R_{Th} = 1/i_t$. Alternatively, we could use a current source as a test source and let $i_t = 1\text{ A}$; then $v_t = (1)R_{Th}$.

A variety of examples is now presented to demonstrate the utility of these theorems. Before beginning our analysis of several examples that will demonstrate the utility of these theorems, the reader is cautioned to remember that these theorems, in addition to an alternate method of attack, often permit us to solve several small problems rather than one large one. They allow us to replace a network, no matter how large, *at a pair of terminals* with a Thévenin or Norton equivalent circuit. In fact, we could represent the entire U.S. power grid at a pair of terminals with one of the equivalent circuits. Once this is done we can quickly analyze the effect of different loads on a network. Thus, these theorems provide us with additional insight into the operation of a specific network.

Circuits Containing Only Independent Sources

EXAMPLE 4.7

Let us use Thévenin's and Norton's theorems to find V_o in the network in Example 4.3.

SOLUTION   The circuit is redrawn in Fig. 4.11a. To determine the Thévenin equivalent, we break the network at the 6-kΩ load as shown in Fig. 4.11b. KVL indicates that the open-circuit voltage, V_{oc}, is equal to 3 V plus the voltage V_1, which is the voltage across the current source. The 2 mA from the current source flows through the two resistors (where else could it possibly go!) and, therefore, $V_1 = (2 \times 10^{-3})(1k + 2k) = 6$ V. Therefore, $V_{oc} = 9$ V. By making both sources zero, we can find the Thévenin equivalent resistance, R_{Th}, using the circuit in Fig. 4.11c. Obviously, $R_{Th} = 3$ kΩ. Now our Thévenin equivalent circuit, consisting of V_{oc} and R_{Th}, is connected back to the original terminals of the load, as shown in Fig. 4.11d. Using a simple voltage divider, we find that $V_o = 6$ V.

To determine the Norton equivalent circuit at the terminals of the load, we must find the short-circuit current as shown in Fig. 4.11e. Note that the short circuit causes the 3-V source to be directly across (i.e., in parallel with) the resistors and the current source. Therefore, $I_1 = 3/(1k + 2k) = 1$ mA. Then, using KCL, $I_{sc} = 3$ mA. We have already determined R_{Th} and, therefore, connecting the Norton equivalent to the load results in the circuit in Fig. 4.11f. Hence, V_o is equal to the source current multiplied by the parallel resistor combination, which is 6 V.   ❏

Consider for a moment some salient features of this example. Note that in applying the theorems there is no point in breaking the network to the left of the 3-V source, since the resistors in parallel with the current source are already a Norton equivalent, which can be immediately changed to a Thévenin equivalent using source transformation! Furthermore, once the network has been simplified using a Thévenin or Norton equivalent, we simply have a new network with which we can apply the theorems again. The following example illustrates this approach.

EXAMPLE 4.8

Let us use Thévenin's theorem to find V_o in the network in Fig. 4.8a, which is redrawn in Fig. 4.12a.

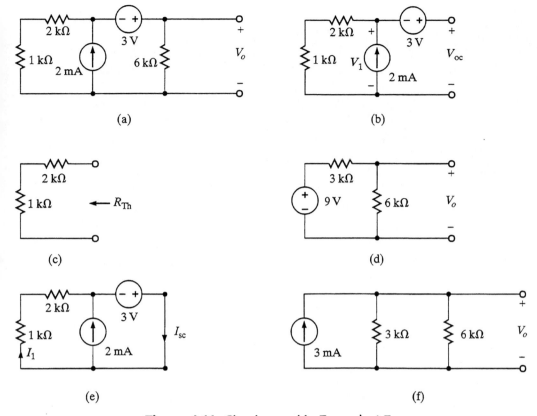

Figure 4.11 Circuits used in Example 4.7.

D4.5 Find V_o in the following
network using Thévenin's theorem.

ANSWER: $V_o = 1$ V

SOLUTION   If we break the network to the left of the current source, the open-circuit voltage V_{oc_1} is as shown in Fig. 4.12b. Since there is no current in the 2-kΩ resistor and therefore no voltage across it, V_{oc_1} is equal to the voltage across the 6-kΩ resistor, which can be determined by voltage division as

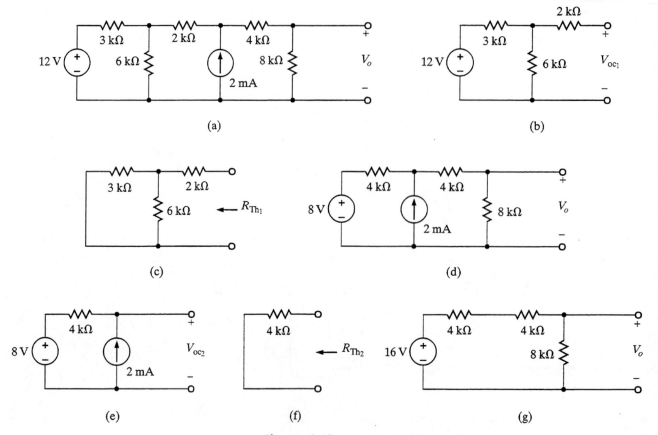

Figure 4.12 Circuits used in Example 4.8.

$$V_{oc_1} = 12\left(\frac{6k}{6k + 3k}\right) = 8 \text{ V}$$

The Thévenin equivalent resistance, R_{Th_1}, is found from Fig. 4.12c as

$$R_{Th_1} = 2k + \frac{(3k)(6k)}{3k + 6k} = 4 \text{ k}\Omega$$

Connecting this Thévenin equivalent back to the original network produces the circuit shown in Fig. 4.12d. We can now apply Thévenin's theorem again, and this time we break the network to the right of the current source as shown in Fig. 4.12e. In this case V_{oc_2} is

$$V_{oc_2} = (2 \times 10^{-3})(4k) + 8 = 16 \text{ V}$$

and R_{Th_2} obtained from Fig. 4.12f is 4 kΩ. Connecting this Thévenin equivalent to the remainder of the network produces the circuit shown in Fig. 4.12g. Simple voltage division applied to this final network yields $V_o = 8$ V. Norton's theorem can be applied in a similar manner to solve this network; however, we save that solution as an exercise for the reader. ❏

EXAMPLE 4.9

It is instructive to examine the use of Thévenin's and Norton's theorems in the solution of the network in Fig. 4.4a, which is redrawn in Fig. 4.13a.

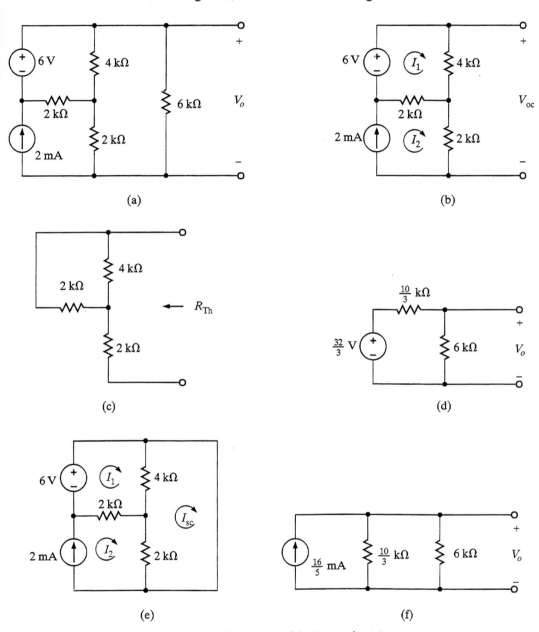

Figure 4.13 Circuits used in Example 4.9.

SOLUTION   If we break the network at the 6-kΩ load, the open-circuit voltage is found from Fig. 4.13b. The equations for the mesh currents are

$$-6 + 4kI_1 + 2k(I_1 - I_2) = 0$$

and

$$I_2 = 2 \times 10^{-3}$$

from which we easily obtain $I_1 = 5/3$ mA. Then, using KVL, V_{oc} is

$$V_{oc} = 4kI_1 + 2kI_2$$

$$= 4k\left(\frac{5}{3} \times 10^{-3}\right) + 2k(2 \times 10^{-3})$$

$$= \frac{32}{3}\text{ V}$$

R_{Th} is derived from Fig. 4.13c and is

$$R_{Th} = (2k//4k) + 2k = \frac{10}{3}\text{ k}\Omega$$

Attaching the Thévenin equivalent to the load produces the network in Fig. 4.13d. Then using voltage division, we obtain

$$V_o = \frac{32}{3}\left(\frac{6k}{6k + \frac{10}{3}k}\right)$$

$$= \frac{48}{7}\text{ V}$$

In applying Norton's theorem to this problem, we must find the short-circuit current shown in Fig. 4.13e. At this point the quick-thinking reader stops immediately! Three mesh equations applied to the original circuit will immediately lead to the solution, but the three mesh equations in the circuit in Fig. 4.13e will provide only part of the answer, specifically the short-circuit current. Sometimes the use of the theorems is more complicated than a straightforward attack using node or loop analysis. This would appear to be one of those situations. Interestingly, it is not. We can find I_{sc} from the network in Fig. 4.13e without using the mesh equations. The technique is simple, but a little tricky, and so we ignore it at this time. Having said all these things, let us now finish what we have started. The mesh equations for the network in Fig. 4.13e are

$$-6 + 4k\left(I_1 - I_{sc}\right) + 2k\left(I_1 - 2 \times 10^{-3}\right) = 0$$

$$2k\left(I_{sc} - 2 \times 10^{-3}\right) + 4k\left(I_{sc} - I_1\right) = 0$$

where we have incorporated the fact that $I_2 = 2 \times 10^{-3}$ A. Solving these equations yields $I_{sc} = 16/5$ mA. R_{Th} has already been determined in the Thévenin analysis. Connecting the Norton equivalent to the load results in the circuit in Fig. 4.13f. Applying Ohm's law to this circuit yields $V_o = 48/7$ V.   ❑

EXTENSION EXERCISES

E4.4   Use Thévenin theorem to find V_o in the network in Fig. E4.4.

ANSWER: $V_o = -3$ V.

Figure E4.4

E4.5   Find V_o in the circuit in Fig. E4.2 using both Thévenin's and Norton's theorems. When deriving the Norton equivalent circuit, break the network to the left of the 4-kΩ resistor. Why?

ANSWER: $V_o = \dfrac{4}{3}$ V.

Circuits Containing Only Dependent Sources

As we have stated earlier, the Thévenin or Norton equivalent of a network containing only dependent sources is R_{Th}. The following examples will serve to illustrate how to determine this Thévenin equivalent resistance.

EXAMPLE 4.10

We wish to determine the Thévenin equivalent of the network in Fig. 4.14a at the terminals A-B.

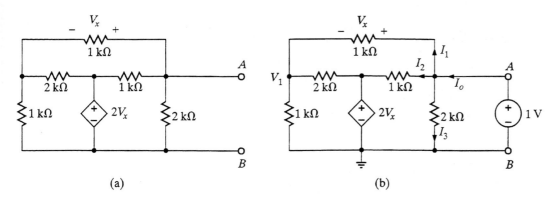

(a)   (b)

Figure 4.14 Networks employed in Example 4.10.

SOLUTION   Our approach to this problem will be to apply a 1-V source at the terminals as shown in Fig. 4.14b and then compute the current I_o and $R_{\text{Th}} = 1/I_o$.

The equations for the network in Fig. 4.14b are as follows. KVL around the outer loop specifies that

$$V_1 + V_x = 1$$

The KCL equation at the node labeled V_1 is

$$\frac{V_1}{1k} + \frac{V_1 - 2v_x}{2k} + \frac{V_1 - 1}{1k} = 0$$

Solving the equations for V_x yields $V_x = 3/7$ V. Knowing V_x, we can compute the currents $I_1, I_2,$ and I_3. Their values are

$$I_1 = \frac{V_x}{1k} = \frac{3}{7} \text{ mA}$$

$$I_2 = \frac{1 - 2V_x}{1k} = \frac{1}{7} \text{ mA}$$

$$I_3 = \frac{1}{2k} = \frac{1}{2} \text{ mA}$$

Therefore,

$$I_o = I_1 + I_2 + I_3$$

$$= \frac{15}{14} \text{ mA}$$

and

$$R_{Th} = \frac{1}{I_o}$$

$$= \frac{14}{15} \text{ k}\Omega$$

EXAMPLE 4.11

Let us determine R_{Th} at the terminals A-B for the network in Fig. 4.15a.

SOLUTION   Our approach to this problem will be to apply a 1-mA current source at the terminals A-B and compute the terminal voltage V_2 as shown in Fig. 4.15b. Then $R_{Th} = V_2/0.001$.

The node equations for the network are

$$\frac{V_1 - 2000I_x}{2k} + \frac{V_1}{1k} + \frac{V_1 - V_2}{3k} = 0$$

$$\frac{V_2 - V_1}{3k} + \frac{V_2}{2k} = 1 \times 10^{-3}$$

and

Figure 4.15 Networks used in Example 4.10.

$$I_x = \frac{V_1}{1\text{k}}$$

Solving these equations yields

$$V_2 = \frac{10}{7}\text{ V}$$

and hence,

$$R_{\text{Th}} = \frac{V_2}{1 \times 10^{-3}}$$

$$= \frac{10}{7}\text{ k}\Omega \qquad \square$$

EXTENSION EXERCISE

E4.6 Find the Thévenin equivalent circuit of the network in Fig. E4.6 at the terminals A-B.

Figure E4.6

ANSWER: $R_{\text{Th}} = \frac{8}{3}\text{ k}\Omega$

Circuits Containing Both Independent and Dependent Sources

In these type of circuits we must calculate both the open-circuit voltage and short-circuit current in order to calculate the Thévenin equivalent resistance. Furthermore, we must remember that we cannot split the dependent source and its controlling variable when we break the network to find the Thévenin or Norton equivalent.

We now present two examples: one with a dependent voltage source and one with a dependent current source.

EXAMPLE 4.12

Let us use Thévenin's theorem to find V_o in the network in Fig. 4.16a.

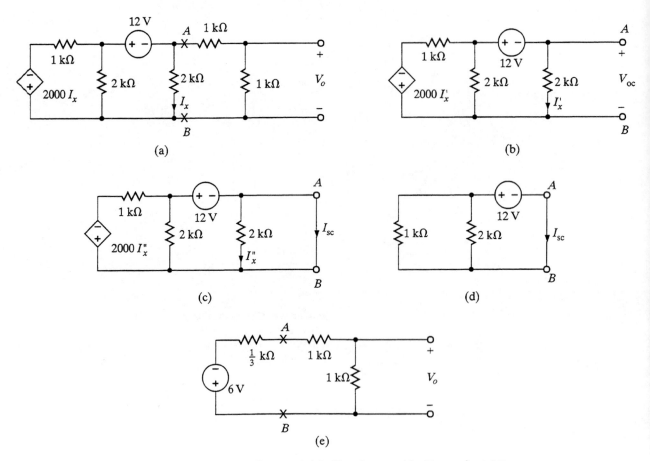

Figure 4.16 Circuits used in Example 4.12.

SOLUTION   To begin, we break the network at points A-B. Could we break it just to the right of the 12-V source? No! Why? The open-circuit voltage is calculated from the network in Fig. 4.16b. Note that we now use the source $2000I'_x$ because this circuit is different from that in Fig. 4.16a. KCL for the supernode around the 12-V source is

$$\frac{(V_{oc} + 12) - (-2000I'_x)}{1k} + \frac{V_{oc} + 12}{2k} + \frac{V_{oc}}{2k} = 0$$

where

$$I'_x = \frac{V_{oc}}{2k}$$

yielding $V_{oc} = -6$ V.

I_{sc} can be calculated from the circuit in Fig. 4.16c. Note that the presence of the short circuit forces I''_x to zero and, therefore, the network is reduced to that shown in Fig. 4.16d.

Therefore,

$$I_{sc} = \frac{-12}{\frac{2}{3}k} = -18 \text{ mA}$$

Then

$$R_{Th} = \frac{V_{oc}}{I_{sc}} = \frac{1}{3} k\Omega$$

Connecting the Thévenin equivalent circuit to the remainder of the network at terminals A-B produces the circuit in Fig. 4.16e. At this point, simple voltage division yields

$$V_o = (-6)\left(\frac{1k}{1k + 1k + \frac{1}{3}k}\right) = \frac{-18}{7} \text{ V} \qquad \square$$

EXAMPLE 4.13

Let us find V_o in the network in Fig. 3.24 using Thévenin's theorem.

SOLUTION   The network is redrawn in Fig. 4.17a. V_{oc} is determined from the network in Fig. 4.17b. Note that

$$I_1 = \frac{V'_x}{2k}$$

$$I_2 = 2 \text{ mA}$$

and

$$V'_x = 4k\left(\frac{V'_x}{2k} - 2 \times 10^{-3}\right)$$

Figure 4.17 Circuits used in Example 4.13.

Solving these equations yields $I_1 = 4$ mA and, hence,

$$V_{oc} = 2kI_1 + 3 = 11 \text{ V}$$

I_{sc} is derived from the circuit in Fig. 4.17c. Note that if we collapse the short circuit, the network is reduced to that in Fig. 4.17d. Although we have temporarily lost sight of I_{sc}, we can easily find the branch currents and they, in turn, will yield I_{sc}. KCL at the node at the bottom left of the network is

$$\frac{V_x''}{4k} = \frac{V_x''}{2000} - 2 \times 10^{-3}$$

or

$$V_x'' = 8 \text{ V}$$

Then since

$$I_3 = \frac{3}{2k} = \frac{3}{2} \text{ mA}$$

as shown in Fig. 4.17c

$$I_{sc} = \frac{V''_x}{2000} + I_3$$

$$= \frac{11}{2} \text{ mA}$$

Then

$$R_{Th} = \frac{V_{oc}}{I_{sc}} = 2 \text{ k}\Omega$$

Connecting the Thévenin equivalent circuit to the remainder of the original network produces the circuit in Fig. 4.17e. Simple voltage division yields

$$V_o = 11\left(\frac{6k}{2k + 6k}\right)$$

$$= \frac{33}{4} \text{ V} \qquad \square$$

Problem-Solving Strategy for Applying Thévenin's Theorem

- Remove the load and find the voltage across the open circuit terminals, V_{oc}. All the circuit analysis techniques presented here can be used to compute this voltage.

- Determine the Thévenin equivalent resistance of the network at the open terminals with the load removed. Three different types of circuits may be encountered in determining the resistance, R_{Th}.
    - If the circuit contains only independent sources, they are made zero by replacing the voltage sources with short circuits and the current sources with open circuits. R_{Th} is then found by computing the resistance of the purely resistive network at the open terminals.
    - If the circuit contains only dependent sources, an independent voltage or current source is applied at the open terminals and the corresponding current or voltage at these terminals is measured. The voltage/current ratio at the terminals is the Thévenin equivalent resistance. Since there is no energy source, the open-circuit voltage is zero in this case.
    - If the circuit contains both independent and dependent sources, the open-circuit terminals are shorted and the short-circuit current between these terminals is determined. The ratio of the open-circuit voltage to the short-circuit current is the resistance R_{Th}.

- If the load is now connected to the Thévenin equivalent circuit, consisting of V_{oc} in series with R_{Th}, the desired solution can be obtained.

The problem-solving strategy for Norton's theorem is essentially the same as that for Thévenin's theorem with the exception that we are dealing with the short-circuit current instead of the open-circuit voltage.

E4.7 Find V_o in the circuit in Fig. E4.7 using Thévenin's theorem.

Figure E4.7

ANSWER: $V_o = \dfrac{36}{13}\,\text{V}$.

At this point it is worthwhile to pause for a moment and reflect about what we have learned; that is, let us compare the use of node or loop analysis with that of the theorems discussed in this chapter. When we examine a network for analysis, one of the first things we should do is count the number of nodes and loops. Next we consider the number of sources. For example, are there a number of voltage sources or current sources present in the network? All these data, together with the information that we expect to glean from the network, give a basis for selecting the simplest approach. With the current level of computational power available to us, we can solve the node or loop equations that define the network in a flash.

With regard to the theorems, we have found that in some cases the theorems do not necessarily simplify the problem and a straightforward attack using node or loop analysis is as good an approach as any. This is a valid point provided that we are simply looking for some particular voltage or current. However, the real value of the theorems is the insight and understanding that they provide about the physical nature of the network. For example, superposition tells us what each source contributes to the quantity under investigation. However, a computer solution of the node or loop equations does not tell us the effect of changing certain parameter values in the circuit. It does not help us understand the concept of loading a network or the ramifications of interconnecting networks or the idea of matching a network for maximum power transfer. The theorems help us to understand the effect of using a transducer at the input of an amplifier with a given input resistance. They help us explain the effect of a load, such as a speaker, at the output of an amplifier. We derive none of this information from a node or loop analysis. In fact, as a simple example, suppose that a network at a specific pair of terminals has a Thévenin equivalent circuit consisting of a voltage source in series with a 2-kΩ resistor. If we connect a 2-Ω resistor to the network at these terminals, the voltage across the 2-Ω resistor will be essentially nothing. This result is fairly obvious using the Thévenin theorem approach; however, a node or loop analysis gives us no clue as to why we have obtained this result.

We have studied networks containing only dependent sources. This is a very important topic because all electronic devices, such as transistors, are modeled in this fashion. Motors in power systems are also modeled in this way. We use these amplification devices for many different purposes, such as the speed control for an automobile.

Finally, it is interesting to note that when we employ source transformation as we did in Example 4.6, we are simply converting back and forth between a Thévenin equivalent circuit and a Norton equivalent circuit.

4.5 MAXIMUM POWER TRANSFER

In circuit analysis we are sometimes interested in determining the maximum power that can be delivered to a load. By employing Thévenin's theorem, we can determine the maximum power that a circuit can supply and the manner in which to adjust the load to effect maximum power transfer.

Suppose that we are given the circuit shown in Fig. 4.18. The power that is delivered to the load is given by the expression

$$P_{load} = i^2 R_L$$
$$= \left(\frac{v}{R + R_L}\right)^2 R_L$$

Figure 4.18 Equivalent circuit for examining maximum power transfer.

We want to determine the value of R_L that maximizes this quantity. Hence, we differentiate this expression with respect to R_L and equate the derivative to zero.

$$\frac{dP_{load}}{dR_L} = \frac{(R + R_L)^2 v^2 - 2v^2 R_L(R + R_L)}{(R + R_L)^4} = 0$$

which yields

$$R_L = R$$

In other words, maximum power transfer takes place when the load resistance $R_L = R$. Although this is a very important result, we have derived it using the simple network in Fig. 4.18. However, we should recall that v and R in Fig. 4.18 could represent the Thévenin equivalent circuit for any linear network.

EXAMPLE 4.14

Let us find the value of R_L for maximum power transfer in the network in Fig. 4.19a and the maximum power that can be transferred to this load.

SOLUTION   To begin, we derive the Thévenin equivalent circuit for the network exclusive of the load. V_{oc} can be calculated from the circuit in Fig. 4.19b. The mesh equations for the network are

$$I_1 = 2 \times 10^{-3}$$
$$3k(I_2 - I_1) + 6kI_2 + 3 = 0$$

DRILL

D4.6 In the network in Fig. 4.18, $v = 12$ V and $R = 2$ kΩ. Determine the R_L for maximum power transfer and the maximum power transferred.

ANSWER:

$$R_L = 2 \text{ k}\Omega$$
$$P_{max} = 18 \text{ mW}$$

Figure 4.19 Circuits used in Example 4.14.

Solving these equations yields $I_2 = 1/3$ mA and, hence,

$$V_{oc} = 4kI_1 + 6kI_2$$

$$= 10 \text{ V}$$

R_{Th}, shown in Fig. 4.19c, is 6 kΩ; therefore, $R_L = R_{Th} = 6$ kΩ for maximum power transfer. The maximum power transferred to the load is

$$P_L = \left(\frac{10}{12k}\right)^2 (6k) = \frac{25}{6} \text{ mW} \qquad \square$$

EXAMPLE 4.15

Let us find R_L for maximum power transfer and the maximum power transferred to this load in the circuit in Fig. 4.20a.

SOLUTION   We wish to reduce the network to the form shown in Fig. 4.18. We could form the Thévenin equivalent circuit by breaking the network at the load. However, a close examination of the network indicates that our analysis will be simpler if we break the network to the left of the 4-kΩ resistor. When we do this, however, we must realize, as shown in Fig. 4.18, that for maximum power transfer $R_L = R_{Th} + 4$ kΩ. V_{oc} can be calculated from the network in Fig. 4.20b. Forming a su-

Figure 4.20 Circuits used in Example 4.15.

pernode around the dependent source and its connecting nodes, the KCL equation for this supernode is

$$\frac{V_{oc} - 2000I'_x}{1k + 3k} + (-4 \times 10^{-3}) + \frac{V_{oc}}{2k} = 0$$

where

$$I'_x = \frac{V_{oc}}{2k}$$

These equations yield $V_{oc} = 8\,\text{V}$. The short-circuit current can be found from the network in Fig. 4.20c. It is here that we find the advantage of breaking the network to the left of the 4-kΩ resistor. The short circuit shorts the 2-kΩ resistor and, therefore, $I''_x = 0$. Hence, the circuit is reduced to that in Fig. 4.20d, where clearly $I_{sc} = 4\,\text{mA}$. Then

$$R_{Th} = \frac{V_{oc}}{I_{sc}} = 2\,\text{k}\Omega$$

Connecting the Thévenin equivalent to the remainder of the original circuit produces the network in Fig. 4.20e. For maximum power transfer $R_L = R_{Th} + 4\,\text{k}\Omega = 6\,\text{k}\Omega$, and the maximum power transferred is

$$P_L = \left(\frac{8}{12k}\right)^2 (6k) = \frac{8}{3}\,\text{mW} \qquad \square$$

E4.8 Given the circuit in Fig. E4.8, find R_L for maximum power transfer and the maximum power transferred.

Figure E4.8

ANSWER: $R_L = 6\,\text{k}\Omega, P_L = 2/3\,\text{mW}$.

4.6 DC SPICE ANALYSIS USING SCHEMATIC CAPTURE

Introduction

Originally, SPICE (Simulation Program with Integrated Circuit Emphasis) was developed at the University of California at Berkeley. It quickly became an industry standard for simulating electronic circuits. With the advent of the PC industry, several companies began selling PC and Macintosh compatible versions of SPICE. One such company, MicroSim, produced a PC compatible version called PSPICE, which we will discuss in some detail in this text. Recently, MicroSim was purchased by OrCAD. The version of PSPICE contained here is 8.0—the last MicroSim version. Version 9.0 is under development by OrCAD and will be introduced in the book's Web site when it becomes available.

In SPICE, circuit information, such as the names and values of resistors and sources as well as how they are interconnected, is input using data statements with a specific format. The particulars for each element must be typed in a precise order. This makes debugging difficult since you must know the proper format to recognize your errors.

PSPICE, as well as other SPICE-based simulators, now employs a feature known as schematic capture. The schematic editor in version 8.0 is called *Schematics*. The *Schematics* editor permits us to bypass the cryptic formatting and simply draw the circuit diagram, assigning element values via dialog boxes. *Schematics* then converts the circuit diagram to an original SPICE format for actual simulation.

When elements such as resistors and voltage sources are given values, it is convenient to use scale factors. PSPICE supports the following list:

T–tera–10^{12}	K–kilo–10^3	N–nano–10^{-9}
G–giga–10^9	M–milli–10^{-3}	P–pico–10^{-12}
MEG–mega–10^6	U–micro–10^{-6}	F–femto–10^{-15}

The component value must be immediately followed by a character—spaces are not allowed. Any text can follow the scale factor. Finally, PSPICE is case insensitive. So there is no difference between 1 MΩ and 1 mΩ in PSPICE.

Figure 4.21 shows the five-step procedure for simulating circuits using *Schematics* and PSPICE. We will demonstrate this procedure in the following example.

Figure 4.21 The five-step procedure for PSPICE simulations.

A dc Simulation Example

The following font conventions will be used throughout this tutorial. Uppercase text refers to PSPICE programs, menus, dialog boxes, and utilities. All boldface text denotes keyboard or mouse inputs. In each instance, boldface text's case matches that in PSPICE. Let's simulate the circuit in Fig. 4.22 using PSPICE. Following the flowchart procedure shown in Fig. 4.21, the first step is to open *Schematics* using **Start/Programs/Design-Lab 8/Schematics** sequence of pop-up menus. When *Schematics* opens, our screen will change to the Schematic Editor window shown in Fig. 4.23.

Figure 4.22 A dc circuit used for simulation.

Step 1: Drawing the Schematic

Next, we obtain and place the required parts—two resistors and a dc voltage source. To get a part, left click on the **Draw** menu and select **Get New Part**, as shown in Fig. 4.24. The **Part Browser** in Fig. 4.25 appears, listing all the parts available in PSPICE. Since we do not know the *Schematics* name for a dc voltage source, select **Libraries**, and the **Library Browser** in Fig. 4.26 appears. This browser lists all the parts libraries available to us. The dc

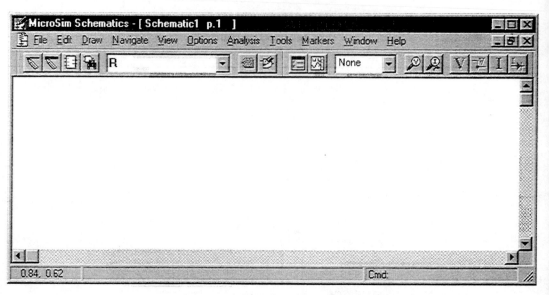

Figure 4.23 The *Schematics* editor window.

voltage source is called VDC and is located in the SOURCE.slb library that was selected and is shown in Fig. 4.26. Thus, we select the part **VDC** and click **OK**. The box in Fig. 4.25 reappears.

Figure 4.24 Getting a new part in *Schematics*.

If we now select **Place & Close**, we revert back to the Schematic Editor in Fig. 4.23 with one difference: the mouse pointer has become a dc voltage source symbol. The source can be positioned within the drawing area by moving the mouse and placed by left clicking <u>once</u>. The result is shown in Fig. 4.27. Note that *Schematics* uses the battery symbol for a dc source. Since the circuit contains only one source, we click the right mouse button to stop placing sources and the mouse pointer returns to its original status.

Next we place the resistors. We repeat the process of **Draw/Get New Part** except that this time, when the **Part Browser Basic** dialog box appears, we type in **R** and select **Place &**

Figure 4.25 The Parts Browser window.

Figure 4.26 The Library Browser window.

Figure 4.27 The *Schematics* editor after placing the dc voltage source.

Figure 4.28 The parts after placement.

Close. The mouse pointer then becomes a resistor. We left click once to place R1, *move the mouse some distance*, and left click again to place R2. To stop adding resistors, we right click once. Moving the mouse between part placements keeps the parts from stacking atop one another in the diagram. In order to make the schematic match the one in Fig. 4.22, we rotate R2 and place it at the right end of the diagram. This rotation is accomplished by first clicking on the **Edit** menu and selecting **Rotate**. This causes R2 to spin 90° counterclockwise. Next, we click on R2 and drag it to the desired location. A diagram similar to that shown in Fig. 4.28 should result. Note that the resistors are automatically assigned default values of 1 kΩ.

The parts can now be connected by going to the **Draw** menu and selecting **Wire**. The mouse pointer will then turn into a symbolic pencil. To connect the top of the source to R1 (i.e., R_a), use the mouse to place the pencil at the end of the wire stub protruding from Vin, click once, and release. Next, we move the mouse up and over to the left end of R1. A line is drawn up and over a 90° angle and appears dashed, as shown in Figure 4.29. *Dashed lines are not yet wires!* We must left click again to complete and cut the connection. The dashed lines become solid and the wiring connection is made. Excess wire fragments (extended dashed lines) can be removed by selecting **Redraw** from the **View** menu.

The two remaining wires can be drawn using a *Schematics* shortcut. To reactivate the wiring pencil, double right click. This shortcut reactivates the most recent mouse use. We simply repeat the steps listed previously to complete the wiring.

Figure 4.29 Wiring the parts.

PSPICE requires that all schematics have a ground, or reference, terminal. The voltage there will be zero and all other node voltages are referenced to it. The required part is the analog ground (AGND) in the PORT.slb library shown in Fig. 4.26. Get this part and place it at the bottom of the schematic, as shown in Fig. 4.30. Make sure that the part touches the bottom wire in the diagram so that the node dot appears in your schematic. The wiring is now completed.

Figure 4.30 The schematic after wiring and adding the AGND part.

Step 2: Changing Component Names and Values

In order to change the voltage source's name, we double click on the text, V1. The **Edit Reference Designator** dialog box appears as shown in Fig. 4.31. Simply type in the new name for the source, in this case **Vin**, and select **OK**. Next we will change the source's voltage value by double clicking on the value, as shown in Fig. 4.30. Now the **Attributes** dialog box in Figure 4.32 appears. Type in **10V** and select **OK**. In a similar manner, the two resistors are renamed Ra and Rb and their values are set at 4 kΩ and 6 kΩ, respectively. The circuit shown in Fig. 4.33 is now ready for simulation.

Figure 4.31 Renaming the voltage source.

Figure 4.32 Setting the source value.

Figure 4.33 The schematic after changing the part names and values.

Step 3: Saving the Schematic

The schematic can be saved by simply going to the **File** menu and selecting **Save**. It is better to save your work in a directory other than the main PSPICE directory. All *Schematics* files are automatically given the extension .sch.

The Netlist

The netlist is actually the old-fashioned PSPICE code listing for the diagram drawn in *Schematics*. To create the netlist, go to the **Analysis** menu in Fig. 4.23 and select **Create Netlist**. What will appear at the bottom of the *Schematics* Editor window is either the message *Netlist Created* or a dialog box informing you of netlist errors. The netlist can be viewed by returning to the **Analysis** menu and selecting **Examine Netlist**, which opens the file shown in Fig. 4.34. All three elements appear along with their proper values. The text elements $N_001 $N_002 etc. are the node numbers that *Schematics* created when it converted the diagram to a netlist.

By tracing through the node numbers, we can be assured that our circuit is properly connected. The source is 10 V positive at node 2 with respect to node 0. AGND is always node number zero. Ra connects node 2 to node 1 and Rb completes the loop between nodes 0 and 1— it is perfect.

How did PSPICE generate these particular node numbers? In PSPICE, the terminals of a part symbol are called pins and are numbered. For example, the pin numbers for the symbol that represents a resistor, dc voltage, and current source symbols are shown in Figure 4.35. None of the parts in Fig. 4.35 have been rotated. When the GET NEW PART sequence generates a part with a horizontal orientation, like the resistor, pin 1 is on the left. All vertically oriented parts (the sources) have pin 1 at the top. In the netlist, the order of the node numbers is always pin 1 and then pin 2 for each component. Furthermore, when a part is rotated, the pin numbers also rotate. The most critical aspect of the pin numbers is current direction. PSPICE always yields the current flowing into pin 1 and out of pin 2. To change the node number order of a part in the netlist, we must rotate that part 180°. This can be done by clicking on the part and selecting **Rotate** in the **Edit** menu twice.

Figure 4.34 The netlist.

Figure 4.35 Pin numbers for resistors, voltage sources and current sources.

Step 4: Simulating the Circuit

The node voltages in the circuit are typically of interest to us, and *Schematics* permits us to identify each node with a unique name. For example, if we wish to call the output node at Rb "Vout", we simply double click on the wiring at the output node and the dialog box in Fig. 4.36 will appear. Then type **Vout** in the space shown and select **OK**. The schematic will then look like that shown in Fig. 4.37.

Figure 4.36 Labeling a node.

Figure 4.37 The schematic after naming the output node.

Version 8.0 has a feature that allows all node voltages and branch currents to be displayed directly on the schematic diagram. Go to the **Analysis** menu and select **Display Results on Schematic**, as shown in Fig. 4.38. To display node voltages, choose **Enable Voltage Display**. Similarly, to display branch currents, choose **Enable Current Display**. The results will be displayed on the schematic when the simulation is performed.

Simulation begins by selecting the type of analysis we wish to perform. This is done by selecting **Setup** from the **Analysis** menu. The SETUP dialog box is shown in Fig. 4.39. A *Schematics* dc analysis is requested by selecting **Bias Point Detail** and then **Close**. Since we enabled both voltage and current displays directly on the schematic, the simulation results will include all node voltages and all branch currents.

Viewing the SETUP box in Fig. 4.39, we see that a number of different analyses could be requested (for example, a DC SWEEP). In this case, *Schematics* will ask for the name of the dc source you wish to sweep, the start and stop values, and the increment. The simulation results will contain all dc node voltages and branch currents as a function of the

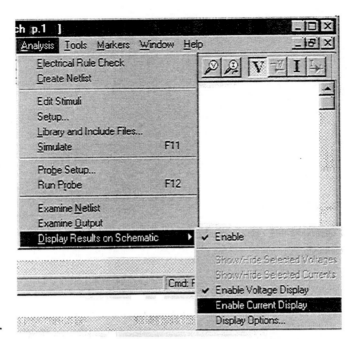

Figure 4.38 Enabling the voltage and current displays.

Figure 4.39 The Analysis Setup Box.

varying source value. Two additional analyses, TRANSIENT and AC SWEEP, will be discussed in subsequent chapters.

After exiting the analysis SETUP Box, we select **Run PSpice** from the **Analysis** menu. Once the simulation has run, your schematic should look much like that shown in Fig. 4.40 with node voltages and branch currents displayed. All node voltages are referenced to the AGND node. Current directions can be found be clicking on the branch current value. A small red arrow will appear, revealing the direction. The simulation results show $V_{\text{out}} = 6\,\text{V}$ and $I = 1\,\text{mA}$.

The use of voltage division and Ohm's law quickly confirms the simulation results; that is,

$$V_{\text{out}} = V_{\text{in}}\left[\frac{R_b}{R_b + R_a}\right] = 10\left[\frac{6\text{k}}{6\text{k} + 4\text{k}}\right] = 6\,\text{V}$$

and

$$I = \frac{10}{4\text{k} + 6\text{k}} = 1\,\text{mA}$$

Figure 4.40 The simulated circuit.

Step 5: Viewing and Printing the Results

As seen in Fig. 4.40, simulation results can be displayed on the schematic. Alternatively, all node voltages, voltage source currents, and total power dissipation are contained in a file called the output file. Let us check the simulation results displayed on the schematic against the output file data. To view this data, select **Examine Output** from the **Analysis** menu. Within the resulting long text file is a section containing the node voltages, voltage source currents, and total power dissipation, as shown in Fig. 4.41. Indeed, V_{out} is 6.00 V and the total power dissipation is 10 mW. The current bears closer inspection. PSPICE says the current through V_{in} is −1 mA. Recall from the netlist discussion that in PSPICE, current flows from a part's pin number 1 to pin 2. As seen in Fig. 4.35, pin 1 is at the positive end of the voltage source. PSPICE is telling us that the current flowing top-down through our source (counterclockwise in Fig. 4.22) is −1 mA. This agrees with both our calculation of 1 mA flowing clockwise around the loop and the current displayed in Fig. 4.40. Note that the output file is displayed in a simple text editor. So the contents of the file can be edited, saved, printed, or copied and pasted to other programs.

Figure 4.41 Simulation results in the output file.

To print the circuit diagram, first go to the **Configure** menu and select **Display Options**. When the **Display Options** dialog box comes up, deselect the **Grid On** checkbox to turn off the grid dots and select **OK**. Next, point the mouse pointer above and left of the upper left-hand corner of the diagram. Click left, and hold and drag the mouse beyond the lower right edge of the drawing. A box will grow as you drag, eventually surrounding the circuit. Go to the **File** menu and select **Print**. The dialog box in Fig. 4.42 will appear. Select the options **Only Print Selected Area** and **User Definable Zoom Factor**. For most of the small schematics you will create, a scale of 125 to 200% will do fine. Other options are self-explanatory. Finally, select **OK** to print.

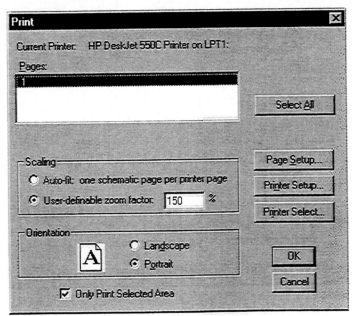

Figure 4.42 The PRINT dialog box.

To incorporate your schematic into a different program, such as Microsoft Word, draw a box around the diagram as described previously, and then under the **Edit** menu, select **Copy to Clipboard**. You can now paste the circuit into other programs.

EXAMPLE 4.16

Using the PSPICE *Schematics* editor, draw the circuit in Fig. 4.43, and use the IPROBE and VIEWPOINT parts to determine the current I and the voltage V_o.

Figure 4.43

SOLUTION The *Schematics* diagram is shown in Fig. 4.44. Note how the AGND part has been judiciously placed so that the VIEWPOINT part value is indeed V_o. From Fig. 4.44, we find $V_o = 5.2348$ V and $I = 4.759$ mA.

Figure 4.44 The schematics diagram for the network in Fig. 4.43. ❑

Let us use PSPICE to find the voltage V_o and the current I_x in the circuit in Fig. 4.45.

Figure 4.45

SOLUTION The PSPICE *Schematics* diagram is shown in Fig. 4.46. From this diagram, we find that $V_o = 150$ V and $I_x = -1.25$ A. ❑

2nd DC Example

Figure 4.46 The schematics diagram for the network in Fig. 4.45.

4.7 APPLICATIONS

EXAMPLE 4.18

On Monday afternoon, Connie suddenly remembers that she has a term paper due Tuesday morning. When she sits at her computer to start typing, she discovers that the computer mouse doesn't work. After disassembly and some inspection, she finds that the mouse contains a printed circuit board that is powered by a 5-V supply contained inside the computer case. Furthermore, the board is found to contain several resistors, some op-amps and one unidentifiable device, which is connected directly to the computer's 5-V supply as shown in Fig. 4.47a. Using a voltmeter to measure the node voltages, Connie confirms that all resistors and op-amps are functioning properly and the power supply voltage reaches the mouse board. However, without knowing the mystery device's function within the circuit, she cannot determine its condition. A phone call to the manufacturer reveals that the device is indeed linear but is also proprietary. With some persuasion, the manufacturer's representative agrees that if Connie can find the Thévenin equivalent circuit for the element at nodes *A-B* with the computer on, he will tell her if it is functioning properly. Armed with a single 1-kΩ resistor and a voltmeter, Connie attacks the problem.

SOLUTION In order to find the Thévenin equivalent for the unknown device, together with the 5-V source, Connie first isolates nodes *A* and *B* from the rest of the devices on the board to measure the open-circuit voltage. The resulting voltmeter reading is $V_{AB} = 2.4$ V. Thus, the Thévenin equivalent voltage is 2.4 V. Then she

Figure 4.47 Networks used in Example 4.18.

connects the 1-kΩ resistor at nodes A-B as shown in Fig. 4.47b. The voltmeter reading is now $V_{AB} = 0.8$ V. Using voltage division to express V_{AB} in term of V_{Th}, R_{Th} and R_{test} in Fig. 4.47b yields the expression

$$0.8 = V_{Th}\left(\frac{1k}{1k + R_{Th}}\right)$$

Solving the equations for R_{Th}, we obtain

$$R_{Th} = 2.0\ k\Omega$$

Therefore, the unknown device and the 5-V source can be represented at the terminals A-B by the Thévenin equivalent circuit shown in Fig. 4.47c. When Connie phones the manufacturer with the data, the representative informs her that the device has indeed failed.

4.8 SUMMARY

- Linearity: This property requires both additivity and homogeneity. Using this property, the voltage or current somewhere in a network can be determined by assuming a specific value for the variable and then determining what source value is required to produce it. The ratio of the specified source value to that computed from the assumed value of the variable, together with the assumed value of the variable, can be used to obtain a solution.

- In a linear network containing multiple independent sources, the principle of superposition allows us to compute any current or voltage in the network as the algebraic sum of the individual contributions of each source acting alone.

- Superposition is a linear property and does not apply to nonlinear functions such as power.

- Source transformation permits us to replace a voltage source V in series with a resistance R by a current source $I = V/R$ in parallel with the resistance R. The reverse is also true.

- Using Thévenin's theorem, we can replace some portion of a network at a pair of terminals with a voltage source V_{oc} in series with a resistor R_{Th}. V_{oc} is the open-circuit voltage at the terminals, and R_{Th} is the Thévenin equivalent resistance obtained by looking into the terminals with all independent sources made zero.

- Using Norton's theorem, we can replace some portion of a network at a pair of terminals with a current source I_{sc} in parallel with a resistor R_{Th}. I_{sc} is the short-circuit current at the terminals and R_{Th} is the Thévenin equivalent resistance.

- Maximum power transfer can be achieved by selecting the load R_L to be equal to R_{Th} found by looking into the network from the load terminals.

- dc PSPICE with Schematic Capture is an effective tool in analyzing dc circuits.

PROBLEMS

Section 4.1

4.1 Find V_o in the network in Fig. P4.1 using linearity and the assumption that $V_o = 1$ mV.

Figure P4.1

Find I_2 using $V_o = 1$ mV and then use the procedure in Example 4.1.

4.2 Find V_o in the network in Fig. P4.2 using linearity and the assumption that $V_o = 1$ mV.

Figure P4.2

Find I_3 and use it to find V_4. Then find I_2. Follow the procedure in Example 4.1.

4.3 Find I_o in the network in Fig. P4.3 using linearity and the assumption that $I_o = 1$ mA.

Find V_2 from $I_o = 1$ mA. Find I_3 and then find I_2. Repeat the procedure until the source current is obtained. Then use the linearity to obtain I_o.

Figure P4.3

4.4 Find I_o in the network in Fig. P4.4 using linearity and the assumption that $I_o = 1$ mA.

Similar to P4.3

Figure P4.4

Section 4.2

4.5 Find V_o in the network in Fig. P4.5 using superposition.

Follow the procedure in Example 4.2.

Figure P4.5

4.6 Find V_o in the network in Fig. P4.6 using superposition.

Similar to P4.5

Figure P4.6

4.7   Find V_o in the network in Fig. P4.7 using superpositon.

Figure P4.7

Similar to P4.5

4.8   Find I_o in the network in Fig. P4.8 using superpositon.

Figure P4.8

Similar to P4.6

4.9   Find V_o in the network in Fig. P4.9 using superpositon.

Figure P4.9

Similar to P4.6.

4.10 Find I_o in the network in Fig. P4.10 using superposition.

Figure P4.10

Similar to P4.6

4.11 Find V_o in the network in Fig. P4.11 using superpositon.

Similar to P4.5

Figure P4.11

4.12 Find I_o in the network in Fig. P4.12 using superpositon.

Similar to P4.6

Figure P4.12

Section 4.3

4.13 Find V_o in the network in Fig. P4.13 using source transformation.

Use the procedure illustrated in Example 4.6.

Figure P4.13

Similar to P4.13

4.14 Find V_o in the network in Fig. P4.11 using source transformation.

Similar to P4.13

4.15 Find V_o in the network in Fig. P4.9 using source transformation.

4.16 Find I_o in the network in Fig. P4.16 using source transformation.

Figure P4.16

Similar to P4.13

4.17 Find V_o in the network in Fig. P4.17 using source transformation.

Figure P4.17

Similar to P4.13

4.18 Use source transformation to find I_o in the network in Fig. P4.18.

Figure P4.18

Similar to P4.13

4.19 Find I_o in the circuit in Fig. P4.18 using source transformation.

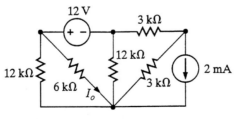

Figure P4.19

Similar to P4.13

4.20 Find V_o in the network in Fig. P4.20 using source transformation.

Similar to P4.13

Figure P4.20

Section 4.4

4.21 Use Thévenin's theorem to find V_o in the network in Fig. P4.21.

Similar to Example 4.7

Figure P4.21

4.22 Find V_o in the network in Fig. P4.22 using Thévenin's theorem.

Similar to P4.21

Figure P4.22

4.23 Find I_o in the network in Fig. P4.23 using Thévenin's theorem.

Similar to P4.21

Figure P4.23

4.24 Find I_o in the network in Fig. P4.24 using Thévenin's theorem.

Figure P4.24

Similar to P4.22

4.25 Find I_o in the network in Fig. P4.25 using Thévenin's theorem.

Figure P4.25

Similar to P4.22

4.26 Find I_o in the network in Fig. P4.26 using Thévenin's theorem.

Figure P4.26

Similar to P4.22

4.27 Find V_o in the network in Fig. P4.27 using Thévenin's theorem.

Figure P4.27

Similar to P4.22

4.28 Find V_o in the network in Fig. P4.28 using Thévenin's theorem.

Similar to P4.22

Figure P4.28

4.29 Find I_o in the network in Fig. P4.29 using Thévenin's theorem.

Similar to P4.22

Figure P4.29

4.30 Find I_o in the network in Fig. P4.30 using Thévenin's theorem.

Similar to P4.22 with the exception that the 6-V source is the load.

Figure P4.30

4.31 Find I_o in the network in Fig. P4.31 using Thévenin's theorem.

Similar to P4.22

Figure P4.31

4.32 Find the Thévenin equivalent of the network in Fig. P4.32 at the terminals A-B.

Figure P4.32

See the procedure in Example 4.10.

4.33 Find the Thévenin equivalent of the network in Fig. P4.33 at the terminals A-B.

Figure P4.33

See the procedure in Example 4.11.

4.34 Find V_o in the network in Fig. P4.34 using Thévenin's theorem.

Figure P4.34

Follow the procedure used in Example 4.12.

4.35 Use Thévenin's theorem to find I_o in the circuit in Fig. P4.35.

Figure P4.35

Similar to P4.34

4.36 Find V_o in the network in Fig. P4.36 using Thévenin's theorem.

Similar to P4.34

Figure P4.36

4.37 Use Thévenin's theorem to find V_o in the circuit in Fig. P4.37.

Use the procedure illustrated in Example 4.13.

Figure P4.37

4.38 Find V_o in the network in Fig. P4.38 using Thévenin's theorem.

Similar to P4.34

Figure P4.38

4.39 Find V_o in the circuit in Fig. P4.39 using Thévenin's theorem.

Similar to P4.37

Figure P4.39

4.40 Find V_o in the network in Fig. P4.40 using Thévenin's theorem.

Figure P4.40

Similar to P4.37

4.41 Find I_o in the network in Fig. P4.41 using Norton's theorem.

Figure P4.41

See Example 4.9

4.42 Find I_o in the network in Fig. P4.42 using Norton's theorem.

Figure P4.42

Similar to P4.41

4.43 Find I_o in the network in Fig. P4.43 using Norton's theorem.

Figure P4.43

Similar to P4.41

4.44 Find V_o in the network in Fig. P4.44 using Norton's theorem.

See Example 4.9.

Figure P4.44

Section 4.5

4.45 Find R_L in the network in Fig. P4.45 in order to achieve maximum power transfer.

Use the procedure outlined in Example 4.14.

Figure P4.45

4.46 In the network in Fig. P4.46, find R_L, for maximum power transfer and the maximum power transferred to this load.

Similar to P4.45

Figure P4.46

4.47 Find R_L for maximum power transfer and the maximum power that can be transferred to the load in Fig. P4.47.

Similar to P4.45

Figure P4.47

4.48 In the network in Fig. P4.48, find R_L for maximum power transfer and the maximum power transferred to this load.

Similar to P4.45

Figure P4.48

4.49 Find R_L for maximum power transfer and the maximum power that can be transferred to the load in Fig. P4.49.

Use the procedure outlined in Example 4.15.

Figure P4.49

Section 4.6

4.50 Using the PSPICE *Schematics* editor, draw the circuit in Fig. P4.50, and use the Enable Voltage and Current Displays feature parts to determine the current I and the voltage V_o.

Figure P4.50

4.51 Using the PSPICE *Schematics* editor, draw the circuit in Fig. P4.51, and use the Enable Voltage and Current Displays feature parts to determine the current I and the voltages V_1 and V_2.

Figure P4.51

Advanced Problems

4.52 Find V_o in the circuit in Fig. P4.52 using linearity and the assumption that $V_o = 1\,\text{V}$.

Figure P4.52

4.53 Find I_o in the network in Fig. P4.53 using superposition.

Figure P4.53

4.54 Use superposition to find V_o in the circuit in Fig. P4.54.

Figure P4.54

4.55 Find V_o in the circuit in Fig. P4.55 using superpositon.

Figure P4.55

4.56 Find I_o the network in Fig. P4.56 using superpositon.

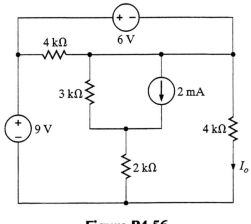

Figure P4.56

4.57 Find V_o in the network in Fig. P4.57 using superposition.

Figure P4.57

4.58 Find I_o the network in Fig. P4.58.

Figure P4.58

4.59 Find V_o in the network in Fig. P4.59.

Figure P4.59

4.60 Find I_o in the network in Fig. P4.60 using source transformation.

Figure P4.60

4.61 Find I_o in the network in Fig. P4.61 using source transformation.

Figure P4.61

4.62 Use source transformation to find V_o in the network in Fig. P4.62.

Figure P4.62

4.63 Find I_o the network in Fig. P4.63 using source transformation.

Figure P4.63

4.64 Find I_o in the network in Fig. P4.64.

Figure P4.64

4.65 Find V_o in the network in Fig. P4.65 using source transformation.

Figure P4.65

4.66 Find V_o in the network in Fig. P4.66 using Thévenin's theorem.

Figure P4.66

4.67 Find V_o in the network in Fig. P4.67 using Thévenin's theorem.

Figure P4.67

4.68 Find V_o in the network in Fig. P4.68 using Thévenin's theorem.

Figure P4.68

4.69 Find V_o in the network in Fig. P4.69 using Thévenin's theorem.

Figure P4.69

4.70 Find I_o in the network in Fig. P4.70 using Thévenin's theorem.

Figure P4.70

4.71 Use a combination of Thévenin's theorem and superposition to find V_o in the circuit in Fig. P4.71.

Figure P4.71

4.72 Find V_o in the network in Fig. P4.72 using Thévenin's theorem.

Figure P4.72

4.73 Use Norton's theorem to find V_o in the network in Fig. P4.73.

Figure P4.73

4.74 Find V_o in the network in Fig. P4.74 using Norton's theorem.

Figure P4.74

4.75 Find V_o in the network in Fig. P4.75 using Norton's theorem.

Figure P4.75

4.76 Find R_L for maximum power transfer and the maximum power that can be transferred in the network in Fig. P4.76.

Figure P4.76

4.77 Find R_L for maximum power transfer and the maximum power that can be transferred in the network in Fig. P4.77.

Figure P4.77

4.78 Find R_L for maximum power transfer and the maximum power that can be transferred in the network in Fig. P4.78.

Figure P4.78

4.79 Find the Thévenin equivalent of the network in Fig. P4.79 at the terminals A-B.

Figure P4.79

4.80 Given the linear circuit in Fig. P4.80, it is known that when a 2-kΩ load is connected to the terminals A-B, the load current is 10 mA. If a 10-kΩ load is connected to the terminals, the load current is 6 mA. Find the current in a 20-kΩ load.

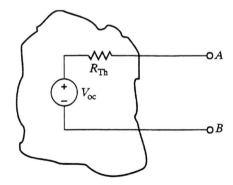

Figure P4.80

4.81 If an 8-kΩ load is connected to the terminals of the network in Fig. P4.81, $V_{AB} = 16$ V. If a 2-kΩ load is connected to the terminals, $V_{AB} = 8$ V. Find V_{AB} if a 20-kΩ load is connected to the terminals.

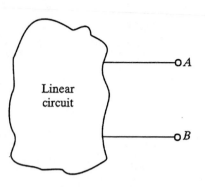

Figure P4.81

4.82 Choose R_L in Fig. P4.82 such that $I = 0.2$ mA.

Figure P4.82

4.83 Choose R_L in Fig. P4.82 for maximum power transfer.

Capacitance and Inductance

5

*H*ave you ever wondered how a tiny camera battery is able to produce a blinding flash or how a hand-held "stun gun" can deliver 50,000 V? The answer is energy storage, and we introduce two elements that possess this property in this chapter: the capacitor and the inductor. Both capacitors and inductors are linear elements; however, unlike the resistor, their terminal characteristics are described by linear differential equations. Another distinctive feature of these elements is their ability to absorb energy from the circuit, store it temporarily, and later return it. Elements that possess this energy storage capability are referred to simply as *storage elements*.

Capacitors are capable of storing energy when a voltage is present across the element. The energy is actually stored in an electric field not unlike that produced by sliding across a car seat on a dry winter day. Conversely, inductors are capable of storing energy when a current is passing through them, causing a magnetic field to form. This phenomenon can be demonstrated by placing a needle compass in the vicinity of a current. The current causes a magnetic field whose energy deflects the compass needle.

A very important circuit, which employs a capacitor in a vital role, is also introduced. This circuit, known as an op-amp integrator, produces an output voltage that is proportional to the integral of the input voltage. The significance of this circuit is that any system (for example, electrical, mechanical, hydraulic, biological, social, economic, and so on) that can be described by a set of linear differential equations with constant coefficients can be modeled by a network consisting of op-amp integrators. Thus, very complex and costly systems can be tested safely and inexpensively prior to construction and implementation.

Finally, we examine some practical circuits where capacitors and inductors are normally found or can be effectively used in circuit design.

5.1 CAPACITORS

Note the use of the passive sign convention.

A *capacitor* is a circuit element that consists of two conducting surfaces separated by a non-conducting, or *dielectric*, material. A simplified capacitor and its electrical symbol are shown in Fig. 5.1.

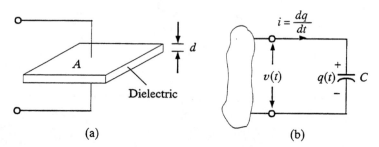

Figure 5.1 Capacitor and its electrical symbol.

There are many different kinds of capacitors, and they are categorized by the type of dielectric material that is used between the conducting plates. Although any good insulator can serve as a dielectric, each type has characteristics that make it more suitable for particular applications.

For general applications in electronic circuits (e.g., coupling between stages of amplification) the dielectric material may be paper impregnated with oil or wax, mylar, polystyrene, mica, glass, or ceramic.

Ceramic dielectric capacitors constructed of barium titanates have a large capacitance-to-volume ratio because of their high dielectric constant. Mica, glass, and ceramic dielectric capacitors will operate satisfactorily at high frequencies.

Aluminum electrolytic capacitors, which consist of a pair of aluminum plates separated by a moistened borax paste electrolyte, can provide high values of capacitance in small volumes. They are typically used for filtering, bypassing, and coupling, and in power supplies and motor-starting applications. Tantalum electrolytic capacitors have lower losses and more stable characteristics than those of aluminum electrolytic capacitors. Figure 5.2 shows a variety of typical capacitors.

In addition to these capacitors, which we deliberately insert in a network for specific applications, stray capacitance is present any time there is a difference in potential between two conducting materials separated by a dielectric. Because this stray capacitance can cause unwanted coupling between circuits, extreme care must be exercised in the layout of electronic systems on printed circuit boards.

Capacitance is measured in coulombs per volt or farads. The unit *farad* (F) is named after Michael Faraday, a famous English physicist. Capacitors may be fixed or variable and typically range from thousands of microfarads (μF) to a few picofarads (pF).

However, capacitor technology, initially driven by the modern interest in electric vehicles, is rapidly changing. For example, the capacitor in the photograph in Fig. 5.3 was designed and built at the Space Power Institute at Auburn University in order to achieve high energy and high power density capacitors for application in electric vehicles and space. This capacitor, which is specially constructed, is rated at 55 F, 500 joules (J). It is interesting to calculate the dimensions of a simple equivalent capacitor consisting of two parallel plates

Figure 5.2 Some typical capacitors (courtesy of Cornell Dubilier).

Figure 5.3 A 55-F, 500-J capacitor and a 10-in ruler for comparison.

each of area A, separated by a distance d as shown in Fig. 5.1. We learned in basic physics that the capacitance of two parallel plates of area A, separated by distance d, is

$$C = \frac{\varepsilon_o A}{d}$$

where ε_o, the permitivity of free space, is 8.85×10^{-12} F/m. If we assume the plates are separated by a distance in air of the thickness of one sheet of oil-impregnated paper, which is about 1.016×10^{-4} m, then

$$55\,\text{F} = \frac{\left(8.85 \times 10^{-12}\right)A}{1.016 \times 10^{-4}}$$

$$A = 6.3141 \times 10^8\,\text{m}^2$$

and since 1 square mile is equal to 2.59×10^6 square meters, the area is

$$A \approx 244 \text{ square miles}$$

which is the area of a medium-sized city! It would now seem that the capacitor in the photograph is much more impressive than it originally appeared. This capacitor is actually constructed using multiple layers of a combination of pressed carbon and metal fibers, which are heat treated to adhere to metal foil. The metal foils are connected to form a stack of 18 cells, within this geometry. There are literally millions of pieces of carbon and metal fibers employed to obtain the required surface area.

Suppose now that a source is connected to the capacitor shown in Fig. 5.1; then positive charges will be transferred to one plate and negative charges to the other. The charge on the capacitor is proportional to the voltage across it such that

$$q = Cv \qquad\qquad\qquad \textbf{5.1}$$

where C is the proportionality factor known as the capacitance of the element in farads.

The charge differential between the plates creates an electric field that stores energy. Because of the presence of the dielectric, the conduction current that flows in the wires that connect the capacitor to the remainder of the circuit cannot flow internally between the plates. However, via electromagnetic field theory it can be shown that this conduction current is equal to the displacement current that flows between the plates of the capacitor and is present any time that an electric field or voltage varies with time.

Our primary interest is in the current–voltage terminal characteristics of the capacitor. Since the current is

$$i = \frac{dq}{dt}$$

then for a capacitor

$$i = \frac{d}{dt} Cv$$

DRILL

D5.1 Write the i-v relationship for the following capacitors.

(a)

(b)

ANSWER:

(a) $i(t) = -C\dfrac{dv(t)}{dt}$

(b) $i(t) = -C\dfrac{dv(t)}{dt}$

which for constant capacitance is

$$i = C\frac{dv}{dt} \qquad \text{5.2}$$

Equation (5.2) can be rewritten as

$$dv = \frac{1}{C} i\, dt$$

Now integrating this expression from $t = -\infty$ to some time t and assuming $v(-\infty) = 0$ yields

$$v(t) = \frac{1}{C} \int_{-\infty}^{t} i(x)\, dx \qquad \text{5.3}$$

where $v(t)$ indicates the time dependence of the voltage. Equation (5.3) can be expressed as two integrals, so that

$$v(t) = \frac{1}{C} \int_{-\infty}^{t_0} i(x)\, dx + \frac{1}{C} \int_{t_0}^{t} i(x)\, dx \qquad \text{5.4}$$

$$= v(t_0) + \frac{1}{C} \int_{t_0}^{t} i(x)\, dx$$

where $v(t_0)$ is the voltage due to the charge that accumulates on the capacitor from time $t = -\infty$ to time $t = t_0$.

The energy stored in the capacitor can be derived from the power that is delivered to the element. This power is given by the expression

$$p(t) = v(t)i(t) = Cv(t)\frac{dv(t)}{dt} \qquad \text{5.5}$$

and hence the energy stored in the electric field is

$$w_C(t) = \int_{-\infty}^{t} Cv(x)\frac{dv(x)}{dx}\,dx = C\int_{-\infty}^{t} v(x)\frac{dv(x)}{dx}\,dx$$

$$= C\int_{v(-\infty)}^{v(t)} v(x)\,dv(x) = \tfrac{1}{2}Cv^2(x)\Big|_{v(-\infty)}^{v(t)}$$

$$= \tfrac{1}{2}Cv^2(t)\ \text{J} \qquad\qquad\qquad \textbf{5.6}$$

since $v(t = -\infty) = 0$. The expression for the energy can also be written using Eq. (5.1) as

$$w_C(t) = \frac{1}{2}\frac{q^2(t)}{C} \qquad\qquad\qquad \textbf{5.7}$$

Equations (5.6) and (5.7) represent the energy stored by the capacitor, which, in turn, is equal to the work done by the source to charge the capacitor.

The polarity of the voltage across a capacitor being charged is shown in Fig. 5.1b. In the ideal case the capacitor will hold the charge for an indefinite period of time if the source is removed. If at some later time an energy-absorbing device (e.g., a flash bulb) is connected across the capacitor, a discharge current will flow from the capacitor and, therefore, the capacitor will supply its stored energy to the device.

EXAMPLE 5.1

If the charge accumulated on two parallel conductors charged to 12 V is 600 pC, what is the capacitance of the parallel conductors?

SOLUTION   Using Eq. (5.1), we find that

$$C = \frac{Q}{V} = \frac{(600)(10^{-12})}{12} = 50\ \text{pF} \qquad \square$$

EXAMPLE 5.2

The voltage across a 5-µF capacitor has the waveform shown in Fig. 5.4a. Determine the current waveform.

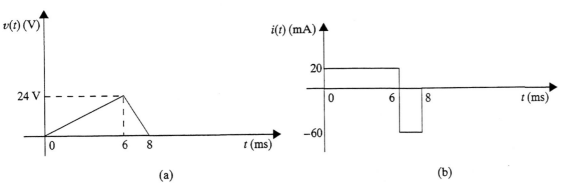

(a)

(b)

Figure 5.4 Voltage and current waveforms for a 5-µF capacitor.

SOLUTION Note that

$$v(t) = \frac{24}{6 \times 10^{-3}} t \qquad\qquad 0 \le t \le 6 \text{ ms}$$

$$= \frac{-24}{2 \times 10^{-3}} t + 96 \qquad 6 \le t < 8 \text{ ms}$$

$$= 0 \qquad\qquad\qquad\quad 8 \text{ ms} \le t$$

Using Eq. (5.2), we find that

$$i(t) = C \frac{dv(t)}{dt}$$

$$= 5 \times 10^{-6}(4 \times 10^3) \qquad 0 \le t \le 6 \text{ ms}$$

$$= 20 \text{ mA} \qquad\qquad\qquad 0 \le t \le 6 \text{ ms}$$

$$i(t) = 5 \times 10^{-6}(-12 \times 10^3) \qquad 6 \le t \le 8 \text{ ms}$$

$$= -60 \text{ mA} \qquad\qquad\qquad 6 \le t < 8 \text{ ms}$$

and

$$i(t) = 0 \qquad\qquad\qquad 8 \text{ ms} \le t$$

Therefore, the current waveform is as shown in Fig. 5.4b and $i(t) = 0$ for $t > 8$ ms.

❏

EXAMPLE 5.3

Determine the energy stored in the electric field of the capacitor in Example 5.2 at $t = 6$ ms.

SOLUTION Using Eq. (5.6), we have

$$w(t) = \tfrac{1}{2} C v^2(t)$$

At $t = 6$ ms,

$$w(6 \text{ ms}) = \tfrac{1}{2}(5 \times 10^{-6})(24)^2$$

$$= 1440 \text{ μJ}$$

❏

EXTENSION EXERCISE

E5.1 A 10-μF capacitor has an accumulated charge of 500 nC. Determine the voltage across the capacitor.

ANSWER: 0.05 V.

EXAMPLE 5.4

The current in an initially uncharged 4-μF capacitor is shown in Fig. 5.5a. Let us derive the waveforms for the voltage, power, and energy and compute the energy stored in the electric field of the capacitor at $t = 2$ ms.

(a)

(b)

(c)

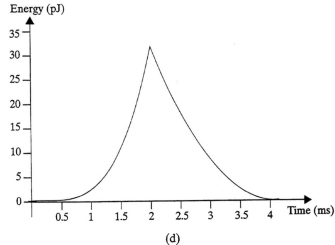

(d)

Figure 5.5 Waveforms used in Example 5.4.

SOLUTION   The equations for the current waveform in the specific time intervals are

$$i(t) = \frac{16 \times 10^{-6}\, t}{2 \times 10^{-3}} \qquad 0 \le t \le 2 \text{ ms}$$

$$= -8 \times 10^{-6} \qquad 2 \text{ ms} \le t \le 4 \text{ ms}$$

$$= 0 \qquad 4 \text{ ms} < t$$

Since $v(0) = 0$, the equation for $v(t)$ in the time interval $0 \le t \le 2$ ms is

$$v(t) = \frac{1}{(4)(10^{-6})} \int_0^t 8(10^{-3})x \, dx = 10^3 t^2$$

and hence,

$$v(2\text{ ms}) = 10^3(2 \times 10^{-3})^2 = 4\text{ mV}$$

In the time interval $2\text{ ms} \le t \le 4\text{ ms}$,

$$v(t) = \frac{1}{(4)(10^{-6})} \int_{2(10^{-3})}^{t} - (8)(10^{-6})\, dx + (4)(10^{-3}) = -2t + 8 \times 10^{-3}$$

The waveform for the voltage is shown in Fig. 5.5b.

Since the power is $p(t) = v(t)i(t)$, the expression for the power in the time interval $0 \le t \le 2\text{ ms}$ is $p(t) = 8t^3$. In the time interval $2\text{ ms} \le t \le 4\text{ ms}$, the equation for the power is

$$p(t) = -(8)(10^{-6})(-2t + 8 \times 10^{-3})$$
$$= 16(10^{-6})t - 64(10^{-9})$$

The power waveform is shown in Fig. 5.5c. Note that during the time interval $0 \le t \le 2$ ms, the capacitor is absorbing energy and during the interval $2\text{ ms} \le t \le 4\text{ ms}$, it is delivering energy.

The energy is given by the expression

$$w(t) = \int_{t_0}^{t} p(x)\, dx + w(t_0)$$

In the time interval $0 \le t \le 2\text{ ms}$,

$$w(t) = \int_{0}^{t} 8x^3\, dx = 2t^4$$

Hence,

$$w(2\text{ ms}) = 32\text{ pJ}$$

In the time interval $2 \le t \le 4\text{ ms}$,

$$w(t) = \int_{2 \times 10^{-3}}^{t} \left[(16 \times 10^{-6})x - (64 \times 10^{-9})\right] dx + 32 \times 10^{-12}$$
$$= \left[(8 \times 10^{-6})x^2 - (64 \times 10^{-9})x\right]_{2 \times 10^{-3}}^{t} + 32 \times 10^{-12}$$
$$= (8 \times 10^{-6})t^2 - (64 \times 10^{-9})t + 128 \times 10^{-12}$$

From this expression we find that $w(2\text{ ms}) = 32$ pJ and $w(4\text{ ms}) = 0$. The energy waveform is shown in Fig. 5.5d. ❑

EXTENSION EXERCISES

E5.2 The voltage across a 2-μF capacitor is shown in Fig. E5.2. Determine the waveform for the capacitor current.

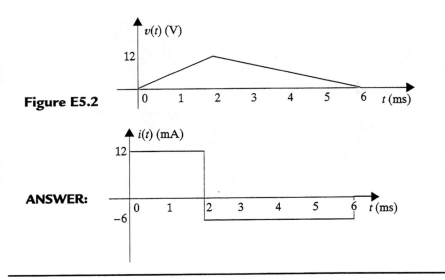

Figure E5.2

ANSWER:

E5.3   Compute the energy stored in the electric field of the capacitor in Extension Exercise E5.2 at $t = 2$ ms.

ANSWER: $w = 144\ \mu\text{J}$.

5.2   INDUCTORS

An *inductor* is a circuit element that consists of a conducting wire usually in the form of a coil. Two typical inductors and their electrical symbols are shown in Fig. 5.6. Inductors are typically categorized by the type of core on which they are wound. For example, the core material may be air or any nonmagnetic material, iron, or ferrite. Inductors made with air or nonmagnetic materials are widely used in radio, television, and filter circuits. Iron-core inductors are used in electrical power supplies and filters. Ferrite-core inductors are widely used in high-frequency applications. Note that in contrast to the magnetic core that confines the flux, as shown in Fig. 5.6b, the flux lines for nonmagnetic inductors extend beyond the inductor itself, as illustrated in Fig. 5.6a. Like stray capacitance, stray inductance can result from any element carrying current surrounded by flux linkages. Figure 5.7 shows a variety of typical inductors.

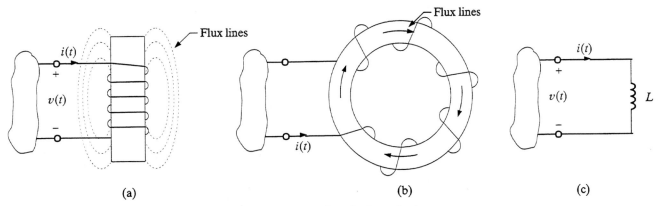

Figure 5.6 Two inductors and their electrical symbols.

Figure 5.7 Some typical inductors (courtesy of Coilcraft).

From a historical standpoint, developments that led to the mathematical model we employ to represent the inductor are as follows. It was first shown that a current-carrying conductor would produce a magnetic field. It was later found that the magnetic field and the current that produced it were linearly related. Finally, it was shown that a changing magnetic field produced a voltage that was proportional to the time rate of change of the current that produced the magnetic field; that is,

$$v(t) = L \frac{di(t)}{dt} \qquad \text{5.8}$$

The constant of proportionality L is called the inductance and is measured in the unit henry, named after the American inventor Joseph Henry, who discovered the relationship. As seen in Eq. (5.8), 1 henry (H) is dimensionally equal to 1 volt-second per ampere.

Following the development of the mathematical equations for the capacitor, we find that the expression for the current in an inductor is

$$i(t) = \frac{1}{L} \int_{-\infty}^{t} v(x)\, dx \qquad \text{5.9}$$

which can also be written as

$$i(t) = i(t_0) + \frac{1}{L} \int_{t_0}^{t} v(x)\, dx \qquad \text{5.10}$$

The power delivered to the inductor can be used to derive the energy stored in the element. This power is equal to

Note the use of the passive sign convention, as illustrated in Fig. 5.6.

DRILL

D5.2   Write the i-v relationship for the following inductors.

(a)

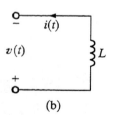

(b)

$$p(t) = v(t)i(t)$$

$$= \left[L \frac{di(t)}{dt} \right] i(t)$$

ANSWER:

5.11   (a) $v(t) = -L \dfrac{di(t)}{dt}$

(b) $v(t) = L \dfrac{di(t)}{dt}$

Therefore, the energy stored in the magnetic field is

$$w_L(t) = \int_{-\infty}^{t} \left[L \frac{di(x)}{dx} \right] i(t) \, dx$$

Following the development of Eq. (5.6), we obtain

$$w_L(t) = \tfrac{1}{2} L i^2(t) \text{ J} \qquad \textbf{5.12}$$

The inductor, like the resistor and capacitor, is a passive element. The polarity of the voltage across the inductor is shown in Fig. 5.6.

Practical inductors typically range from a few microhenrys to tens of henrys. From a circuit design standpoint it is important to note that inductors cannot be easily fabricated on an integrated circuit chip, and therefore chip designs typically employ only active electronic devices, resistors and capacitors, that can be easily fabricated in microcircuit form.

EXAMPLE 5.5

The current in a 10-mH inductor has the waveform shown in Fig. 5.8a. Determine the voltage waveform.

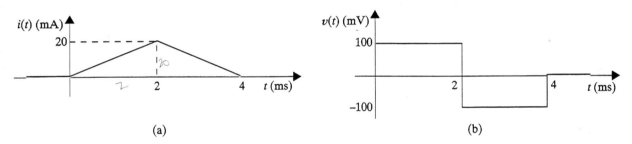

(a)   (b)

Figure 5.8 Current and voltage waveforms for a 10-mH inductor.

SOLUTION   Using Eq. (5.8) and noting that

$$i(t) = \frac{20 \times 10^{-3} t}{2 \times 10^{-3}} \qquad 0 \le t \le 2 \text{ ms}$$

$$i(t) = \frac{-20 \times 10^{-3} t}{2 \times 10^{-3}} + 40 \times 10^{-3} \quad 2 \le t \le 4 \text{ ms}$$

and

$$i(t) = 0 \qquad 4 \text{ ms} < t$$

we find that

$$v(t) = \left(10 \times 10^{-3} \right) \frac{20 \times 10^{-3}}{2 \times 10^{-3}} \quad 0 \le t \le 2 \text{ ms}$$

$$= 100 \text{ mV}$$

and

$$v(t) = (10 \times 10^{-3}) \frac{-20 \times 10^{-3}}{2 \times 10^{-3}} \quad 2 \leq t \leq 4 \text{ ms}$$

$$= -100 \text{ mV}$$

and $v(t) = 0$ for $t > 4$ ms. Therefore, the voltage waveform is shown in Fig. 5.8b.

❑

EXAMPLE 5.6

The current in a 2-mH inductor is

$$i(t) = 2 \sin 377t \text{ A}$$

Determine the voltage across the inductor and the energy stored in the inductor.

SOLUTION   From Eq. (5.8), we have

$$v(t) = L \frac{di(t)}{dt}$$

$$= (2 \times 10^{-3}) \frac{d}{dt} (2 \sin 377t)$$

$$= 1.508 \cos 377t \text{ V}$$

and from Eq. (5.12),

$$w_L(t) = \tfrac{1}{2} L i^2(t)$$

$$= \tfrac{1}{2}(2 \times 10^{-3})(2 \sin 377t)^2$$

$$= 0.004 \sin^2 377t \text{ J}$$

❑

EXAMPLE 5.7

The voltage across a 200-mH inductor is given by the expression

$$v(t) = (1 - 3t)e^{-3t} \text{ mV} \quad t \geq 0$$

$$= 0 \qquad\qquad\quad t < 0$$

Let us derive the waveforms for the current, energy, and power.

SOLUTION   The waveform for the voltage is shown in Fig. 5.9a. The current is derived from Eq. (5.10) as

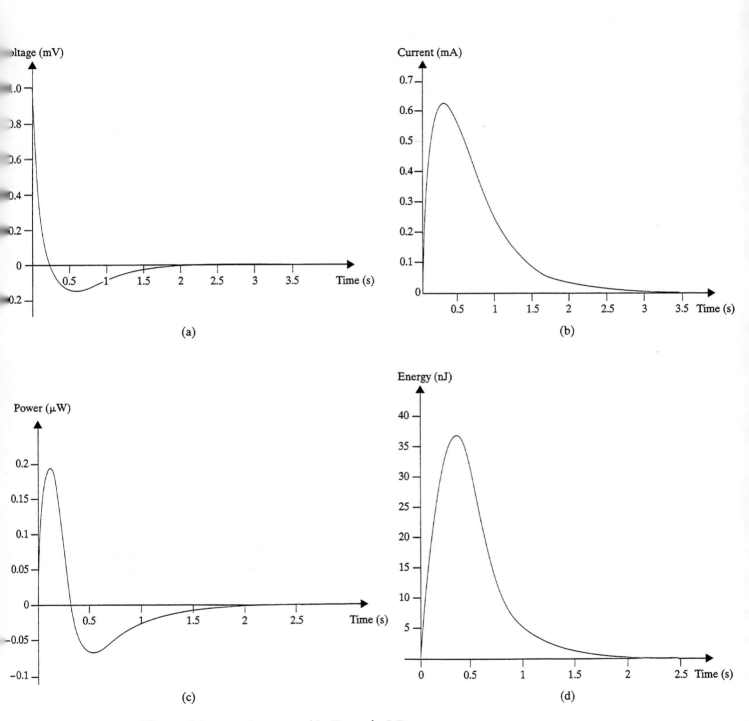

Figure 5.9 Waveforms used in Example 5.7.

$$i(t) = \frac{10^3}{200}\int_0^t (1 - 3x)e^{-3x}\, dx$$

$$= 5\left\{\int_0^t e^{-3x}\, dx - 3\int_0^t xe^{-3x}\, dx\right\}$$

$$= 5\left\{\frac{e^{-3x}}{-3}\bigg|_0^t - 3\left[-\frac{e^{-3x}}{9}(3x + 1)\right]_0^t\right\}$$

$$= 5te^{-3t}\,\text{mA} \qquad t \geq 0$$

$$= 0 \qquad t < 0$$

A plot of the current waveform is shown in Fig. 5.9b.
The power is given by the expression

$$p(t) = v(t)i(t)$$

$$= 5t(1 - 3t)e^{-6t}\,\mu\text{W} \qquad t \geq 0$$

$$= 0 \qquad t < 0$$

The equation for the power is plotted in Fig. 5.9c.
The expression for the energy is

$$w(t) = \tfrac{1}{2}Li^2(t)$$

$$= 2.5t^2 e^{-6t}\,\mu\text{J} \qquad t \geq 0$$

$$= 0 \qquad t < 0$$

This equation is plotted in Fig. 5.9d.   ❏

EXTENSION EXERCISES

E5.4 The current in a 5-mH inductor has the waveform shown in Fig. E5.4.
Compute the waveform for the inductor voltage.

Figure E5.4

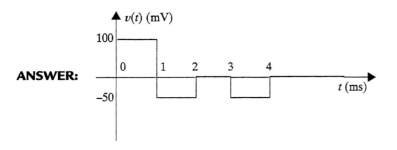

ANSWER:

E5.5   Compute the energy stored in the magnetic field of the inductor in Extension Exercise E5.4 at $t = 1.5$ ms.

ANSWER: $W = 562.5$ nJ.

5.3 FUNDAMENTAL CAPACITOR/INDUCTOR CHARACTERISTICS

Capacitors and inductors have a number of extremely important characteristics. One of the first things that the reader may have noticed is the similarity of the defining equations. Capacitors and inductors have a dual relationship; that is, their defining equations are identical if we interchange C with L and i with v, and vice versa, as shown in Table 5.1. Because of this relationship, the elements have a number of complementary features. For example, if the voltage across a capacitor is constant (i.e., non-time varying), the current through it is zero and, therefore, the capacitor looks like an open circuit to dc. Similarly, if the current in an inductor is constant, the voltage across it is zero and, therefore, the inductor looks like a short circuit to dc. An instantaneous jump in the voltage across a capacitor is not physically realizable because it requires the movement of a finite amount of charge in zero time, which is an infinite current. Similarly, an instantaneous change in the current in an inductor would require an infinite voltage. Therefore, it is not possible to change instantaneously the current in an inductor.

Ideal capacitors and inductors only store energy; they do not dissipate energy as a resistor does. Although we model these elements as ideal devices, in practice there is some

Table 5.1 The Dual Relationship for Capacitors and Inductors

Capacitor	Inductor
$i(t) = C \dfrac{dv(t)}{dt}$	$v(t) = L \dfrac{di(t)}{dt}$
$v(t) = \dfrac{1}{C} \displaystyle\int_{t_0}^{t} i(x)\, dx + v(t_0)$	$i(t) = \dfrac{1}{L} \displaystyle\int_{t_0}^{t} v(x)\, dx + i(t_0)$
$p(t) = Cv(t) \dfrac{dv(t)}{dt}$	$p(t) = Li(t) \dfrac{di(t)}{dt}$
$w(t) = \frac{1}{2} Cv(t)^2$	$w(t) = \frac{1}{2} Li^2(t)$

DRILL

E5.3 Write the i-v relationship for the practical capacitor shown below:

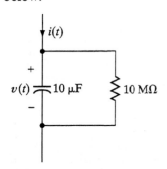

ANSWER: $i(t) =$

$$10^{-5}\frac{dv(t)}{dt} + 10^{-7}v(t)$$

Figure 5.10 (a) Ideal elements and (b) practical elements.

leakage resistance associated with each element, as shown in Fig. 5.10. In the case of the capacitor, there is normally a very large leakage resistance in parallel, which provides a conduction path between the plates. It is through this parallel resistance that the realistic capacitor slowly discharges itself. The physical coil, which forms the inductor, has significant winding resistance that quickly dissipates energy. Thus, the practical inductor is modeled as a resistor in series with an ideal inductor. Of the two elements, the capacitor is by far the better energy-storage device.

5.4 CAPACITOR AND INDUCTOR COMBINATIONS

Series Capacitors

If a number of capacitors are connected in series, their equivalent capacitance can be calculated using KVL. Consider the circuit shown in Fig. 5.11a. For this circuit

$$v(t) = v_1(t) + v_2(t) + v_3(t) + \cdots + v_N(t) \qquad \textbf{5.13}$$

but

$$v_i(t) = \frac{1}{C_i}\int_{t_0}^{t} i(t)\, dt + v_i(t_0) \qquad \textbf{5.14}$$

Therefore, Eq. (5.13) can be written as follows using Eq. (5.14):

Figure 5.11 Equivalent circuit for N series-connected capacitors.

$$v(t) = \left(\sum_{i=1}^{N} \frac{1}{C_i} \right) \int_{t_0}^{t} i(t)\, dt + \sum_{i=1}^{N} v_i(t_0) \qquad\qquad \textbf{5.15}$$

$$= \frac{1}{C_S} \int_{t_0}^{t} i(t)\, dt + v(t_0) \qquad\qquad \textbf{5.16}$$

where

$$v(t_0) = \sum_{i=1}^{N} v_i(t_0)$$

and

$$\frac{1}{C_S} = \sum_{i=1}^{N} \frac{1}{C_i} = \frac{1}{C_1} + \frac{1}{C_2} + \cdots + \frac{1}{C_N} \qquad\qquad \textbf{5.17}$$

Capacitors in series combine like resistors in parallel.

Thus, the circuit in Fig. 5.11b is equivalent to that in Fig. 5.11a under the conditions stated previously.

It is also important to note that since the same current flows in each of the series capacitors, each capacitor gains the same charge in the same time period. The voltage across each capacitor will depend on this charge and the capacitance of the element.

EXAMPLE 5.8

Determine the equivalent capacitance and the initial voltage for the circuit shown in Fig. 5.12.

SOLUTION   Note that these capacitors must have been charged before they were connected in series or else the charge of each would be equal and the voltages would be in the same direction.

The equivalent capacitance is

$$\frac{1}{C_S} = \frac{1}{2} + \frac{1}{3} + \frac{1}{6}$$

where all capacitance values are in microfarads.

Therefore, $C_S = 1\ \mu F$ and, as seen from the figure, $v(t_0) = -3$ V. Note that the total energy stored in the circuit is

$$w(t_0) = \tfrac{1}{2}\left[2 \times 10^{-6}(2)^2 + 3 \times 10^{-6}(-4)^2 + 6 \times 10^{-6}(-1)^2\right]$$

$$= 31\ \mu J$$

However, the energy recoverable at the terminals is

$$w_C(t_0) = \tfrac{1}{2} C_S v^2(t)$$

$$= \tfrac{1}{2}\left[1 \times 10^{-6}(-3)^2\right]$$

$$= 4.5\ \mu J \qquad\qquad \square$$

Figure 5.12 Circuit containing multiple capacitors with initial voltages.

EXAMPLE 5.9

Two previously uncharged capacitors are connected in series and then charged with a 12-V source. One capacitor is 30 μF and the other is unknown. If the voltage across the 30-μF capacitor is 8 V, find the capacitance of the unknown capacitor.

SOLUTION   The charge on the 30-μF capacitor is

$$Q = CV = (30 \ \mu F)(8 \ V) = 240 \ \mu C$$

Since the same current flows in each of the series capacitors, each capacitor gains the same charge in the same time period.

$$C = \frac{Q}{V} = \frac{240 \mu C}{4 V} = 60 \mu F \qquad \square$$

Parallel Capacitors

To determine the equivalent capacitance of N capacitors connected in parallel, we employ KCL. As can be seen from Fig. 5.13a,

$$i(t) = i_1(t) + i_2(t) + i_3(t) + \cdots + i_N(t) \qquad \textbf{5.18}$$

$$= C_1 \frac{dv(t)}{dt} + C_2 \frac{dv(t)}{dt} + C_3 \frac{dv(t)}{dt} + \cdots + C_N \frac{dv(t)}{dt}$$

$$= \left(\sum_{i=1}^{N} C_i \right) \frac{dv(t)}{dt}$$

$$= C_p \frac{dv(t)}{dt} \qquad \textbf{5.19}$$

where

Capacitors in parallel combine like resistors in series.

$$C_p = C_1 + C_2 + C_3 + \cdots + C_N \qquad \textbf{5.20}$$

(a)   (b)

Figure 5.13 Equivalent circuit for N capacitors connected in parallel.

EXAMPLE 5.10

Determine the equivalent capacitance at terminals A-B of the circuit shown in Fig. 5.14.

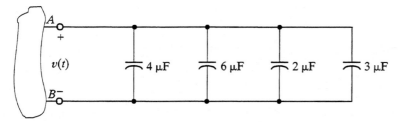

Figure 5.14 Circuit containing multiple capacitors in parallel.

SOLUTION
$$C_p = 4\mu\text{F} + 6\mu\text{F} + 2\mu\text{F} + 3\mu\text{F}$$
$$= 15\mu\text{F}$$

$\square$

EXTENSION EXERCISES

E5.6 Two initially uncharged capacitors are connected as shown in Fig. E5.6. Determine the value of C_1.

ANSWER: $C_1 = 4\ \mu\text{F}.$

Figure E5.6

E5.7 Compute the equivalent capacitance of the network in Fig. E5.7.

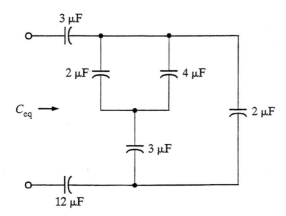

Figure E5.7

ANSWER: $C_{\text{eq}} = 1.5\ \mu\text{F}.$

E5.8 Compute the equivalent capacitance of the network in Fig. E5.8 if all the capacitors are 4 μF.

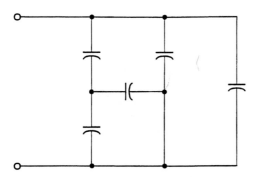

Figure E5.8

ANSWER: $C_{eq} = \dfrac{32}{3}$ μF.

Series Inductors

If N inductors are connected in series, the equivalent inductance of the combination can be determined as follows. Referring to Fig. 5.15a and using KVL, we see that

$$v(t) = v_1(t) + v_2(t) + v_3(t) + \cdots + v_N(t) \tag{5.21}$$

and therefore,

$$v(t) = L_1\frac{di(t)}{dt} + L_2\frac{di(t)}{dt} + L_3\frac{di(t)}{dt} + \cdots + L_N\frac{di(t)}{dt} \tag{5.22}$$

$$= \left(\sum_{i=1}^{N} L_i\right)\frac{di(t)}{dt}$$

$$= L_S\frac{di(t)}{dt} \tag{5.23}$$

where

$$L_S = \sum_{i=1}^{N} L_i \tag{5.24}$$

Inductors in series combine like resistors in series.

Therefore, under this condition the network in Fig. 5.15b is equivalent to that in Fig. 5.15a.

Figure 5.15 Equivalent circuit for N series-connected inductors.

EXAMPLE 5.11

Find the equivalent inductance of the circuit shown in Fig. 5.16.

Figure 5.16 Circuit containing multiple inductors.

SOLUTION   The equivalent inductance of the circuit shown in Fig. 5.16 is

$$L_S = 1H + 2H + 4H$$
$$= 7H$$ ❑

Parallel Inductors

Consider the circuit shown in Fig. 5.17a, which contains N parallel inductors. Using KCL, we can write

$$i(t) = i_1(t) + i_2(t) + i_3(t) + \cdots + i_N(t) \tag{5.25}$$

However,

$$i_j(t) = \frac{1}{L_j} \int_{t_0}^{t} v(x)\, dx + i_j(t_0) \tag{5.26}$$

Substituting this expression into Eq. (5.25) yields

$$i(t) = \left(\sum_{j=1}^{N} \frac{1}{L_j} \right) \int_{t_0}^{t} v(x)\, dx + \sum_{j=1}^{N} i_j(t_0) \tag{5.27}$$

$$= \frac{1}{L_P} \int_{t_0}^{t} v(x)\, dx + i(t_0) \tag{5.28}$$

where

$$\frac{1}{L_p} = \frac{1}{L_1} + \frac{1}{L_2} + \frac{1}{L_3} + \cdots + \frac{1}{L_N} \tag{5.29}$$

Inductors in parallel combine like resistors in parallel.

and $i(t_0)$ is equal to the current in L_P at $t = t_0$. Thus, the circuit in Fig. 5.17b is equivalent to that in Fig. 5.17a under the conditions stated previously.

(a)   (b)

Figure 5.17 Equivalent circuits for N inductors connected in parallel.

EXAMPLE 5.12

Determine the equivalent inductance and the initial current for the circuit shown in Fig. 5.18.

SOLUTION   The equivalent inductance is

$$\frac{1}{L_P} = \frac{1}{12} + \frac{1}{6} + \frac{1}{4}$$

Figure 5.18 Circuit containing multiple inductors with initial currents.

where all inductance values are in millihenrys

$$L_P = 2 \text{ mH}$$

and the initial current is $i(t_0) = -1 \ A$.   ❏

The previous material indicates that capacitors combine like conductances, whereas inductances combine like resistances.

EXTENSION EXERCISES

E5.9 Compute the equivalent inductance of the network in Fig. E5.9 if all inductors are 4 mH.

ANSWER: 4.4 mH.

E5.10 Determine the equivalent inductance of the network in Fig. E5.10 if all inductors are 6 mH.

L_{eq}

Figure E5.9

L_{eq}

Figure E5.10

ANSWER: 9.429 mH.

5.5 *RC* OPERATIONAL AMPLIFIER CIRCUITS

The properties of the ideal op-amp are $v_+ = v_-$ and $i_+ = v_- = 0$.

Two very important *RC* op-amp circuits are the differentiator and the integrator. These circuits are derived from the circuit for an inverting op-amp by replacing the resistors R_1 and R_2, respectively, by a capacitor. Consider, for example, the circuit shown in Fig. 5.19a. The circuit equations are

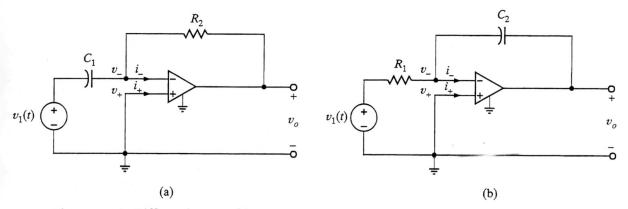

Figure 5.19 Differentiator and integrator operational amplifier circuits.

$$C_1 \frac{d}{dt}(v_1 - v_-) + \frac{v_o - v_-}{R_2} = i_-$$

However, $v_- = 0$ and $i_- = 0$. Therefore,

$$v_o(t) = -R_2 C_1 \frac{dv_1(t)}{dt} \qquad\qquad 5.30$$

Thus, the output of the op-amp circuit is proportional to the derivative of the input.

The circuit equations for the op-amp configuration in Fig. 5.19b are

$$\frac{v_1 - v_-}{R_1} + C_2 \frac{d}{dt}(v_o - v_-) = i_-$$

but since $v_- = 0$ and $i_- = 0$, the equation reduces to

$$\frac{v_1}{R_1} = -C_2 \frac{dv_o}{dt}$$

or

$$v_o(t) = \frac{-1}{R_1 C_2} \int_{-\infty}^{t} v_1(x)\, dx$$

$$= \frac{-1}{R_1 C_2} \int_{0}^{t} v_1(x)\, dx + v_o(0) \qquad\qquad 5.31$$

If the capacitor is initially discharged, then $v_o(0) = 0$; hence,

$$v_o(t) = \frac{-1}{R_1 C_2} \int_{0}^{t} v_1(x)\, dx \qquad\qquad 5.32$$

Thus, the output voltage of the op-amp circuit is proportional to the integral of the input voltage.

EXAMPLE 5.13

The waveform in Fig. 5.20a is applied at the input of the differentiator circuit shown in Fig. 5.19a. If $R_2 = 1\ \text{k}\Omega$ and $C_1 = 2\ \mu\text{F}$, determine the waveform at the output of the op-amp.

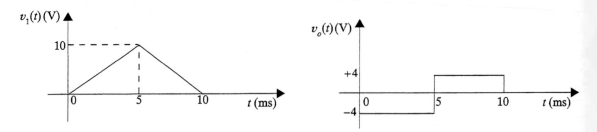

Figure 5.20 Input and output waveforms for a differentiator circuit.

SOLUTION Using Eq. (5.30), we find that the op-amp output is

$$v_o(t) = -R_2 C_1 \frac{dv_1(t)}{dt}$$

$$= -(2)10^{-3} \frac{dv_1(t)}{dt}$$

$dv_1(t)/dt = (2)10^3$ for $0 \le t < 5$ ms, and therefore,

$$v_o(t) = -4 \text{ V} \qquad 0 \le t < 5 \text{ ms}$$

$dv_1(t)/dt = -(2)10^3$ for $5 \le t < 10$ ms, and therefore,

$$v_o(t) = 4 \text{ V} \qquad 5 \le t < 10 \text{ ms}$$

Hence, the output waveform of the differentiator is shown in Fig. 5.20b. ❏

EXAMPLE 5.14

If the integrator shown in Fig. 5.19b has the parameters $R_1 = 5 \text{ k}\Omega$ and $C_2 = 0.2 \text{ }\mu\text{F}$, determine the waveform at the op-amp output if the input waveform is given as in Fig. 5.21a and the capacitor is initially discharged.

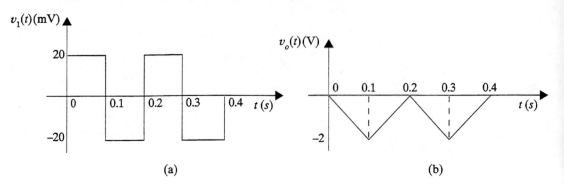

(a)   (b)

Figure 5.21 Input and output waveforms for an integrator circuit.

SOLUTION The integrator output is given by the expression

$$v_o(t) = \frac{-1}{R_1 C_2} \int_0^t v_1(x) \, dx$$

which with the given circuit parameters is

$$v_o(t) = -10^3 \int_0^t v_1(x)\, dx$$

In the interval $0 \le t < 0.1$ s, $v_1(t) = 20$ mV. Hence,

$$v_o(t) = -10^3(20)10^{-3}t \qquad 0 \le t < 0.1 \text{ s}$$
$$= -20t$$

At $t = 0.1$ s, $v_o(t) = -2$ V. In the interval from 0.1 to 0.2 s, the integrator produces a positive slope output of $20t$ from $v_o(0.1) = -2$ V to $v_o(0.2) = 0$ V. This waveform from $t = 0$ to $t = 0.2$ s is repeated in the interval $t = 0.2$ to $t = 0.4$ s, and therefore, the output waveform is shown in Fig. 5.21b.   ❑

EXTENSION EXERCISE

E5.11   The waveform in Fig. E5.11 is applied to the input terminals of the op-amp differentiator circuit. Determine the differentiator output waveform if the op-amp circuit parameters are $C_1 = 2$ F and $R_2 = 2\ \Omega$.

Figure E5.11

ANSWER:

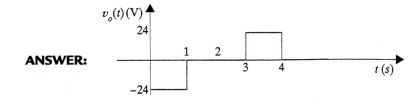

5.6 APPLICATIONS

We now have two elements that have been added to our repertoire of components that can be employed in circuit applications and design—capacitors and inductors. For example, we have just found that when the capacitors are used in conjunction with op-amps, we can construct circuits that will integrate or differentiate a signal. In the remainder of the text, we will find that these two components are an integral part of our circuit analysis and design.

EXAMPLE 5.15

Integrated circuit chips, or ICs, are rectangular pieces of silicon that range in size from about 10/1000 to 200/1000 inch on each side. Chip manufacturers make electrical contact to the silicon with fine gold wire, called wire bonds, which stretch from the silicon surface to metal pins, as shown in Fig. 5.22a. The chip is eventually coated in plastic to protect it from physical damage and the pins become the access terminals to the chip.

In general, wires are not perfect conductors. They have resistance and inductance. In most cases, wiring resistance and inductance are so small that they have no measurable effect on circuit performance. However, in cases where the current changes rapidly, wire inductance can play a significant role. Let us examine the effect of the inductance of a wire bond in a high-speed microprocessor.

SOLUTION Figure 5.22b shows an equivalent network for the situation of interest. The wire bond is modeled by the 10-nH inductor. The current source, $i(t)$, represents the current being used by the microprocessor. Finally, the 5-V source represents the chip supply voltage. As the microprocessor performs various functions, the current demand, $i(t)$, changes as shown in Fig. 5.22c. Using the element values in Fig. 5.22b, let us plot $V_{chip}(t)$ and find its minimum and maximum values.

Applying KVL in Fig. 5.22b, the chip voltage can be expressed as

$$v_{chip}(t) = 5 - L\frac{di(t)}{dt}$$

From Fig. 5.22c for $0 < t < 4$ ns,

$$\frac{di(t)}{dt} = \frac{\Delta i(t)}{\Delta t} = \frac{1}{4n} = 250 \times 10^6$$

Similarly, in the range 12 ns $< t <$ 20 ns,

$$\frac{di(t)}{dt} = -125 \times 10^6$$

Based on these calculations, the resulting chip voltage, shown in Fig. 5.22d, varies widely from the 5-V requirement. The minimum value is 2.5 V and the maximum value is 6.25 V. If we desire a more stable chip voltage, we can simply put several wire bonds in parallel, as shown in Fig. 5.22e. Now $i(t)$ is divided between five different paths and so $di(t)/dt$ and the change in $V_{chip}(t)$ are reduced by a factor of 5. The resulting chip voltage, shown in Fig. 5.22f, has minimum and maximum values of 4.5 V and 5.25 V. This simple technique is employed by most IC manufacturers. ❏

5.7 CIRCUIT DESIGN

EXAMPLE 5.16

During the dyeing process at a textile mill, dye is pumped through a pipe into a vat. As shown in Fig. 5.23a, a flowmeter mounted on the pipe monitors the flow rate

Figure 5.22 Integrated circuit wire bond example.

and outputs 1 volt at v_F for each gallon/second of flow. We want to design a circuit with an output voltage of 1 V for every 10 gallons used. Let us design a network to perform the required conversion using the inverting integrator in Fig. 5.23b, where the voltage source V_F represents the flowmeter output.

(a)

(b)

Figure 5.23 Flowmeter/volume meter conversion.

SOLUTION The relationship between v_x and v_F in Fig. 5.23b is

$$v_x = \frac{-1}{RC} \int v_F \, dt$$

The integral of v_F will produce 1 V-s for each gallon used. Since we want 1 V at v_x for every 10 gallons, RC must be 10. If we arbitrarily choose $C = 1$ μF, then $R = 10$ MΩ. Since v_x is negative, we add the simple inverter in Fig. 5.23b to produce the voltage v_o, which has all the required characteristics. The switch around the capacitor can be closed to initialize the integrator to 0 V at v_x. Finally, since the op-amp output voltage cannot exceed its dc power supplies, as v_o reaches this limit, we will need to log the output value (normally done electronically), close the switch to reset v_o to zero, and then open the switch to begin a new integration cycle. ❑

EXAMPLE 5.17

A traffic survey is to be conducted at an intersection to determine the best traffic light timing sequence. In particular, we need to find the difference in the number of eastbound versus southbound vehicles over a 24-hour period. Sensors in the roadway as shown in Fig. 5.24a produce the voltage waveform in Fig. 5.24b each time a vehicle passes. We want to design a network that will produce a voltage propor-

Figure 5.24 Traffic light survey.

tional to this difference where 1 V corresponds to 10 cars. Let us use an inverting integrator to produce a voltage that meets the requirements and verify its performance with PSPICE using our op-amp model with $R_I = 1\ M\Omega$, $R_o = 50\ \Omega$, and $A = 10^5$.

SOLUTION The required network must do three things. First, it must perform a subtraction of the two sensor voltages. Second, it must integrate the difference to keep a running total of the traffic situation. And third, it must scale the voltage to meet the specification of 1 V per 10 vehicles. The circuit in Fig. 5.24c, where the voltage sources represent the sensor outputs, will meet all requirements.

KCL at the inverting op-amp input yields v_o in terms of v_E and v_S.

$$\frac{v_S}{R_S} - \frac{v_E}{R_E} = -C\left(\frac{dv_o}{dt}\right)$$

Solving for v_o, we obtain

$$v_o = \int \left[\frac{1}{R_E C} v_E - \frac{1}{R_S C} v_S\right] dt$$

Based on the sensor waveform in Fig. 5.24b, the integral of the sensor output is 1.5 V-s for each vehicle that passes. If the RC products in the preceding equation are 15, then one vehicle will produce 0.1 V at v_o as required. If we arbitrarily choose $C = 1.5\ \mu F$, then R_E and R_S are 10 MΩ. As in Example 5.16, as the integrator output voltage nears the op-amp's power supply values, the data must be logged and the capacitor discharged through a switch to reset the integrator. At the end of a 24-hour period, the recorded values of v_o can be summed to yield the daily vehicle count.

The pspice *Schematics* circuit is shown in Fig. 5.24d. To provide an inverting gain for the integrator, the gain of the dependent voltage source has been set to $-100,000$ V/V. Two VPULSE voltage sources are used to model the traffic sensors. For simulation purposes, let one eastbound vehicle pass every second and one southbound vehicle every 2.5 seconds. The required attributes for the VPULSE sources are listed in the following table.

Source	v_{east}	v_{south}
dc voltage	0	0
ac voltage	0	0
Start voltage	0	0
Pulse voltage	10	10
Delay time	0.85	2.35
Rise time	0	0
Fall time	0	0
Pulse width	0.15	0.15
Period	1	2.5

After 10 s, 10 eastbound vehicles have passed versus only four southbound. The difference of six should produce 0.6 V at v_o. From the PROBE plots of v_{east}, v_{south}, and v_o shown in Fig. 5.24e, the output voltage at 10 s is 0.608 V, which is within 1.4% of our ideal op-amp calculations.   ❑

5.8 SUMMARY

- The important (dual) relationships for capacitors and inductors are as follows:

$$q = Cv$$

$$i(t) = C\frac{dv(t)}{dt} \qquad\qquad v(t) = L\frac{di(t)}{dt}$$

$$v(t) = \frac{1}{C}\int_{-\infty}^{t} i(x)\,dx \qquad\qquad i(t) = \frac{1}{L}\int_{-\infty}^{t} v(x)\,dx$$

$$p(t) = Cv(t)\frac{dv(t)}{dt} \qquad\qquad p(t) = Li(t)\frac{di(t)}{dt}$$

$$W_C(t) = 1/2Cv^2(t) \qquad\qquad W_L(t) = 1/2Li^2(t)$$

- The passive sign convention is used with capacitors and inductors.
- In dc steady state a capacitor looks like an open circuit and an inductor looks like a short circuit.
- Leakage resistance is present in practical capacitors and inductors.
- When capacitors are interconnected, their equivalent capacitance is determined as follows: Capacitors in series combine like resistors in parallel and capacitors in parallel combine like resistors in series.
- When inductors are interconnected, their equivalent inductance is determined as follows: Inductors in series combine like resistors in series and inductors in parallel combine like resistors in parallel.
- *RC* operational amplifier circuits can be used to differentiate or integrate an electrical signal.

PROBLEMS

Section 5.1

5.1 A 6-μF capacitor was charged to 12 V. Find the charge accumulated in the capacitor.

5.2 A capacitor has an accumulated charge of 600 μC with 5 V across it. What is the value of capacitance?

5.3 A 12-μF capacitor has an accumulated charge of 480 μC. Determine the voltage across the capacitor.

5.4 A 10-μF capacitor is charged by a constant current source and its voltage is increased to 2 V in 5 s. Find the value of the constant current source.

$$i = C\frac{dv}{dt}$$

Similar to P5.4

5.5 A capacitor is charged by a constant current of 2 mA and results in a voltage increase of 12 V in a 10-s interval. What is the value of the capacitance?

Similar to P5.4

5.6 An uncharged 100-μF capacitor is charged by a constant current of 1 mA. Find the voltage across the capacitor after 4 s.

$i(t) = C \dfrac{dv(t)}{dt}$ and

$W_C(t) = 1/2 C v^2(t)$

5.7 The voltage across a 100-μF capacitor is given by the expression $v(t) = 120 \sin 377t$ V. Find (a) the current in the capacitor and (b) the expression for the energy stored in the element.

$v(t) = \dfrac{1}{C} \displaystyle\int_{t_o}^{t} i(x)\, dx$

$p(t) = v(t)i(t)$

5.8 An uncharged 10-μF capacitor is charged by the current $i(t) = 10 \cos 377t$ mA. Find (a) the expression for the voltage across the capacitor and (b) the expression for the power.

5.9 The voltage across a 6-μF capacitor is shown in Fig. P5.9. Compute the waveform for the current in the capacitor.

Similar to P5.6

Figure P5.9

5.10 The voltage across a 100-μF capacitor is shown in Fig P5.10. Compute the waveform for the current in the capacitor.

Similar to P5.9

Figure P5.10

5.11 The current in a 100-μF capacitor is shown in Fig. P5.11. Determine the waveform for the voltage across the capacitor if it is initially uncharged.

Similar to P5.8

Figure P5.11

5.12 The voltage across a 2-μF capacitor is given by the waveform in Fig. P5.12. Compute the current waveform.

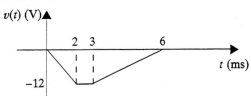

Similar to P5.9

Figure P5.12

5.13 The voltage across a 0.1-F capacitor is given by the waveform in Fig. P5.13. Find the waveform for the current in the capacitor.

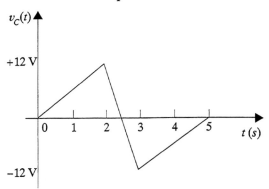

Similar to P5.9

Figure P5.13

5.14 The voltage across a 50-μF capacitor is shown in Fig. P5.14. Determine the current waveform.

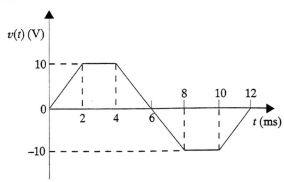

Similar to P5.9

Figure P5.14

5.15 Draw the waveform for the current in a 12-μF capacitor when the capacitor voltage is as described in Fig. P5.15.

Similar to P5.9

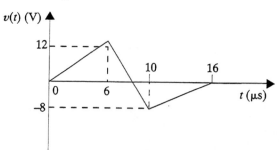

Figure P5.15

5.16 Draw the waveform for the current in a 3-μF capacitor when the voltage across the capacitor is give in Fig. P5.16.

Similar to P5.9

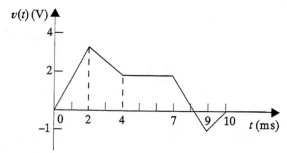

Figure P5.16

5.17 The voltage across a 6-μF capacitor is given by the waveform in Fig. P5.17. Plot the waveform for the capacitor current.

Similar to P5.9

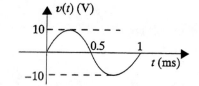

Figure P5.17

5.18 The waveform for the current in a 200-μF capacitor is shown in Fig. P5.18. Determine the waveform for the capacitor voltage.

Similar to P5.8

Figure P5.18

5.19 The waveform for the current in a 100-μF initially uncharged capacitor is shown in Fig. P5.19. Determine the waveform of the voltage.

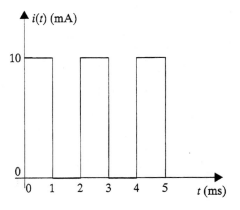

Figure P5.19

Similar to P5.8

5.20 The waveform for the current in a 100-μF initially uncharged capacitor is shown in Fig. P5.20. Determine the waveform for the capacitor's voltage.

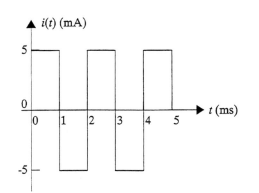

Figure P5.20

Similar to P5.8

Section 5.2

5.21 A 10-mH inductor has a sudden current change from 200 mA to 100 mA in 1 ms. Find the induced voltage.

$$v(t) = L\frac{di(t)}{dt}$$

5.22 The current in an inductor changes from 0 to 200 mA in 4 ms and induces a voltage of 100 mV. What is the value of the inductor?

Similar to P5.21

5.23 If the current $i(t) = 1.5t$ A flows through a 2-H inductor, find the energy stored at $t = 2$ s.

$$W_L(t) = 1/2 Li(t)^2$$

5.24 The current in a 100-mH inductor is $i(t) = 2 \sin 377t$ A. Find (a) the voltage across the inductor and (b) the expression for the energy stored in the element.

Similar to P5.21 and P5.23

$$i(t) = \frac{1}{L} \int_{t_o}^{t} v(x)\, dx$$

$$p(t) = v(t)i(t)$$

5.25 The induced voltage across a 10-mH inductor is $v(t) = 120 \cos 377t$ V. Find (a) the expression for the inductor current and (b) the expression for the power.

5.26 The current in a 10-mH inductor is shown in Fig. P5.26. Determine the waveform for the voltage across the inductor.

Similar to P5.21

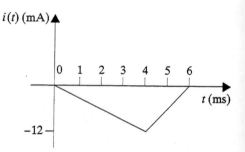

Figure P5.26

5.27 The current in a 50-mH inductor is given in Fig. P5.27. Sketch the inductor voltage.

Similar to P5.21

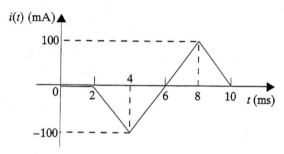

Figure P5.27

5.28 The current in a 25-mH inductor is given by the expressions

$$i(t) = 0 \qquad\qquad t < 0$$
$$i(t) = 10(1 - e^{-t})\text{mA} \qquad t > 0$$

Similar to P5.24

Find (a) the voltage across the inductor and (b) the expression for the energy stored in it.

5.29 Given the data in the previous problem, find the voltage across the inductor and the energy stored in it after 1 s.

5.30 The current in a 50-mH inductor is specified as follows:

$$i(t) = 0 \qquad\qquad t < 0$$
$$i(t) = 2te^{-4t}\,\text{A} \qquad t > 0$$

Similar to P5.28

Find (a) the voltage across the inductor, (b) the time at which the current is a maximum, and (c) the time at which the voltage is a minimum.

5.31 The current in a 10-mH inductor is shown in Fig. P5.31. Find the voltage across the inductor.

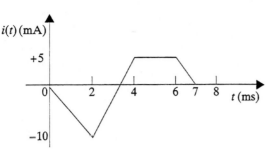

Figure P5.31

Similar to P5.21

5.32 Draw the waveform for the voltage across a 10-mH inductor when the inductor current is given by the waveform shown in Fig. P5.32.

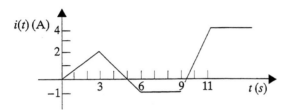

Figure P5.32

Similar to P5.21

5.33 The current in a 16-mH inductor is given by the waveform in Fig. P5.33. Find the waveform for the voltage across the inductor.

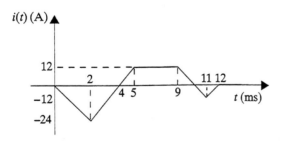

Figure P5.33

Similar to P5.21

5.34 The voltage across a 2-H inductor is given by the waveform shown in Fig. P5.34. Find the waveform for the current in the inductor.

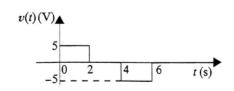

Figure P5.34

Similar to P5.25

5.35 The waveform for the voltage across a 20-mH inductor is shown in Fig. P5.35. Compute the waveform for the inductor current.

Similar to P5.25

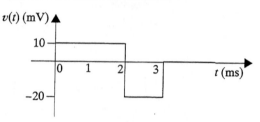

Figure P5.35

5.36 The voltage across a 10-mH inductor is shown in Fig. P5.36. Determine the waveform for the inductor current.

Similar to P5.25

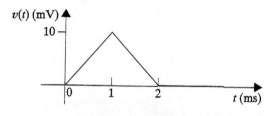

Figure P5.36

5.37 The voltage across a 2-H inductor is given by the waveform shown in Fig. P5.37. Find the waveform for the current in the inductor.

Similar to P5.25

Figure P5.37

5.38 The current in a 4-mH inductor is given by the waveform in Fig. P5.38. Plot the voltage across the inductor.

Similar to P5.21

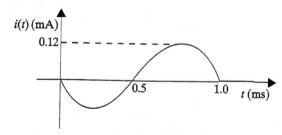

Figure P5.38

Section 5.3

5.39 A practical inductor, as shown in Fig. P5.39, has a current $i(t) = 10\cos 377t$ A. Find the voltage across this inductor.

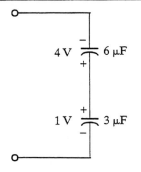

10 mH   0.1 mΩ

Figure P5.39

$$v(t) = L\frac{di(t)}{dt} + Ri(t)$$

Section 5.4

5.40 A 3-μF capacitor and 6-μF capacitor are connected in parallel and charged to 12 V. Find (a) the charge stored by each capacitor and (b) the total energy stored.

1. Find C_{eq}
2. Find total voltage V
3. $W = \frac{1}{2}C_{eq}v^2$

5.41 The two capacitors in Fig. P5.41 were charged and then connected as shown. Determine the equivalent capacitance, the initial voltage at the terminals, and the total energy stored in the network.

4 V — 6 μF

1 V — 3 μF

Figure P5.41

Similar to P5.40 except that
$$\frac{1}{C_{total}} = \frac{1}{C_1} + \frac{1}{C_2}$$

5.42 Two capacitors are connected in series as shown in Fig. P5.42. Find V_o.

V_o — 12 μF

24 V — 6 μF

Figure P5.42

$$v = \frac{1}{C}\int i(t)\,dt$$

$v \propto \dfrac{1}{C}$ since $\int i(t)\,dt$ is the same for both capacitors

5.43 Three capacitors are connected as shown in Fig. P5.43. Find V_1 and V_2.

V_1 — 8 μF

V_2 — 4 μF

12 V

Figure P5.43

Similar to P5.42

Use the series and parallel relationships for combining capacitors.

5.44 What values of capacitance can be obtained by interconnecting a 4-μF capacitor, a 6-μF capacitor, and a 12-μF capacitor?

Similar to P5.44

5.45 Given a 1-, 3-, and 4-μF capacitor, can they be interconnected to obtain an equivalent 2-μF capacitor?

Similar to P5.44

5.46 Given four 2-μF capacitors, find the maximum value and minimum value that can be obtained by interconnecting the capacitors in series/parallel combinations.

5.47 Select the value of C to produce the desired total capacitance of $C_T = 2$ μF in the circuit in Fig. P5.47.

Find the expression for the total capacitance in terms of C and set it equal to 2 μF.

Figure P5.47

See the figure in Example 5.8

5.48 A 20-mH inductor and a 12-mH inductor are connected in series with a 1-A current source. Find (a) the equivalent inductance and (b) the total energy stored.

$i = \dfrac{1}{L} \displaystyle\int v(t)\, dx$ and thus

$i \propto \dfrac{1}{L}$

5.49 Two inductors are connected in parallel, as shown in Fig. P5.49. Find i.

Figure P5.49

5.50 Select the value of L that produces a total inductance of $L_T = 10$ mH in the circuit in Fig. P5.50.

Similar to P5.47 with the exception that the elements are inductors.

Figure P5.50

Section 5.5

5.51 For the network in Fig. P5.51 choose C such that

$$v_o = -10 \int v_S \, dt$$

Assume a perfect op-amp and derive the relationship between v_o and v_S. Compare this equation to the derived relationship.

Figure P5.51

5.52 For the network in Fig. P5.52, $v_S(t) = 120 \cos 377t$ V. Find $v_o(t)$

Similar to P5.51

Figure P5.52

5.53 For the network in Fig. P5.53, $v_S(t) = 115 \sin 377t$ V. Find $v_o(t)$.

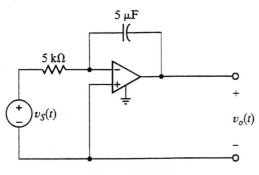

Similar to P5.52

Figure P5.53

Advanced Problems

5.54 Determine the total capacitance of the network in Fig. P5.54.

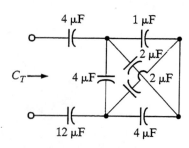

Figure P5.54

5.55 Find the equivalent capacitance at terminals *A-B* in Fig. P5.55.

Figure P5.55

5.56 Find the total capacitance C_T of the network in Fig. P5.56.

Figure P5.56

5.57 Find the total capacitance C_T of the network in Fig. P5.57.

Figure P5.57

5.58 In the network in Fig. P5.58, find the capacitance C_T if (a) the switch is open and (b) the switch is closed.

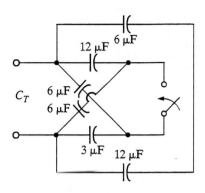

Figure P5.58

5.59 Determine the inductance at terminals A-B in the network in Fig. P5.59.

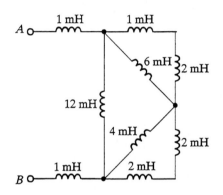

Figure P5.59

5.60 Determine the inductance at terminals A-B in the network in Fig P5.60.

Figure P5.60

5.61 Find the total inductance at the terminals A-B shown in the network in Fig. P5.61.

Figure P5.61

5.62 Given the network shown in Fig. P5.62, find (a) the equivalent inductance at terminals A-B with terminals C-D short circuited; and (b) the equivalent inductance at terminals C-D with terminals A-B, open circuited.

Figure P5.62

5.63 Find the total inductance at the terminals of the network in Fig. P5.63.

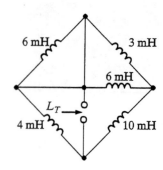

Figure P5.63

5.64 Find L_T in the network in Fig. P5.64 (a) with the switch open and (b) with the switch closed. All inductors are 12 mH.

L_T

Figure P5.64

5.65 Find the value of L in the network in Fig. P5.65 so that the total inductance L_T will be 2 mH.

L_T

4 mH   2 mH

L

6 mH

Figure P5.65

5.66 Design an op-amp circuit that will produce the output voltage

$$v_o = \int v_S \, dt - 10v_S$$

5.67 For the network in Fig. P5.67, $v_{S_1}(t) = 80 \cos 377t$ V and $v_{S_2}(t) = 40 \cos 377t$ V. Find $v_o(t)$.

1 μF

20 kΩ   10 kΩ

$v_{S_1}(t)$ $v_{S_2}(t)$ $v_o(t)$

Figure P5.67

First-Order Transient Circuits

6

*I*n this chapter we perform what is normally referred to as a first-order transient analysis because the networks that we examine contain only a single storage element. When only a single storage element is present in the network, the network can be described by a first-order differential equation.

Our analysis involves an examination and description of the behavior of a circuit as a function of time after a sudden change in the network occurs due to switches opening or closing. Because of the presence of the storage element, the circuit response to a sudden change will go through a transition period prior to settling down to a steady-state value. It is this transition period that we will examine carefully in our transient analysis.

One of the important parameters that we will examine in our transient analysis is the circuit's time constant. This is a very important network parameter because it tells us how fast the circuit will respond to changes. We can contrast two very different systems to obtain a feel for the parameter. For example, consider the model for a room air-conditioning system and the model for a single-transistor stage of amplification in a computer chip. If we change the setting for the air conditioner from 70 degrees to 60 degrees, the unit will come on and the room will begin to cool. However, the temperature measured by a thermometer in the room will fall very slowly and, thus, the time required to reach the desired temperature is long. However, if we send a trigger signal to a transistor to change state, the action may take only a few nanoseconds. These two systems will have vastly different time constants.

Two analysis techniques are presented for performing a transient analysis: the differential equation approach, in which a differential equation is written and solved for each network, and a step-by-step approach, which takes advantage of the known form of the solution in every case. Finally, a number of application-oriented examples are presented and discussed.

6.1 INTRODUCTION

In Chapter 5 we found that capacitors and inductors were capable of storing electric energy. In the case of a charged capacitor, the energy is stored in the electric field that exists between the positively and negatively charged plates. This stored energy can be released if a circuit is somehow connected across the capacitor that provides a path through which the negative charges move to the positive charges. As we know, this movement of charge constitutes a current. The rate at which the energy is discharged is a direct function of the parameters in the circuit that is connected across the capacitor's plates.

As an example, consider the flash circuit in a camera. Recall that the operation of the flash circuit, from a user standpoint, involves depressing the push button on the camera that triggers both the shutter and the flash and then waiting a few seconds before repeating the process to take the next picture. This operation can be modeled using the circuit in Fig. 6.1a. The voltage source and resistor R_S model the batteries that power the camera and flash. The capacitor models the energy storage, the switch models the push button, and finally the resistor R models the xenon flash lamp. Thus, if the capacitor is charged, when the switch is closed, the capacitor voltage drops and energy is released through the xenon lamp, producing the flash. In practice this energy release takes about a millisecond and the discharge time is a function of the elements in the circuit. When the push button is released and the switch is then opened, the battery begins to recharge the capacitor. Once again, the time required to charge the capacitor is a function of the circuit elements. The discharge and charge cycles are graphically illustrated in Fig. 6.1b. While the discharge time is very fast, it is not instantaneous. In order to provide further insight into this phenomenon, consider what we might call a *free-body diagram* of the right half of the network in Fig. 6.1a as shown in Fig. 6.1c (that is, a charged capacitor that is discharged through a resistor). When the switch is closed, KCL for the circuit is

$$C \frac{dv_c(t)}{dt} + \frac{v_c(t)}{R} = 0$$

or

$$\frac{dv_c(t)}{dt} + \frac{1}{RC} v_c(t) = 0$$

We will demonstrate in the next section that the solution of this equation is

$$v_c(t) = V_o e^{-t/RC}$$

Note that this function is a decaying exponential and the rate at which it decays is a function of the values of R and C. The product RC is a very important parameter and we will give it a special name in the following discussions.

Figure 6.1 Diagrams used to describe a camera's flash circuit.

6.2 GENERAL FORM OF THE RESPONSE EQUATIONS

In our study of first-order transient circuits we will show that the solution of these circuits (i.e., finding a voltage or current) requires us to solve a first-order differential equation of the form

$$\frac{dx(t)}{dt} + ax(t) = f(t) \qquad \textbf{6.1}$$

Although there are a number of techniques for solving an equation of this type, we will obtain a general solution that we will then employ in two different approaches to transient analysis.

A fundamental theorem of differential equations states that if $x(t) = x_p(t)$ is any solution to Eq. (6.1), and $x(t) = x_c(t)$ is any solution to the homogeneous equation

$$\frac{dx(t)}{dt} + ax(t) = 0 \qquad \textbf{6.2}$$

then

$$x(t) = x_p(t) + x_c(t) \qquad \textbf{6.3}$$

is a solution to the original Eq. (6.1). The term $x_p(t)$ is called the *particular integral solution*, or forced response, and $x_c(t)$ is called the *complementary solution*, or natural response.

At the present time we confine ourselves to the situation in which $f(t) = A$ (i.e., some constant). The general solution of the differential equation then consists of two parts that are obtained by solving the two equations

$$\frac{dx_p(t)}{dt} + ax_p(t) = A \qquad\qquad 6.4$$

$$\frac{dx_c(t)}{dt} + ax_c(t) = 0 \qquad\qquad 6.5$$

Since the right-hand side of Eq. (6.4) is a constant, it is reasonable to assume that the solution $x_p(t)$ must also be a constant. Therefore, we assume that

$$x_p(t) = K_1 \qquad\qquad 6.6$$

Substituting this constant into Eq. (6.4) yields

$$K_1 = \frac{A}{a} \qquad\qquad 6.7$$

Examining Eq. (6.5), we note that

$$\frac{dx_c(t)/dt}{x_c(t)} = -a \qquad\qquad 6.8$$

This equation is equivalent to

$$\frac{d}{dt}\left[\ln x_c(t)\right] = -a$$

Hence,

$$\ln x_c(t) = -at + c$$

and therefore,

$$x_c(t) = K_2 e^{-at} \qquad\qquad 6.9$$

Therefore, a solution of Eq. (6.1) is

$$x(t) = x_p(t) + x_c(t)$$

$$= \frac{A}{a} + K_2 e^{-at} \qquad\qquad 6.10$$

The constant K_2 can be found if the value of the independent variable $x(t)$ is known at one instant of time.

Equation (6.10) can be expressed in general in the form

$$x(t) = K_1 + K_2 e^{-t/\tau} \qquad\qquad 6.11$$

Once the solution in Eq. (6.11) is obtained, certain elements of the equation are given names that are commonly employed in electrical engineering. For example, the term K_1 is referred to as the *steady-state solution*: the value of the variable $x(t)$ as $t \to \infty$ when the second term becomes negligible. The constant τ is called the *time constant* of the circuit.

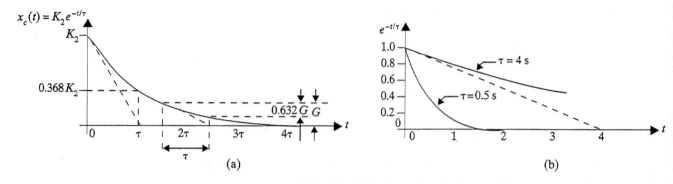

Figure 6.2 Time-constant illustrations.

Note that the second term in Eq. (6.11) is a decaying exponential that has a value, if $\tau > 0$, of K_2 for $t = 0$ and a value of 0 for $t = \infty$. The rate at which this exponential decays is determined by the time constant τ. A graphical picture of this effect is shown in Fig. 6.2a. As can be seen from the figure, the value of $x_c(t)$ has fallen from K_2 to a value of $0.368K_2$ in one time constant, a drop of 63.2%. In two time constants the value of $x_c(t)$ has fallen to $0.135K_2$, a drop of 63.2% from the value at time $t = \tau$. This means that the gap between a point on the curve and the final value of the curve is closed by 63.2% each time constant. Finally, after five time constants, $x_c(t) = 0.0067K_2$, which is less than 1%.

An interesting property of the exponential function shown in Fig. 6.2a is that the initial slope of the curve intersects the time axis at a value of $t = \tau$. In fact, we can take any point on the curve, not just the initial value, and find the time constant by finding the time required to close the gap by 63.2%. Finally, the difference between a small time constant (i.e., fast response) and a large time constant (i.e., slow response) is shown in Fig. 6.2b. These curves indicate that if the circuit has a small time constant, it settles down quickly to a steady-state value. Conversely, if the time constant is large, more time is required for the circuit to settle down or reach steady state. In any case, note that the circuit response essentially reaches steady state within five time constants (i.e., 5τ).

Note that the previous discussion has been very general in that no particular form of the circuit has been assumed—only that it results in a first-order differential equation.

6.3 ANALYSIS TECHNIQUES

The Differential Equation Approach

Equation (6.11) defines the general form of the solution of first-order transient circuits; that is, it represents the solution of the differential equation that describes an unknown current or voltage *anywhere in the network*. One of the ways that we can arrive at this solution is to solve the equations that describe the network behavior using what is often called the *state-variable approach*. In this technique we write the equation for the voltage across the capacitor and/or the equation for the current through the inductor. Recall from Chapter 5 that these quantities cannot change instantaneously. Let us first illustrate this technique in the general sense and then examine two specific examples.

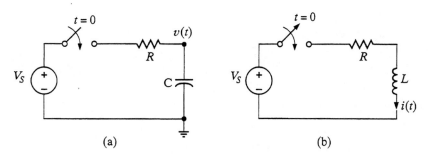

Figure 6.3 *RC* and *RL* circuits.

Consider the circuit shown in Fig. 6.3a. At time $t = 0$, the switch closes. The KCL equation that describes the capacitor voltage for time $t > 0$ is

$$C\frac{dv(t)}{dt} + \frac{v(t) - V_S}{R} = 0$$

or

$$\frac{dv(t)}{dt} + \frac{v(t)}{RC} = \frac{V_S}{RC}$$

From our previous development, we assume that the solution of this first-order differential equation is of the form

$$v(t) = K_1 + K_2 e^{-t/\tau}$$

Substituting this solution into the differential equation yields

$$-\frac{K_2}{\tau}e^{-t/\tau} + \frac{K_1}{RC} + \frac{K_2}{RC}e^{-t/\tau} = \frac{V_S}{RC}$$

Equating the constant and exponential terms, we obtain

$$K_1 = V_S$$
$$\tau = RC$$

Therefore,

$$v(t) = V_S + K_2 e^{-t/RC}$$

where V_S is the steady-state value and RC is the network's time constant. K_2 is determined by the initial condition of the capacitor. For example, if the capacitor is initially uncharged (that is, the voltage across the capacitor is zero at $t = 0$), then

$$0 = V_S + K_2$$

or

$$K_2 = -V_S$$

Hence, the complete solution for the voltage $v(t)$ is

$$v(t) = V_S - V_S e^{-t/RC}$$

The circuit in Fig. 6.3b can be examined in a similar manner. The KVL equation that describes the inductor current for $t > 0$ is

$$L\frac{di(t)}{dt} + Ri(t) = V_S$$

A development identical to that just used yields

$$i(t) = \frac{V_S}{R} + K_2 e^{-\left(\frac{R}{L}\right)t}$$

where V_S/R is the steady-state value and L/R is the circuit's time constant. If there is no initial current in the inductor, then at $t = 0$

$$0 = \frac{V_S}{R} + K_2$$

and

$$K_2 = \frac{-V_S}{R}$$

Hence,

$$i(t) = \frac{V_S}{R} - \frac{V_S}{R}e^{-\frac{R}{L}t}$$

is the complete solution. Note that if we wish to calculate the voltage across the resistor, then

$$v_R(t) = Ri(t)$$

$$= V_S\left(1 - e^{-\frac{R}{L}t}\right)$$

EXAMPLE 6.1

Consider the circuit shown in Fig. 6.4a. Assuming that the switch has been in position 1 for a long time, at time $t = 0$ the switch is moved to position 2. We wish to calculate the current $i(t)$ for $t > 0$.

SOLUTION   At $t = 0-$ the capacitor is fully charged and conducts no current since the capacitor acts like an open circuit to dc. The initial voltage across the capacitor can be found using voltage division. As shown in Fig. 6.4b,

$$v_c(0-) = 12\left(\frac{3k}{6k + 3k}\right) = 4\ V$$

The network for $t > 0$ is shown in Fig. 6.4c. The KCL equation for the voltage across the capacitor is

$$\frac{v(t)}{R_1} + C\frac{dv(t)}{dt} + \frac{v(t)}{R_2} = 0$$

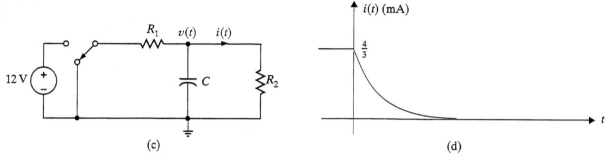

Figure 6.4 Analysis of RC circuits.

Using the component values, the equation becomes

$$\frac{dv(t)}{dt} + 5v(t) = 0$$

The form of the solution to this homogeneous equation is

$$v(t) = K_2 e^{-t/\tau}$$

If we substitute this solution into the differential equation, we find that $\tau = 0.2$ s. Thus,

$$v(t) = K_2 e^{-t/0.2} \text{ V}$$

Using the initial condition $v_c(0-) = v_c(0+) = 4$ V, we find that the complete solution is

$$v(t) = 4e^{-t/0.2} \text{ V}$$

Then $i(t)$ is simply

$$i(t) = \frac{v(t)}{R_2}$$

or

$$i(t) = \frac{4}{3} e^{-t/0.2} \text{ mA} \qquad \square$$

EXAMPLE 6.2

The switch in the network in Fig. 6.5a opens at $t = 0$. Let us find the output voltage $v_o(t)$ for $t > 0$.

Figure 6.5 Analysis of an *RL* circuit.

SOLUTION At $t = 0-$ the circuit is in steady state and the inductor acts like a short circuit. The initial current through the inductor can be found in many ways; however, we will form a Thévenin equivalent for the part of the network to the left of the inductor, as shown in Fig. 6.5b. From this network we find that $I_1 = 4$ A and $V_{oc} = 4$ V. In addition, $R_{Th} = 1$ Ω. Hence, $i_L(0-)$ obtained from Fig. 6.5c is $i_L(0-) = 4/3$ A.

The network for $t > 0$ is shown in Fig. 6.5d. The KVL equation for the circuit is

$$-V_{S_1} + R_1 i(t) + L \frac{di(t)}{dt} + R_3 i(t) = 0$$

which with the component values reduces to

$$\frac{di(t)}{dt} + 2i(t) = 6$$

The solution to this equation is of the form

$$i(t) = K_1 + K_2 e^{-t/\tau}$$

which when substituted into the differential equation yields

$$K_1 = 3$$
$$\tau = 1/2$$

Therefore,

$$i(t) = \left(3 + K_2 e^{-2t}\right) A$$

Evaluating this function at the initial condition, which is

$$i_L(0-) = i_L(0+) = i(0) = 4/3 \, A$$

we find that

$$K_2 = \frac{-5}{3}$$

Hence,

$$i(t) = \left(3 - \frac{5}{3} e^{-2t}\right) A$$

and then

$$v_o(t) = 6 - \frac{10}{3} e^{-2t} \, V$$

A plot of the voltage $v_o(t)$ is shown in Fig. 6.5e.   ❑

EXTENSION EXERCISES

E6.1   Find $v_C(t)$ for $t > 0$ in the circuit shown in Fig. E6.1.

Figure E6.1

ANSWER: $v_C(t) = 8e^{-t/0.6} \, V$.

E6.2   In the circuit shown in Fig. E6.2, the switch opens at $t = 0$. Find $i_1(t)$ for $t > 0$.

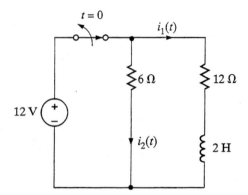

Figure E6.2

ANSWER: $i_1(t) = 1e^{-9t} \, \text{A}$.

The Step-by-Step Approach

In the previous analysis technique we derived the differential equation for the capacitor voltage or inductor current, solved the differential equation, and used the solution to find the unknown variable in the network. In the very methodical technique that we will now describe we will use the fact that Eq. (6.11) is the form of the solution and employ circuit analysis to determine the constants K_1, K_2, and τ.

From Eq. (6.11) we note that as $t \to \infty$, $e^{-at} \to 0$ and $x(t) = K_1$. Therefore, if the circuit is solved for the variable $x(t)$ in steady state (i.e., $t \to \infty$) with the capacitor replaced by an open circuit [v is constant and therefore $i = C(dv/dt) = 0$] or the inductor replaced by a short circuit [i is constant and therefore $v = L(di/dt) = 0$], then the variable $x(t) = K_1$. Note that since the capacitor or inductor has been removed, the circuit is a dc circuit with constant sources and resistors, and therefore only dc analysis is required in the steady-state solution.

The constant K_2 in Eq. (6.11) can also be obtained via the solution of a dc circuit in which a capacitor is replaced by a voltage source or an inductor is replaced by a current source. The value of the voltage source for the capacitor or the current source for the inductor is a known value at one instant of time. In general, we will use the initial condition value since it is generally the one known, but the value at any instant could be used. This value can be obtained in numerous ways and is often specified as input data in a statement of the problem. However, a more likely situation is one in which a switch is thrown in the circuit and the initial value of the capacitor voltage or inductor current is determined from the previous circuit (i.e., the circuit before the switch is thrown). It is normally assumed that the previous circuit has reached steady state, and therefore the voltage across the capacitor or the current through the inductor can be found in exactly the same manner as was used to find K_1.

Finally, the value of the time constant can be found by determining the Thévenin equivalent resistance at the terminals of the storage element. Then $\tau = R_{\text{Th}}C$ for an RC circuit, and $\tau = L/R_{\text{Th}}$ for an RL circuit.

Let us now reiterate this procedure in a step-by-step fashion.

Problem-Solving Strategy for Using the Step-by-Step Approach

Step 1. We assume a solution for the variable $x(t)$ of the form $x(t) = K_1 + K_2 e^{-t/\tau}$.

Step 2. Assuming that the original circuit has reached steady state before a switch was thrown (thereby producing a new circuit), draw this previous circuit with the capacitor replaced by an open circuit or the inductor replaced by a short circuit. Solve for the voltage across the capacitor, $v_C(0-)$, or the current through the inductor, $i_L(0-)$, prior to switch action.

Step 3. Assuming that the energy in the storage element cannot change in zero time, draw the circuit, valid only at $t = 0+$. The switches are in their new positions and the capacitor is replaced by a voltage source with a value of $v_C(0+) = v_C(0-)$ or the inductor is replaced by a current source with value $i_L(0+) = i_L(0-)$. Solve for the initial value of the variable $x(0+)$.

Step 4. Assuming that steady state has been reached after the switches are thrown, draw the equivalent circuit, valid for $t > 5\tau$, by replacing the capacitor by an open circuit or the inductor by a short circuit. Solve for the steady-state value of the variable

$$x(t)|_{t > 5\tau} \doteq x(\infty)$$

Step 5. Since the time constant for all voltages and currents in the circuit will be the same, it can be obtained by reducing the entire circuit to a simple series circuit containing a voltage source, resistor, and a storage element (i.e., capacitor or inductor) by forming a simple Thévenin equivalent circuit at the terminals of the storage element. This Thévenin equivalent circuit is obtained by looking into the circuit from the terminals of the storage element. The time constant for a circuit containing a capacitor is $\tau = R_{Th}C$, and for a circuit containing an inductor it is $\tau = L/R_{Th}$.

Step 6. Using the results of steps 3, 4, and 5, we can evaluate the constants in step 1 as

$$x(0+) = K_1 + K_2$$
$$x(\infty) = K_1$$

and therefore, $K_1 = x(\infty)$, $K_2 = x(0+) - x(\infty)$, and hence the solution is

$$x(t) = x(\infty) + [x(0+) - x(\infty)]e^{-t/\tau}$$

Keep in mind that this solution form applies only to a first-order circuit having constant, dc sources. If the sources are not dc, the forced response will be different. Generally, the forced response is of the same form as the forcing functions (sources) and their derivatives.

EXAMPLE 6.3

Consider the circuit shown in Fig. 6.6a. The circuit is in steady state prior to time $t = 0$, when the switch is closed. Let us calculate the current $i(t)$ for $t > 0$.

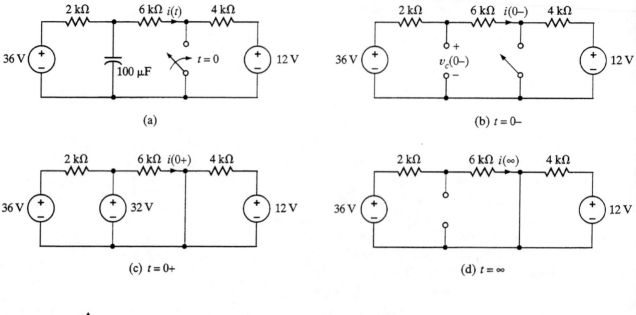

Figure 6.6 Analysis of an *RC* transient circuit with a constant forcing function.

SOLUTION

Step 1. $i(t)$ is of the form $K_1 + K_2 e^{-t/\tau}$.

Step 2.   The initial voltage across the capacitor is calculated from Fig. 6.6b as

$$v_C(0-) = 36 - (2)(2)$$

$$= 32 \text{ V}$$

Step 3.   The new circuit, valid only for $t = 0+$, is shown in Fig. 6.6c. The value of the voltage source that replaces the capacitor is $v_C(0-) = v_C(0+) = 32$ V. Hence,

$$i(0+) = \frac{32}{6k} = \frac{16}{3} \text{ mA}$$

Step 4. The equivalent circuit, valid for $t > 5\tau$, is shown in Fig. 6.6d. The current $i(\infty)$ caused by the 36-V source is

$$i(\infty) = \frac{36}{2k + 6k} = \frac{36}{8}\,\text{mA}$$

Step 5. The Thévenin equivalent resistance, obtained by looking into the open-circuit terminals of the capacitor in Fig. 6.6d, is

$$R_{\text{Th}} = \frac{(2k)(6k)}{2k + 6k} = \frac{3}{2}\,k\Omega$$

Therefore, the circuit time constant is

$$\tau = R_{\text{Th}}\,C$$

$$= \left(\frac{3}{2}\right)(10^3)(100)(10^{-6})$$

$$= 0.15\,\text{s}$$

Step 6.

$$K_1 = i(\infty) = \frac{36}{8}\,\text{mA}$$

$$K_2 = i(0+) - i(\infty) = i(0+) - K_1$$

$$= \frac{16}{3} - \frac{36}{8}$$

$$= \frac{5}{6}\,\text{mA}$$

Therefore,

$$i(t) = \frac{36}{8} + \frac{5}{6}\,e^{-t/0.15}\,\text{mA}$$

Once again we see that although the voltage across the capacitor is continuous at $t = 0$, the current $i(t)$ in the 6-kΩ resistor jumps at $t = 0$ from 2 mA to $5\frac{1}{3}$ mA, and finally decays to $4\frac{1}{2}$ mA. ❏

EXAMPLE 6.4

The circuit shown in Fig. 6.7a is assumed to have been in a steady-state condition prior to switch closure at $t = 0$. We wish to calculate the voltage $v(t)$ for $t > 0$.

SOLUTION

Step 1. $v(t)$ is of the form $K_1 + K_2 e^{-t/\tau}$.

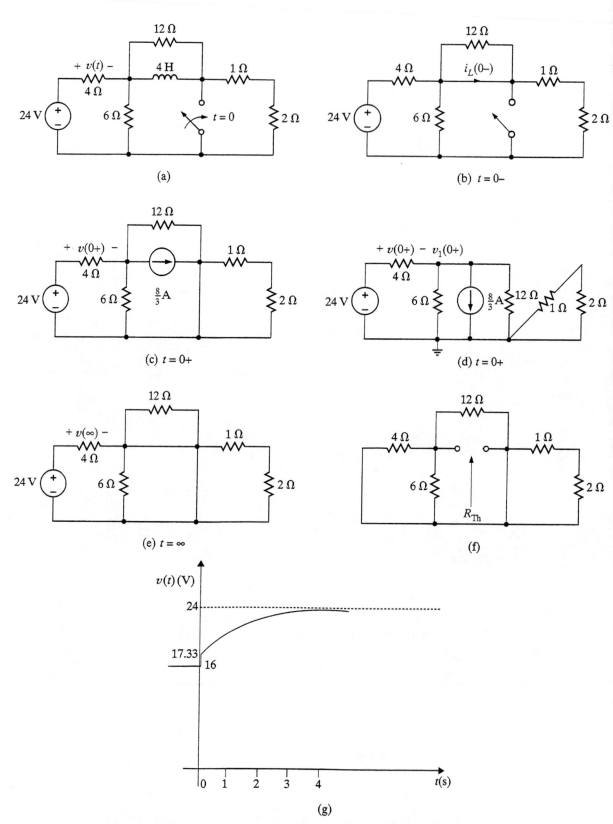

Figure 6.7 Analysis of an *RL* transient circuit with a constant forcing function.

Step 2.   In Fig. 6.7b we see that

$$i_L(0-) = \frac{24}{4 + \dfrac{(6)(3)}{6 + 3}} \left(\frac{6}{6 + 3}\right)$$

$$= \frac{8}{3}\,\text{A}$$

Step 3.   The new circuit, valid only for $t = 0+$, is shown in Fig. 6.7c, which is equivalent to the circuit shown in Fig. 6.7d. The value of the current source that replaces the inductor is $i_L(0-) = i_L(0+) = \frac{8}{3}$ A. The node voltage $v_1(0+)$ can be determined from the circuit in Fig. 6.7d using a single-node equation, and $v(0+)$ is equal to the difference between the source voltage and $v_1(0+)$. The equation for $v_1(0+)$ is

$$\frac{v_1(0+) - 24}{4} + \frac{v_1(0+)}{6} + \frac{8}{3} + \frac{v_1(0+)}{12} = 0$$

or

$$v_1(0+) = \frac{20}{3}\,\text{V}$$

Then

$$v(0+) = 24 - v_1(0+)$$

$$= \frac{52}{3}\,\text{V}$$

Step 4.   The equivalent circuit for the steady-state condition after switch closure is given in Fig. 6.7e. Note that the 6-, 12-, 1-, and 2-Ω resistors are shorted, and therefore $v(\infty) = 24$ V.

Step 5.   The Thévenin equivalent resistance is found by looking into the circuit from the inductor terminals. This circuit is shown in Fig. 6.7f. Note carefully that R_{Th} is equal to the 4-, 6-, and 12-Ω resistors in parallel. Therefore, $R_{\text{Th}} = 2\ \Omega$, and the circuit time constant is

$$\tau = \frac{L}{R_{\text{Th}}} = \frac{4}{2} = 2\ \text{s}$$

Step 6.   From the previous analysis we find that

$$K_1 = v(\infty) = 24$$

$$K_2 = v(0+) - v(\infty) = -\frac{20}{3}$$

and hence that

$$v(t) = 24 - \frac{20}{3}\,e^{-t/2}\,\text{V}$$

From Fig. 6.7b we see that the value of $v(t)$ before switch closure is 16 V. Therefore, the circuit response $v(t)$, as a function of time, is shown in Fig. 6.7g.   ❑

EXTENSION EXERCISES

E6.3 Consider the network in Fig. E6.3. The switch opens at $t = 0$. Find $v_o(t)$ for $t > 0$.

Figure E6.3

ANSWER: $v_o(t) = \dfrac{24}{5} + \dfrac{1}{5}e^{-(5/8)t}$ V.

E6.4 Consider the network in Fig. E6.4. If the switch opens at $t = 0$, find the output voltage $v_o(t)$ for $t > 0$.

Figure E6.4

ANSWER: $v_o(t) = 6 - \dfrac{10}{3}e^{-2t}$ V.

EXAMPLE 6.5

The circuit shown in Fig. 6.8a has reached steady state with the switch in position 1. At time $t = 0$ the switch moves from position 1 to position 2. We want to calculate $v_o(t)$ for $t > 0$.

SOLUTION

Step 1. $v_o(t)$ is of the form $K_1 + K_2 e^{-t/\tau}$.

Step 2.   Using the circuit in Fig. 6.8b, we can calculate $i_L(0-)$

$$i_A = \frac{12}{4} = 3 \text{ A}$$

Then

$$i_L(0-) = \frac{12 + 2i_A}{6} = \frac{18}{6} = 3 \text{ A}$$

Step 3.   The new circuit, valid only for $t = 0+$, is shown in Fig. 6.8c. The value of the current source that replaces the inductor is $i_L(0-) = i_L(0+) = 3$ A. Because of the current source

$$v_o(0+) = (3)(6) = 18 \text{ V}$$

Step 4.   The equivalent circuit, for the steady-state condition after switch closure, is given in Fig. 6.8d. Using the voltages and currents defined in the figure, we can compute $v_o(\infty)$ in a variety of ways. For example, using node equations we can find $v_o(\infty)$ from

$$\frac{v_B - 36}{2} + \frac{v_B}{4} + \frac{v_B + 2i'_A}{6} = 0$$

$$i'_A = \frac{v_B}{4}$$

$$v_o(\infty) = v_B + 2i'_A$$

or, using loop equations,

$$36 = 2(i_1 + i_2) + 4i_1$$
$$36 = 2(i_1 + i_2) + 6i_2 - 2i_1$$
$$v_o(\infty) = 6i_2$$

Using either approach, we find that $v_o(\infty) = 27$ V.

Step 5.   The Thévenin equivalent resistance can be obtained via v_{oc} and i_{sc} because of the presence of the dependent source. From Fig. 6.8e we note that

$$i''_A = \frac{36}{2 + 4} = 6 \text{ A}$$

Figure 6.8 Analysis of an *RL* transient circuit containing a dependent source.

Therefore,

$$v_{oc} = (4)(6) + 2(6)$$
$$= 36 \text{ V}$$

From Fig. 6.8f we can write the following loop equations:

$$36 = 2(i_A''' + i_{sc}) + 4i_A'''$$
$$36 = 2(i_A''' + i_{sc}) + 6i_{sc} - 2i_A'''$$

Solving these equations for i_{sc} yields

$$i_{sc} = \frac{36}{8} \text{ A}$$

Therefore,

$$R_{Th} = \frac{v_{oc}}{i_{sc}} = \frac{36}{36/8} = 8 \ \Omega$$

Hence, the circuit time constant is

$$\tau = \frac{L}{R_{Th}} = \frac{3}{8} \text{ s}$$

Step 6.   Using the information just computed, we can derive the final equation for $v_o(t)$:

$$K_1 = v_o(\infty) = 27$$
$$K_2 = v_o(0+) - v_o(\infty) = 18 - 27 = -9$$

Therefore,

$$v_o(t) = 27 - 9e^{-t/(3/8)} \text{ V} \qquad \square$$

EXTENSION EXERCISE

E6.5   If the switch in the network in Fig. E6.5 closes at $t = 0$, find $v_o(t)$ for $t > 0$.

Figure E6.5

ANSWER: $v_o(t) = 24 + 36e^{-(t/12)} \text{ V}$.

At this point it is appropriate to state that not all switch action will always occur at time $t = 0$. It may occur at any time t_0. In this case the results of the step-by-step analysis yield the following equations:

$$x(t_0) = K_1 + K_2$$
$$x(\infty) = K_1$$

and

$$x(t) = x(\infty) + \left[x(t_0) - x(\infty)\right]e^{-(t-t_0)/\tau} \qquad t > t_0$$

The function is essentially time shifted by t_0 seconds.

Finally, it should be noted that if more than one independent source is present in the network, we can simply employ superposition to obtain the total response.

Thus far we have considered only sources that were constant. Although this represents a very important type of problem, it is by no means the only type nor is it necessarily the most important.

Returning to Eq. (6.1), we ask, "What is the solution if $f(t)$ is not a constant and the network equation is actually as follows?"

$$\frac{dx(t)}{dt} + ax(t) = f(t) \qquad\qquad\qquad \textbf{6.12}$$

Recall that $x(t)$ consists of two parts: the natural response (complementary solution) and the forced response (particular solution). That is,

$$x(t) = x_c(t) + x_p(t) \qquad\qquad\qquad \textbf{6.13}$$

where $x_c(t)$ and $x_p(t)$ must satisfy the equations

$$\frac{dx_c(t)}{dt} + ax_c(t) = 0 \qquad\qquad\qquad \textbf{6.14}$$

and

$$\frac{dx_p(t)}{dt} + ax_p(t) = f(t) \qquad\qquad\qquad \textbf{6.15}$$

Thus $x_c(t)$ must still be of the form

$$x_c(t) = K_2 e^{-at} \qquad\qquad\qquad \textbf{6.16}$$

However, since $f(t)$ is no longer a constant, $x_p(t)$ will, in general, not be one either. Although there are many different mathematical methods for determining the appropriate form for $x_p(t)$, we will employ a deductive approach here. A method that is systematic and often more efficient will be introduced in Chapter 14.

A careful inspection of Eq. (6.15) suggests that $x_p(t)$ must consist of functional forms such as $f(t)$ and its first derivative.

EXAMPLE 6.6

Consider the circuit in Fig. 6.9. The capacitor has an initial charge and an exponentially decaying source is applied at $t = 0$. We wish to determine both $v_o(t)$ and $i_o(t)$ for $t > 0$.

Figure 6.9 *RC* circuit with a nonconstant forcing function.

SOLUTION   A nodal analysis yields the equation

$$C\frac{dv_o(t)}{dt} + \frac{v_o(t) - v_S(t)}{R} = 0 \qquad t > 0$$

or

$$\frac{dv_o(t)}{dt} + 8v_o(t) = 48e^{-2t}$$

Now the natural response (complementary solution) must, as before, satisfy the homogenous equation

$$\frac{dv_{oc}(t)}{dt} + 8v_{oc}(t) = 0$$

which leads to

$$v_{oc}(t) = K_2 e^{-8t}$$

The forced response (particular solution) must satisfy the equation

$$\frac{dv_{op}(t)}{dt} + 8v_{op}(t) = 48e^{-2t}$$

We deduce that $v_{op}(t)$ must be of the form of the forcing function, e^{-2t}, and its derivative, which in this case is also e^{-2t}. Thus, we assume that

$$v_{op}(t) = K_1 e^{-2t}$$

and by substituting this assumed solution back into the differential equation we can determine the required value of K_1.

$$\frac{d}{dt}\left(K_1 e^{-2t}\right) + 8\left(K_1 e^{-2t}\right) = 48e^{-2t}$$

$$(-2 + 8)K_1 e^{-2t} = 48e^{-2t}$$

$$K_1 = 8$$

Therefore, the total solution is

$$v_o(t) = 8e^{-2t} + K_2 e^{-8t}$$

However, since $v_o(0) = -18$, K_2 can be obtained from the expression for $v_o(t)$ as

$$v_o(0) = -18 = 8e^{-2(0)} + K_2 e^{-8(0)}$$

or

$$K_2 = -26$$

Then

$$v_o(t) = 8e^{-2t} - 26e^{-8t}\,\text{V} \qquad t > 0$$

To determine $i(t)$, we note that

$$i(t) = C\frac{dv_o(t)}{dt}$$

$$= 25 \times 10^{-6}(-16e^{-2t} + 208e^{-8t})$$

$$= -0.4e^{-2t} + 5.2e^{-8t}\,\text{mA} \quad t > 0 \qquad \square$$

EXAMPLE 6.7

Consider the network in Fig. 6.10a. The initial inductor current is zero and a sinusoidal voltage is applied at $t = 0$. Let us find both $i_L(t)$ and $v_o(t)$ for $t > 0$.

(a)

(b)

(c)

Figure 6.10 *RL* circuit with a sinusoidal forcing function.

SOLUTION   To simplify the network, we will first find the Thévenin equivalent circuit looking toward the source at nodes *A-B*, as shown in Fig. 6.10b. The Thévenin equivalent voltage is

$$v_{\text{Th}}(t) = v_S(t)\left(\frac{300}{300 + 150}\right) = 10\sin(10^4t)\,\text{V}$$

The equivalent resistance is the parallel combination of 150 Ω with 300 Ω or $R_{Th} = 100\ \Omega$. Now the original circuit is reduced to that in Fig. 6.10c. A loop analysis yields the expression

$$v_S(t) = (R_{Th} + R_3)i(t) + L\frac{di(t)}{dt}$$

or

$$\frac{di(t)}{dt} + 2(10^4)i(t) = 10^3 \sin(10^4 t)$$

The complementary solution must satisfy the homogeneous equation

$$\frac{di_c(t)}{dt} + 2(10^4)i_c(t) = 0$$

which requires a solution of the form

$$i_c(t) = K_1 e^{-2(10^4)t}$$

The forced response must satisfy the equation

$$\frac{di_p(t)}{dt} + 2(10^4)i_p(t) = 10^3 \sin(10^4 t)$$

The solution must be of the form of the forcing function $v_S(t)$ and its derivative. Therefore, we can write

$$i_p(t) = K_2 \sin(10^4 t) + K_3 \cos(10^4 t)$$

and by substituting this assumed solution into the differential equation we can determine K_2 and K_3.

$$\frac{d}{dt}\left[K_2 \sin(10^4 t) + K_3 \cos(10^4 t)\right] + 2(10^4)\left[K_2 \sin(10^4 t) + K_3 \cos(10^4 t)\right]$$
$$= 10^3 \sin(10^4 t)$$

Grouping all sine terms separately from the cosine terms produces two equations:

$$(2(10)^4 K_2 - 10^4 K_3)\sin(10^4 t) = 10^3 \sin(10^4 t)$$
$$(10^4 K_2 + 2(10^4 K_3)\cos(10^4 t) = 0$$

Solving for K_2 and K_3 simultaneously yields

$$K_2 = 0.040$$
$$K_3 = 0.020$$

The final expression for $i_p(t)$ is

$$i_p(t) = 40 \sin(10^4 t) - 20 \cos(10^4 t)\ \text{mA}$$

The total current is the sum of the complementary and forced solutions

$$i(t) = 0.04 \sin(10^4 t) - 0.02 \cos(10^4 t) + K_1 e^{-2(10^4)t}\ \text{A}$$

The value of K_1 can be found using the initial value of $i(t)$. At $t = 0$

$$i(0) = -0.02 + K_1 = 0$$

and

$$K_1 = 0.020$$

Therefore,

$$i(t) = 0.04 \sin(10^4 t) - 0.02 \cos(10^4 t) + 0.02 e^{-2(10^4)t} \text{ A}$$

and

$$v_o(t) = 4 \sin(10^4 t) - 2 \cos(10^4 t) + 2 e^{-2(10^4)t} \text{ V}$$ □

6.4 PULSE RESPONSE

Thus far we have examined networks in which a voltage or current source is suddenly applied. As a result of this sudden application of a source, voltages or currents in the circuit are forced to change abruptly. A forcing function whose value changes in a discontinuous manner or has a discontinuous derivative is called a *singular function*. Two such singular functions that are very important in circuit analysis are the unit impulse function and the unit step function. We will defer a discussion of the former until a later chapter and concentrate on the latter.

The *unit step function* is defined by the following mathematical relationship:

$$u(t) = \begin{cases} 0 & t < 0 \\ 1 & t > 0 \end{cases}$$

In other words, this function, which is dimensionless, is equal to zero for negative values of the argument and equal to 1 for positive values of the argument. It is undefined for a zero argument where the function is discontinuous. A graph of the unit step is shown in Fig. 6.11a. The unit step is dimensionless, and therefore a voltage step of V_o volts or a current step of I_o amperes is written as $V_o u(t)$ and $I_o u(t)$, respectively. Equivalent circuits for a voltage step are shown in Figs. 6.11b and c. Equivalent circuits for a current step are shown in Figs. 6.11d and e. If we use the definition of the unit step, it is easy to generalize this function by replacing the argument t by $t - t_0$. In this case

$$u(t - t_0) = \begin{cases} 0 & t < t_0 \\ 1 & t > t_0 \end{cases}$$

A graph of this function is shown in Fig. 6.11f. Note that $u(t = t_0)$ is equivalent to delaying $u(t)$ by t_0 seconds, so that the abrupt change occurs at time $t = t_0$.

Step functions can be used to construct one or more pulses. For example, the voltage pulse shown in Fig. 6.12a can be formulated by initiating a unit step at $t = 0$ and subtracting one that starts at $t = T$, as shown in Fig. 6.12b. The equation for the pulse is

$$v(t) = A[u(t) - u(t - T)]$$

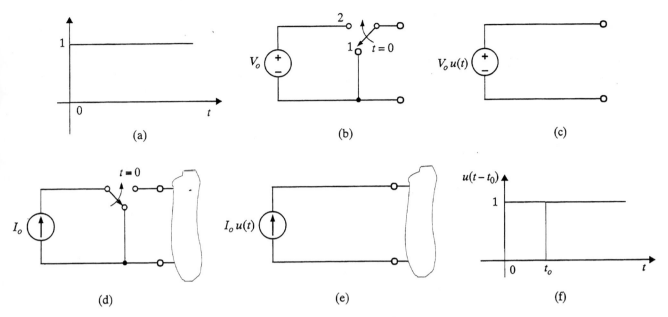

Figure 6.11 Graphs and models of the unit step function.

If the pulse is to start at $t = t_0$ and have width T, the equation would be

$$v(t) = A\{u(t - t_0) - u[t - (t_0 + T)]\}$$

Using this approach, we can write the equation for a pulse starting at any time and ending at any time. Similarly, using this approach, we could write the equation for a series of pulses, called a *pulse train*, by simply forming a summation of pulses constructed in the manner illustrated previously.

The following example will serve to illustrate many of the concepts we have just presented.

EXAMPLE 6.8

Consider the circuit shown in Fig. 6.13a. The input function is the voltage pulse shown in Fig. 6.13b. Since the source is zero for all negative time, the initial conditions for the network are zero [i.e., $v_C(0-) = 0$]. The response $v_o(t)$ for $0 < t < 0.3$ s is due to the application of the constant source at $t = 0$ and is not influenced by any source changes that will occur later. At $t = 0.3$ s the forcing function becomes zero and therefore $v_o(t)$ for $t > 0.3$ s is the source-free or natural response of the network.

Let us determine the expression for the voltage $v_0(t)$.

SOLUTION   Since the output voltage $v_o(t)$ is a voltage division of the capacitor voltage, and the initial voltage across the capacitor is zero, we know that $v_0(0+) = 0$.

If no changes were made in the source after $t = 0$, the steady-state value of $v_o(t)$ [i.e., $v_o(\infty)$] due to the application of the unit step at $t = 0$ would be

$$v_o(\infty) = \frac{9}{6k + 4k + 8k}(8k)$$
$$= 4 \text{ V}$$

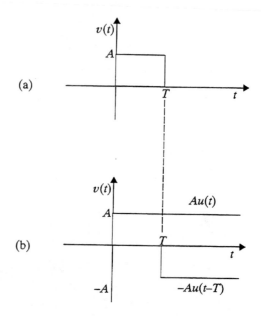

Figure 6.12 Construction of a pulse via two step functions.

DRILL

D6.3 Plot following functions:

(a) $i(t) = 10[u(t) - u(t - 0.01)]$ mA

(b) $v(t) = -10[u(t - 1) - u(t - 2)]$ V

ANSWER: (a)   (b)

The Thévenin equivalent resistance is

$$R_{Th} = \frac{(6k)(12k)}{6k + 12k}$$

$$= 4 \text{ k}\Omega$$

Therefore, the circuit time constant τ is

$$\tau = R_{Th}C$$

$$= (4)(10^3)(100)(10^{-6})$$

$$= 0.4 \text{ s}$$

Therefore, the response $v_o(t)$ for the period $0 < t < 0.3$ s is

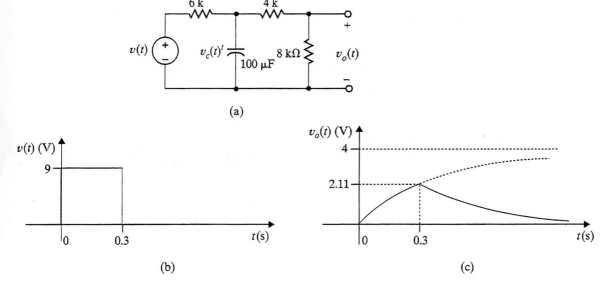

Figure 6.13 Pulse response of a network.

$$v_o(t) = 4 - 4e^{-t/0.4}\,\text{V} \qquad 0 < t < 0.3\,\text{s}$$

The capacitor voltage can be calculated by realizing that using voltage division, $v_o(t) = \frac{2}{3}v_C(t)$. Therefore,

$$v_C(t) = \frac{3}{2}\left(4 - 4e^{-t/0.4}\right)\,\text{V}$$

Since the capacitor voltage is continuous,

$$v_C(0.3-) = v_C(0.3+)$$

and therefore,

$$\begin{aligned}
v_o(0.3+) &= \tfrac{2}{3}v_C(0.3-)\\
&= 4\left(1 - e^{-0.3/0.4}\right)\\
&= 2.11\,\text{V}
\end{aligned}$$

Since the source is zero for $t > 0.3$ s, the final value for $v_o(t)$ as $t \to \infty$ is zero. Therefore, the expression for $v_o(t)$ for $t > 0.3$ s is

$$v_o(t) = 2.11e^{-(t-0.3)/0.4}\,\text{V} \qquad t > 0.3\,\text{s}$$

The term $e^{-(t-0.3)/0.4}$ indicates that the exponential decay starts at $t = 0.3$ s. The complete solution can be written by means of superposition as

$$v_o(t) = 4\left(1 - e^{-t/0.4}\right)u(t) - 4\left(1 - e^{-(t-0.3)/0.4}\right)u(t - 0.3)\,\text{V}$$

or, equivalently, the complete solution is

$$v_o(t) = \begin{cases} 0 & t < 0 \\ 4\left(1 - e^{-t/0.4}\right)\text{V} & 0 < t < 0.3\,\text{s} \\ 2.11e^{-(t-0.3)/0.4}\,\text{V} & 0.3\,\text{s} < t \end{cases}$$

which in mathematical form is

$$v_o(t) = 4\left(1 - e^{-t/0.4}\right)\left[u(t) - u(t - 0.3)\right] + 2.11e^{-(t-0.3)/0.4}u(t - 0.3) \text{ V}$$

Note that the term $[u(t) - u(t - 0.3)]$ acts like a gating function that captures only the part of the step response that exists in the time interval $0 < t < 0.3$ s. The output as a function of time is shown in Fig. 6.13c. ❑

EXTENSION EXERCISE

E6.6 The voltage source in the network in Fig. E6.6a is shown in Fig. E6.6b. The initial current in the inductor must be zero. (Why?) Determine the output voltage $v_0(t)$ for $t > 0$.

(a) (b)

Figure E6.6

ANSWER: $v_o(t) = 0$ for $t < 0, 4\left(1 - e^{-(3/2)t}\right)$ V for $0 \leq t \leq 1$, and $3.11e^{-(3/2)(t-1)}$ V for $1 < t$.

6.5 APPLICATIONS

EXAMPLE 6.9

The network in Fig. 6.14 models the circuit for a flashing neon light. The capacitor charges toward the source voltage of 9 V, through the resistor. The neon light is designed to turn on when the capacitor voltage reaches 8 V. When the light turns on, it discharges the capacitor completely, turns off, and the cycle begins again. This is one form of what is called a relaxation oscillator. If $C = 10 \ \mu\text{F}$, let us find R so that the light flashes once per second.

Figure 6.14 Circuit used in Example 6.9.

SOLUTION   We must first find an expression for the capacitor voltage assuming the neon light never turns on. An analogous situation exists in the network in Fig. 6.3a, where the voltage across the capacitor is

$$v_c(t) = V_S - V_S e^{-t/RC} = 9 - 9e^{-t/RC}$$

At $t = 1$ s, the capacitor voltage should be 8 V. Therefore,

$$v(1) = 8 = 9 - 9e^{-1/RC}$$

Solving for RC yields

$$RC = 0.455 \text{ s}$$

Finally,

$$R = 0.455/C = 45.5 \text{ k}\Omega \qquad \square$$

EXAMPLE 6.10

A heart pacemaker circuit is shown in Fig. 6.15. The SCR (silicon-controlled rectifier) is a solid-state device that has two distinct modes of operation. When the voltage across the SCR is increasing but less than 5 V, the SCR behaves like an open circuit, as shown in Fig. 6.16a. Once the voltage across the SCR reaches 5 V, the device functions like a current source, as shown in Fig. 6.16b. This behavior will continue as long as the SCR voltage remains above 0.2 V. At this voltage, the SCR shuts off and again becomes an open circuit.

Assume that at $t = 0$, $v_c(t)$ is 0 V and the 1-μF capacitor begins to charge toward the 6-V source voltage. Find the resistor value such that $v_c(t)$ will equal 5 V (the SCR firing voltage) at 1 s. At $t = 1$ s, the SCR fires and begins discharging the capacitor. Find the time required for $v_c(t)$ to drop from 5 V to 0.2 V. Finally, plot $v_c(t)$ for the three cycles.

Figure 6.15 Heart pacemaker equivalent circuit.

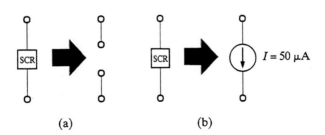

Figure 6.16 Equivalent circuits for silicon-controlled rectifier.

SOLUTION For $t < 1$ s, the equivalent circuit for the pacemaker is shown in Fig. 6.17. As indicated in the previous example, the capacitor voltage has the form

$$v_c(t) = 6 - 6e^{-t/RC}$$

Figure 6.17 Pacemaker equivalent network during capacitor charge cycle.

A voltage of 0.2 V occurs at

$$t_1 = 0.034RC$$

while a voltage of 5 V occurs at

$$t_2 = 1.792RC$$

We desire that $t_2 - t_1 = 1$ s. Therefore,

$$t_2 - t_1 = 1.758RC = 1 \text{ s}$$

and

$$RC = 0.569 \text{ s and } R = 569 \text{ k}\Omega$$

At $t = 1$ s the SCR fires and the pacemaker is modeled by the circuit in Fig. 6.18. The form of the discharge waveform is

$$v(t) = K_1 + K_2 e^{-(t-1)/RC}$$

The term $(t - 1)$ appears in the exponential to shift the function 1 s, since during that time the capacitor was charging. Just after the SCR fires at $t = 1^+$ s, $v_c(t)$ is still 5 V, while at $t = \infty$, $v_c(t) = 6 - IR$. Therefore,

$$K_1 + K_2 = 5 \text{ and } K_1 = 6 - IR$$

Our solution, then, is of the form

$$v_c(t) = 6 - IR + (IR - 1)e^{-(t-1)/RC}$$

Let T be the time beyond 1 s necessary for $v(t)$ to drop to 0.2 V. We write

$$v_c(T + 1) = 6 - IR + (IR - 1)e^{-T/RC} = 0.2$$

Figure 6.18 Pacemaker equivalent network during capacitor discharge cycle.

Substituting for I, R, and C, we find

$$T = 0.11 \text{ s}$$

The output waveform is shown in Fig. 6.19.   ❑

6.6 CIRCUIT DESIGN

EXAMPLE 6.11

Consider the flashing neon light discussed in Example 6.9. Let us generate the design curve for resistance value versus frequency for flash rates in the range from 0.1 to 5 flashes per second. If f represents the flash frequency, determine the value of R for flash rates of $f = 0.5$ and $f = 5$ flashes per second.

SOLUTION   From Example 6.9, we know that

$$v_c(t) = 9 - 9e^{-t/RC}$$

Let the period between flashes be denoted by the variable T. A flash should occur every time $v_c(t)$ charges to 8 V. Therefore,

$$v_c(T) = 8 = 9 - 9e^{-T/RC}$$

Solving for R yields

$$R = 0.455T/C = 45.5 \times 10^3 T$$

Since the period T is $1/f$,

$$R = 45.5 \times 10^3/f$$

As a quick check, if $f = 1$ flash/second, as is the case in Example 6.9, the equation for R yields a value of 45.5 kΩ, which matches the value from Example 6.9. The plot

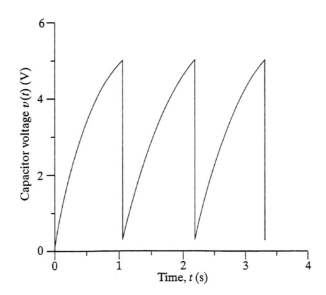

Figure 6.19 Heart pacemaker output voltage waveform.

of R versus f is shown in Fig. 6.20. For 0.5 flashes/second, $R = 91.0$ kΩ, and for 5 flashes/second, $R = 9.1$ kΩ.

Figure 6.20 Resistor value versus flash rate for the circuit in Example 6.9.

6.7 SUMMARY

- An RC or RL transient circuit is said to be first order if it contains only a single capacitor or single inductor. The voltage or current anywhere in the network can be obtained by solving a first-order differential equation.

- The form of a first-order differential equation with a constant forcing function is

$$\frac{dx(t)}{dt} + \frac{x(t)}{\tau} = A$$

and the solution is

$$x(t) = A\tau + K_2 e^{-t/\tau}$$

where $A\tau$ is referred to as the steady-state solution and τ is called the time constant.

- The function $e^{-t/\tau}$ decays to a value that is less than 1% of its initial value after a period of 5τ. Therefore, the time constant, τ, determines the time required for the circuit to reach steady state.

- The time constant for an RC circuit is $R_{Th}\,C$ and for an RL circuit is L/R_{Th}, where R_{Th} is the Thévenin equivalent resistance looking into the circuit at the terminals of the storage element (i.e., capacitor or inductor).

- The two approaches proposed for solving first-order transient circuits are the differential equation approach and the step-by-step method. In the former case, the differential equation that describes the dynamic behavior of the circuit is solved to determine the desired solution. In the latter case, the initial conditions and the steady-state value of the voltage across the capacitor or current in the inductor are used in conjunction with the circuit's time constant and the known form of the desired variable to obtain a solution.

- The response of a first-order transient circuit to an input pulse can be obtained by treating the pulse as a combination of two step-function inputs.

PROBLEMS

Section 6.3

6.1 Use the differential equation approach to find $v_o(t)$ for $t > 0$ in the circuit in Fig. P6.1 and plot the response including the time interval just prior to switch action.

Follow the technique used in Example 6.1.

Figure P6.1

6.2 Use the differential equation approach to find $v_c(t)$ for $t > 0$ in the circuit in Fig. P6.2 and plot the response including the time interval just prior to opening the switch.

Similar to P6.1

Figure P6.2

6.3 Use the differential equation approach to find $v_c(t)$ for $t > 0$ in the circuit in Fig. P6.3 and plot the response including the time interval just prior to closing the switch.

Similar to P6.1

Figure P6.3

6.4 Use the differential equation approach to find $v_o(t)$ for $t > 0$ in the circuit in Fig. P6.4 and plot the response including the time interval just prior to switch action.

Similar to P6.1

Figure P6.4

6.5 Use the differential equation approach to find $i(t)$ for $t > 0$ in the circuit in Fig. P6.5 and plot the response including the time interval just prior to switch movement.

Follow the technique outlined in Example 6.2

Figure P6.5

6.6 Use the differential equation approach to find $i(t)$ for $t > 0$ in the circuit in Fig. P6.6 and plot the response including the time interval just prior to opening the switch.

Similar to P6.5

Figure P6.6

6.7 Use the differential equation approach to find $i_L(t)$ for $t > 0$ in the circuit in Fig. P6.7 and plot the response including the time interval just prior to opening the switch.

Figure P6.7

Similar to P6.5

6.8 Use the differential equation approach to find $i(t)$ for $t > 0$ in the circuit in Fig. P6.8 and plot the response including the time interval just prior to switch movement.

Figure P6.8

Similar to P6.5

6.9 Use the differential equation approach to find $v_o(t)$ for $t > 0$ in the circuit in Fig. P6.9 and plot the response including the time interval just prior to opening the switch.

Figure P6.9

Similar to P6.1

6.10 Use the differential equation approach to find $i_o(t)$ for $t > 0$ in the circuit in Fig. P6.10 and plot the response including the time interval just prior to closing the switch.

Examine the effect of the switch on the capacitor voltage.

Figure P6.10

6.11 Use the differential equation approach to find $v_o(t)$ for $t > 0$ in the circuit in Fig. P6.11 and plot the response including the time interval just prior to closing the switch.

The initial condition of the capacitor must include the influence from both sources. Then the problem is similar to P6.1.

Figure P6.11

6.12 Use the differential equation approach to find $i_o(t)$ for $t > 0$ in the circuit in Fig. P6.12 and plot the response including the time interval just prior to opening the switch.

Similar to P6.1

Figure P6.12

6.13 Use the differential equation approach to find $i_o(t)$ for $t > 0$ in the circuit in Fig. P6.13 and plot the response including the time interval just prior to opening the switch.

Figure P6.13

Similar to P6.5

6.14 Use the differential equation approach to find $i_o(t)$ for $t > 0$ in the circuit in Fig. P6.14 and plot the response including the time interval just prior to opening the switch.

Figure P6.14

Similar to P6.1

6.15 Use the differential equation approach to find $i(t)$ for $t > 0$ in the circuit in Fig. P6.15 and plot the response including the time interval just prior to opening the switch.

Figure P6.15

Similar to P6.5

6.16 Using the differential equation approach find $i_o(t)$ for $t > 0$ in the network in Fig. P6.16 and plot the response including the time interval just prior to opening the switch.

Similar to P6.5

Figure P6.16

6.17 Use the differential equation approach to find $v_o(t)$ for $t > 0$ in the circuit in Fig. P6.17 and plot the response including the time interval just prior to opening the switch.

Similar to P6.1

100 μF

12 V

12 kΩ

24 V

$t = 0$

12 kΩ $v_o(t)$

12 kΩ

Figure P6.17

6.18 Use the differential equation approach to find $i_o(t)$ for $t > 0$ in the circuit in Fig. P6.18 and plot the response including the time interval just prior to opening the switch.

Similar to P6.1

Figure P6.18

6.19 Use the differential equation approach to find $v_o(t)$ for $t > 0$ in the circuit in Fig. P6.19 and plot the response including the time interval just prior to opening the switch.

Figure P6.19

Similar to P6.5

6.20 Use the differential equation approach to find $v_o(t)$ for $t > 0$ in the circuit in Fig. P6.20 and plot the response including the time interval just prior to opening the switch.

Figure P6.20

Similar to P6.1

6.21 Use the step-by step technique to find $i_o(t)$ for $t > 0$ in the network in Fig. P6.21.

Figure P6.21

Follow the technique outlined in Example 6.3.

6.22 Use the step-by step method to find $v_o(t)$ for $t > 0$ in the network in Fig. P6.22.

Similar to P6.21

Figure P6.22

6.23 Use the step-by step method to find $i_o(t)$ for $t > 0$ in the circuit in Fig. P6.23.

Similar to P6.21

Figure P6.23

Similar to P6.21

6.24 Find $v_o(t)$ for $t > 0$ in the network in Fig. P6.23 using the step-by-step technique.

6.25 Use the step-by step method to find $i_o(t)$ for $t > 0$ in the network in Fig. P6.25.

Similar to P6.21

Figure P6.25

Similar to P6.21

6.26 Find $v_o(t)$ for $t > 0$ in the circuit in Fig. P6.25 using the step-by-step method.

6.27 Use the step-by step technique to find $i_o(t)$ for $t > 0$ in the network in Fig. P6.27.

Figure P6.27

Similar to P6.21

6.28 Find $i_o(t)$ for $t > 0$ in the network in Fig. P6.28 using the step-by-step method.

Figure P6.28

Similar to P6.21

6.29 Find $i_o(t)$ for $t > 0$ in the network in Fig. P6.29 using the step-by-step method.

Figure P6.29

Follow the technique outlined in Example 6.4.

6.30 Find $i_o(t)$ for $t > 0$ in the network in Fig. P6.30 using the step-by-step method.

Figure P6.30

Similar to P6.29.

6.31 Find $i_o(t)$ for $t > 0$ in the network in Fig. P6.31 using the step-by-step method.

Examine the effect of opening the switch.

Figure P6.31

6.32 Find $v_o(t)$ for $t > 0$ in the network in Fig. P6.32 using the step-by-step method.

Similar to P6.29

Figure P6.32

6.33 Find $v_o(t)$ for $t > 0$ in the network in Fig. P6.33 using the step-by-step technique.

Similar to P6.29

Figure P6.33

6.34 Use the step-by step method to find $v_o(t)$ for $t > 0$ in the network in Fig. P6.34.

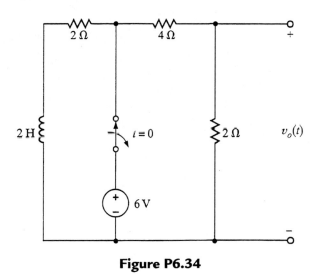

Similar to P6.29

Figure P6.34

6.35 Use the step-by step technique to find $i_o(t)$ for $t > 0$ in the circuit in Fig. P6.35.

Similar to P6.29

Figure P6.35

6.36 Use the step-by step method to find $i_o(t)$ for $t > 0$ in the circuit in Fig. P6.36.

First find the equivalent inductance. Then the problem is similar to P6.29.

Figure P6.36

6.37 Find $i_o(t)$ for $t > 0$ in the circuit in Fig. P6.37 using the step-by-step method.

Similar to P6.29

Figure P6.37

6.38 Find $v_o(t)$ for $t > 0$ in the network in Fig. P6.38 using the step-by-step technique.

Similar to P6.29

Figure P6.38

6.39 Find $i_o(t)$ for $t > 0$ in the network in Fig. P6.39 using the step-by-step method.

Follow the technique outlined in Example 6.5.

Figure P6.39

6.40 Use the step-by step method to find $v_o(t)$ for $t > 0$ in the circuit in Fig. P6.40.

Figure P6.40

Similar to P6.39

6.41 Use the step-by step method to find $v_o(t)$ for $t > 0$ in the network in Fig. P6.41.

Figure P6.41

Follow the technique outlined in Example 6.6.

6.42 Find $v_o(t)$ for $t > 0$ in the circuit in Fig. P6.42 using the step-by-step technique.

Figure P6.42

Similar to P6.41

6.43 Find $v_o(t)$ for $t > 0$ in the network in Fig. P6.43.

Figure P6.43

Follow the technique outlined in Example 6.7.

Section 6.4

6.44 The current source in the network in Fig. P6.44a is defined in Fig. P6.44b. The initial voltage across the capacitor must be zero. (Why?) Determine the current $i_o(t)$ for $t > 0$

Follow the technique outlined in Example 6.8.

 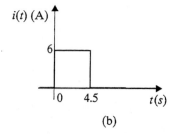

(a)   (b)

Figure P6.44

6.45 Determine the equation for the voltage $v_o(t)$ for $t > 0$, in Fig. P6.45a when subjected to the input pulse shown in Fig. P6.45b.

Similar to P6.44

 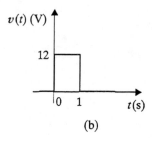

(a)   (b)

Figure P6.45

6.46 Find the output voltage $v_o(t)$ in the network in Fig. P6.46 if the input voltage is

$$v_i(t) = 5(u(t) - u(t - 0.05)) \, \text{V}.$$

Similar to P6.44

Figure P6.46

6.47 Find the output voltage $v_o(t)$ in the network in Fig. P6.47 if the input voltage is

$$v_i(t) = 6(u(t) - u(t - 0.05)) \, \text{V}.$$

Similar to P6.44

Figure P6.47

Advanced Problems

6.48 Using the differential equation approach, find $i_o(t)$ for $t > 0$ in the circuit in Fig. P6.48 and plot the response including the time interval just prior to opening the switch.

Figure P6.48

6.49 Voltages inside the picture tube of a television set can reach several thousand volts. This produces a serious hazard to technicians. Suppose a 1.2μF capacitor is connected to the tube and is charged to 20,000 V. The technician wants to discharge the capacitor through a 500-kΩ resistor. How long will it take to discharge the capacitor by 99%?

6.50 How long would it take to discharage the capacitor in Problem 6.49 to only 20 V? What is the maximum current through the resistor during discharge? What is the maximum power dissipated by the resistor during discharge?

6.51 Use the step-by-step technique to find $i_o(t)$ for $t > 0$ in the network in Fig. P6.51.

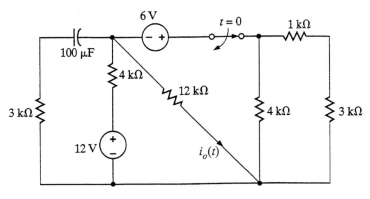

Figure P6.51

6.52 Find $v_o(t)$ for $t > 0$ in the network in Fig. P6.52 using the step-by-step method.

Figure P6.52

6.53 Find $i_o(t)$ for $t > 0$ in the circuit in Fig. P6.53 using the step-by-step technique.

Figure P6.53

6.54 Use the step-by-step technique to find $i_o(t)$ for $t > 0$ in the network in Fig. P6.54.

Figure P6.54

6.55 Find $v_o(t)$ for $t > 0$ in the network in Fig. P6.55 using the step-by-step method.

Figure P6.55

6.56 Find $v_o(t)$ for $t > 0$ in the network in Fig. P6.56 using the step-by-step method.

Figure P6.56

6.57 Use the step-by-step technique to find $v_o(t)$ for $t > 0$ in the circuit in Fig. P6.57.

Figure P6.57

6.58 Use the step-by-step method to find $i_o(t)$ for $t > 0$ in the network in Fig. P6.58.

Figure P6.58

6.59 Find $v_o(t)$ for $t > 0$ in the circuit in Fig. P6.59 using the step-by-step method.

Figure P6.59

6.60 Use the step-by-step method to find $v_o(t)$ for $t > 0$ in the network in Fig. P6.60.

Figure P6.60

6.61 Use the step-by-step technique to find $v_o(t)$ for $t > 0$ in the circuit in Fig. P6.61.

Figure P6.61

6.62 Find $v_o(t)$ for $t > 0$ in the circuit in Fig. P6.62 using the step-by-step method.

Figure P6.62

6.63 Find $v_o(t)$ for $t > 0$ in the network in Fig. P6.63 using the step-by-step technique.

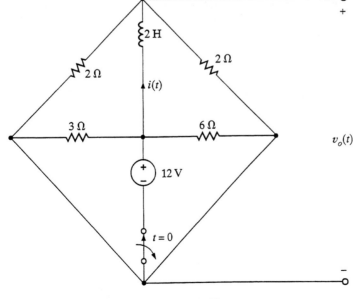

Figure P6.63

6.64 Find $v_o(t)$ for $t > 0$ in the circuit in Fig. P6.64 using the step-by-step method.

Figure P6.64

6.65 Use the step-by-step method to find $v_o(t)$ for $t > 0$ in the circuit in Fig. P6.65.

Figure P6.65

6.66 Use the step-by-step technique to find $i_o(t)$ for $t > 0$ in the network in Fig. P6.66.

Figure P6.66

6.67 Find $i_o(t)$ for $t > 0$ in the circuit in Fig. P6.67 using the step-by-step method.

Figure P6.67

6.68 Find $i_o(t)$ for $t > 0$ in the circuit in Fig. P6.68 using the step-by-step technique.

Figure P6.68

6.69 Use the step-by-step method to find $i_o(t)$ for $t > 0$ in the network in Fig. P6.69.

Figure P6.69

6.70 Use the step-by-step technique to find $v_o(t)$ for $t > 0$ in the network in Fig. P6.70.

Figure P6.70

6.71 Find $v_o(t)$ for $t > 0$ in the network in Fig. P6.71 using the step-by-step method.

Figure P6.71

6.72 Find $v_o(t)$ for $t > 0$ in the network in Fig. P6.72 using the step-by-step method.

Figure P6.72

Second-Order Transient Circuits

7

*I*n this chapter we extend our analysis of circuits that contain only a single storage element to the case where an inductor and a capacitor are present simultaneously. Although the *RLC* circuits are more complicated than those we have analyzed in Chapter 6, we will follow a development similar to that used earlier in obtaining a solution.

Our presentation will deal only with very simple circuits, since the analysis can quickly become complicated for networks that contain more than one loop or one nonreference node. Once we have presented the mathematical development of the response equations, we will analyze several networks in which the parameters have been chosen to illustrate the different types of response. In addition, we will extend our PSPICE analysis techniques. Finally, we will provide a number of practical applications that indicate the importance of these second-order transient circuits.

A more efficient method for providing a general mathematical solution for *RLC* networks will be presented in Chapter 14.

7.1 THE BASIC CIRCUIT EQUATION

To begin our development, let us consider the two basic RLC circuits shown in Fig. 7.1. We assume that energy may be initially stored in both the inductor and capacitor. The node equation for the parallel RLC circuit is

$$\frac{v}{R} + \frac{1}{L} \int_{t_0}^{t} v(x)\,dx + i_L(t_0) + C\frac{dv}{dt} = i_S(t)$$

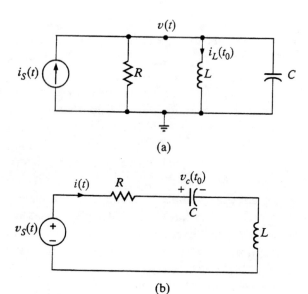

(a)

(b)

Figure 7.1 Parallel and series
RLC series.

Similarly, the loop equation for the series RLC circuit is

$$Ri + \frac{1}{C}\int_{t_0}^{t} i(x)\,dx + v_C(t_0) + L\frac{di}{dt} = v_S(t)$$

Note that the equation for the node voltage in the parallel circuit is of the same form as that for the loop current in the series circuit. Therefore, the solution of these two circuits is dependent on solving one equation. If the two preceding equations are differentiated with respect to time, we obtain

$$C\frac{d^2v}{dt^2} + \frac{1}{R}\frac{dv}{dt} + \frac{v}{L} = \frac{di_S}{dt}$$

and

$$L\frac{d^2i}{dt^2} + R\frac{di}{dt} + \frac{i}{C} = \frac{dv_S}{dt}$$

Since both circuits lead to a second-order differential equation with constant coefficients, we will concentrate our analysis on this type of equation.

D7.1   Write the differential equation that describes the node voltage in (a) and the loop current in (b).

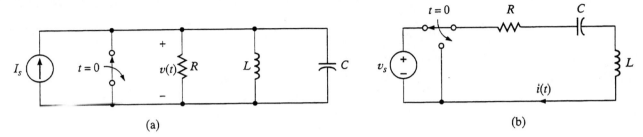

(a)   (b)

ANSWER: (a) $\dfrac{d^2 v}{dt^2} + \dfrac{1}{RC}\dfrac{dv}{dt} + \dfrac{v}{LC} = 0$

(b) $L\dfrac{d^2 i}{dt^2} + R\dfrac{di}{dt} + \dfrac{i}{C} = 0$

7.2 MATHEMATICAL DEVELOPMENT OF THE RESPONSE EQUATIONS

In concert with our development of the solution of a first-order differential equation that results from the analysis of either an RL or an RC circuit as outlined in Chapter 6, we will now employ the same approach here to obtain the solution of a second-order differential equation that results from the analysis of RLC circuits. As a general rule, for this case we are confronted with an equation of the form

$$\frac{d^2 x(t)}{dt^2} + a_1 \frac{dx(t)}{dt} + a_2 x(t) = f(t) \qquad 7.1$$

Once again we use the fact that if $x(t) = x_p(t)$ is a solution to Eq. (7.1), and if $x(t) = x_c(t)$ is a solution to the homogeneous equation

$$\frac{d^2 x(t)}{dt^2} + a_1 \frac{dx(t)}{dt} + a_2 x(t) = 0$$

then

$$x(t) = x_p(t) + x_c(t)$$

is a solution to the original Eq. (7.1). If we again confine ourselves to a constant forcing function [i.e., $f(t) = A$], the development at the beginning of Chapter 6 shows that the solution of Eq. (7.1) will be of the form

$$x(t) = \frac{A}{a_2} + x_c(t) \qquad 7.2$$

Let us now turn our attention to the solution of the homogeneous equation

$$\frac{d^2 x(t)}{dt^2} + a_1 \frac{dx(t)}{dt}\, a_2 x(t) = 0$$

D7.2   Given the homogenous differential equation

$$4\frac{d^2 x(t)}{dt^2} + 8\frac{dx(t)}{dt}$$
$$+ 16x(t) = 0$$

determine the characteristic equation, the damping ratio, and the undamped resonant frequency.

ANSWER: Characteristic equation:

$$s^2 + 2s + 4 = 0$$
$$\zeta = \tfrac{1}{2}$$
$$\omega_o = 2$$

where a_1 and a_2 are constants. For simplicity we will rewrite the equation in the form

$$\frac{d^2 x(t)}{dt^2} + 2\zeta\omega_o \frac{dx(t)}{dt} + \omega_o^2 x(t) = 0 \tag{7.3}$$

where we have made the following simple substitutions for the constants $a_1 = 2\zeta\omega_o$ and $a_2 = \omega_o^2$.

Following the development of a solution for the first-order homogeneous differential equation in Chapter 6, the solution of Eq. (7.3) must be a function whose first- and second-order derivatives have the same form, so that the left-hand side of Eq. (7.3) will become identically zero for all t. Again we assume that

$$x(t) = Ke^{st}$$

Substituting this expression into Eq. (7.3) yields

$$s^2 Ke^{st} + 2\zeta\omega_o sKe^{st} + \omega_o^2 ke^{st} = 0$$

Dividing both sides of the equation by Ke^{st} yields

$$s^2 + 2\zeta\omega_o s + \omega_o^2 = 0 \tag{7.4}$$

This equation is commonly called the *characteristic equation*; ζ is called the exponential *damping ratio*, and ω_o is referred to as the *undamped resonant frequency*. The importance of this terminology will become clear as we proceed with the development. If this equation is satisfied, our assumed solution $x(t) = Ke^{st}$ is correct. Employing the quadratic formula, we find that Eq. (7.4) is satisfied if

$$s = \frac{-2\zeta\omega_o \pm \sqrt{4\zeta^2 \omega_o^2 - 4\omega_o^2}}{2}$$

$$= -\zeta\omega_o \pm \omega_o \sqrt{\zeta^2 - 1} \tag{7.5}$$

Therefore, there are two values of s, s_1 and s_2, that satisfy Eq. (7.4):

$$s_1 = -\zeta\omega_o + \omega_o \sqrt{\zeta^2 - 1}$$
$$s_2 = -\zeta\omega_o - \omega_o \sqrt{\zeta^2 - 1} \tag{7.6}$$

This means that $x_1(t) = K_1 e^{s_1 t}$ is a solution of Eq. (7.3) and that $x_2(t) = K_2 e^{s_2 t}$ is also a solution of Eq. (7.3); that is,

$$\frac{d^2}{dt^2}\left(K_1 e^{s_1 t}\right) + 2\zeta\omega_o \frac{d}{dt}\left(K_1 e^{s_1 t}\right) + \omega_o^2 K_1 e^{s_1 t} = 0$$

and

$$\frac{d^2}{dt^2}\left(K_2 e^{s_2 t}\right) + 2\zeta\omega_o \frac{d}{dt}\left(K_2 e^{s_2 t}\right) + \omega_o^2 K_2 e^{s_2 t} = 0$$

The addition of these two equations produces the equality

$$\frac{d^2}{dt^2}\left(K_1 e^{s_1 t} + K_2 e^{s_2 t}\right) + 2\zeta\omega_o \frac{d}{dt}\left(K_1 e^{s_1 t} + K_2 e^{s_2 t}\right) + \omega_o^2\left(K_1 e^{s_1 t} + K_2 e^{s_2 t}\right) = 0$$

Note that the sum of the two solutions is also a solution. Therefore, in general, the complementary solution of Eq. (7.3) is of the form

$$x_c(t) = K_1 e^{s_1 t} + K_2 e^{s_2 t} \qquad\qquad \textbf{7.7}$$

K_1 and K_2 are constants that can be evaluated via the initial conditions $x(0)$ and $dx(0)/dt$. For example, since

$$x(t) = K_1 e^{s_1 t} + K_2 e^{s_2 t}$$

then

$$x(0) = K_1 + K_2$$

and

$$\left. \frac{dx(t)}{dt} \right|_{t=0} = \frac{dx(0)}{dt} = s_1 K_1 + s_2 K_2$$

Hence, $x(0)$ and $dx(0)/dt$ produce two simultaneous equations, which when solved yield the constants K_1 and K_2.

Close examination of Eqs. (7.6) and (7.7) indicates that the form of the solution of the homogeneous equation is dependent upon the value ζ. For example, if $\zeta > 1$, the roots of the characteristic equation, s_1 and s_2, also called the *natural frequencies* because they determine the natural (unforced) response of the network, are real and unequal; if $\zeta < 1$, the roots are complex numbers; and finally, if $\zeta = 1$, the roots are real and equal. Each of these cases is very important; hence, we will now examine each one in some detail.

Case 1, $\zeta > 1$. This case is commonly called *overdamped*. The natural frequencies s_1 and s_2 are real and unequal, and therefore the natural response of the network described by the second-order differential equation is of the form

$$x_c(t) = K_1 e^{-(\zeta\omega_o - \omega_o \sqrt{\zeta^2 - 1})t} + K_2 e^{-(\zeta\omega_o + \omega_o \sqrt{\zeta^2 - 1})t} \qquad\qquad \textbf{7.8}$$

where K_1 and K_2 are found from the initial conditions. This indicates that the natural response is the sum of two decaying exponentials.

Case 2, $\zeta < 1$. This case is called *underdamped*. Since $\zeta < 1$, the roots of the characteristic equation given in Eq. (7.6) can be written as

$$s_1 = -\zeta\omega_o + j\omega_o \sqrt{1 - \zeta^2} = -\sigma + j\omega_d$$
$$s_2 = -\zeta\omega_o - j\omega_o \sqrt{1 - \zeta^2} = -\sigma - j\omega_d$$

where $j = \sqrt{-1}$, $\sigma = \zeta\omega_o$ and $\omega_d = \omega_o \sqrt{1 - \zeta^2}$. Thus, the natural frequencies are complex numbers (briefly discussed in Appendix B). The natural response is then

$$x_c(t) = K_1 e^{-(\sigma - j\omega_d)t} + K_2 e^{-(\sigma + j\omega_d)t}$$
$$x_c(t) = e^{-\sigma t} \left(K_1 e^{j\omega_d t} + K_2 e^{-j\omega_d t} \right)$$

Using Euler's identities

$$e^{\pm j\theta} = \cos\theta \pm j\sin\theta$$

we obtain

$$x_c(t) = e^{-\sigma t}\left[K_1(\cos \omega_d t + j \sin \omega_d t) + K_2(\cos \omega_d t - j \sin \omega_d t)\right]$$
$$= e^{-\sigma t}\left[(K_1 + K_2) \cos \omega_d t + (jK_1 - jK_2) \sin \omega_d t\right]$$

$$x_c(t) = e^{-\sigma t}(A_1 \cos \omega_d t + A_2 \sin \omega_d t) \tag{7.9}$$

where A_1 and A_2, like K_1 and K_2, are constants, which are evaluated using the initial conditions $x(0)$ and $dx(0)/dt$. If $x_c(t)$ is real, K_1 and K_2 will be complex and $K_2 = K_1^*$. $A_1 = K_1 + K_2$ is therefore 2 times the real part of K_1, and $A_2 = jK_1 - jK_2$ is -2 times the imaginary part of K_1. A_1 and A_2 are the real numbers. This illustrates that the natural response is an exponentially damped oscillatory response.

Case 3, $\zeta = 1$.   This case, called *critically damped*, results in

$$s_1 = s_2 = -\zeta\omega_o$$

as shown in Eq. (7.6). Therefore, Eq. (7.7) reduces to

$$x_c(t) = K_3 e^{-\zeta\omega_o t}$$

where $K_3 = K_1 + K_2$. However, this cannot be a solution to the second-order differential Eq. (7.3) because in general it is not possible to satisfy the two initial conditions $x(0)$ and $dx(0)/dt$ with the single constant K_3.

In the case where the characteristic equation has repeated roots, a solution can be obtained in the following manner. If $x_1(t)$ is known to be a solution of the second-order homogeneous equation, then via the substitution $x(t) = x_1(t)y(t)$ we can transform the given differential equation into a first-order equation in $dy(t)/dt$. Since this resulting equation is only a function of $y(t)$, it can be solved to find the general solution $x(t) = x_1(t)y(t)$.

For the present case, $s_1 = s_2 = -\zeta\omega_o$. For simplicity we let $\alpha = \zeta\omega_o$ and, hence, the basic equation is

$$\frac{d^2 x(t)}{dt^2} + 2\alpha \frac{dx(t)}{dt} + \alpha^2 x(t) = 0 \tag{7.10}$$

and one known solution is

$$x_1(t) = K_3 e^{-\alpha t}$$

By employing the substitution

$$x_2(t) = x_1(t)y(t) = K_3 e^{-\alpha t} y(t)$$

Equation (7.10) becomes

$$\frac{d^2}{dt^2}\left[K_3 e^{-\alpha t} y(t)\right] + 2\alpha \frac{d}{dt}\left[K_3 e^{-\alpha t} y(t)\right] + \alpha^2 K_3 e^{-\alpha t} y(t) = 0$$

Evaluating the derivatives, we obtain

$$\frac{d}{dt}\left[K_3 e^{-\alpha t} y(t)\right] = -K_3 \alpha e^{-\alpha t} y(t) + K_3 e^{-\alpha t}\frac{dy(t)}{dt}$$

$$\frac{d^2}{dt^2}\left[K_3 e^{-\alpha t} y(t)\right] = K_3 \alpha^2 e^{-\alpha t} y(t) - 2K_3 \alpha e^{-\alpha t}\frac{dy(t)}{dt} + K_3 e^{-\alpha t}\frac{d^2 y(t)}{dt^2}$$

Substituting these expressions into the preceding equation yields

$$K_3 e^{-\alpha t}\frac{d^2 y(t)}{dt^2} = 0$$

Therefore,

$$\frac{d^2 y(t)}{dt^2} = 0$$

and hence,

$$y(t) = A_1 + A_2 t$$

Therefore, the general solution is

$$x_2(t) = x_1(t)y(t)$$
$$= K_3 e^{-\alpha t}(A_1 + A_2 t)$$

which can be written as

$$x_2(t) = B_1 e^{-\zeta \omega_0 t} + B_2 t e^{-\zeta \omega_0 t} \qquad \textbf{7.11}$$

where B_1 and B_2 are constants derived from the initial conditions.

It is informative to sketch the natural response for the three cases we have discussed: overdamped, Eq. (7.8); underdamped, Eq. (7.9); and critically damped, Eq. (7.11). Figure 7.2 graphically illustrates the three cases for the situations in which $x_c(0) = 0$. Note that the critically damped response peaks and decays faster than the overdamped response. The underdamped response is an exponentially damped sinusoid whose rate of decay is dependent on the factor ζ. Actually, the terms $\pm e^{-\zeta \omega_0 t}$ define what is called the *envelope* of the response, and the damped oscillations (i.e., the oscillations of decreasing amplitude) exhibited by the waveform in Fig. 7.2b are called *ringing*.

D7.4   Determine the general form of the solution of the equation

$$\frac{d^2 x(t)}{dt^2} + \frac{4 \, dx(t)}{dt} + 4 \, x(t) = 0$$

ANSWER:

$$x_c(t) = B_1 e^{-2t} + B_2 t e^{-2t}$$

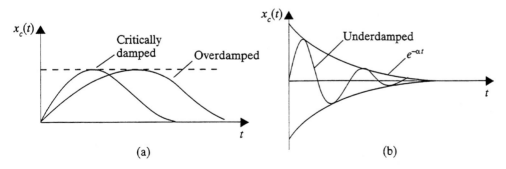

Figure 7.2 Comparison of overdamped, critically damped, and underdamped responses.

EXTENSION EXERCISES

E7.1 A parallel RLC circuit has the following circuit parameters: $R = 1\,\Omega$, $L = 2\,\text{H}$, and $C = 2\,\text{F}$. Compute the damping ratio and the undamped natural frequency of this network.

ANSWER: $\zeta = 0.5, \omega_n = 0.5$ rad/s.

E7.2 A series RLC circuit consists of $R = 2\,\Omega$, $L = 1\,\text{H}$, and a capacitor. Determine the type of response exhibited by the network if (a) $C = \frac{1}{2}\text{F}$, (b) $C = 1\,\text{F}$, and (c) $C = 2\,\text{F}$.

ANSWER: (a) underdamped, (b) critically damped, (c) overdamped.

7.3 THE NETWORK RESPONSE

We will now analyze a number of simple RLC networks that contain both nonzero initial conditions and constant forcing functions. Circuits that exhibit overdamped, underdamped, and critically damped responses will be considered. The approach will illustrate not only the differential equation techniques, but also the manner in which equivalent circuits, valid at $t = 0+$ and $t = \infty$, are used in the analysis.

Problem-Solving Strategy for Second-Order Transient Circuits

- Write the differential equation that describes the circuit.
- Derive the characteristic equation, which can be written in the form $s^2 + 2\omega_o s + \omega_o^2 = 0$, where ζ is the damping ratio and ω_o is the undamped resonant frequency.
- The two roots of the characteristic equation will determine the type of response. If the roots are real and unequal (i.e., $\zeta > 1$), the network response is overdamped. If the roots are real and equal (i.e., $\zeta = 1$), the network response is critically damped. If the roots are complex (i.e., $\zeta < 1$), the network response is underdamped.
- The damping condition and corresponding response for the aforementioned three cases outlined are as follows:

  Overdamped: $x(t) = K_1 e^{-(\zeta\omega_o - \omega_o\sqrt{\zeta^2-1})t} + K_2 e^{-(\zeta\omega_o + \omega_o\sqrt{\zeta^2-1})t}$

  Critically damped: $x(t) = B_1 e^{-\zeta\omega_o t} + B_2 t e^{-\zeta\omega_o t}$

  Underdamped: $x(t) = e^{-\sigma t}(A_1 \cos \omega_d t + A_2 \sin \omega_d t)$, where $\sigma = \zeta\omega_o$, and $\omega_d = \omega_o \sqrt{1 - \zeta^2}$

- Two initial conditions, either given or derived, are required to obtain the two unknown coefficients in the response equation.

The following examples will serve to demonstrate the analysis techniques.

EXAMPLE 7.1

Consider the parallel RLC circuit shown in Fig. 7.3. The second-order differential equation that describes the voltage $v(t)$ is

$$\frac{d^2 v}{dt^2} + \frac{1}{RC}\frac{dv}{dt} + \frac{v}{LC} = 0$$

Figure 7.3 Parallel RLC circuit.

A comparison of this equation with Eqs. (7.3) and (7.4) indicates that for the parallel RLC circuit the damping term is $1/2RC$ and the resonant frequency is $1/\sqrt{LC}$. If the circuit parameters are $R = 2\,\Omega$, $C = \frac{1}{5}\,\text{F}$, and $L = 5\,\text{H}$, the equation becomes

$$\frac{d^2 v}{dt^2} + 2.5\frac{dv}{dt} + v = 0$$

Let us assume that the initial conditions on the storage elements are $i_L(0) = -1\,\text{A}$ and $v_C(0) = 4\,\text{V}$.

Let us find the node voltage $v(t)$ and the inductor current.

SOLUTION   The characteristic equation for the network is

$$s^2 + 2.5s + 1 = 0$$

and the roots are

$$s_1 = -2$$
$$s_2 = -0.5$$

Since the roots are real and unequal, the circuit is overdamped, and $v(t)$ is of the form

$$v(t) = K_1 e^{-2t} + K_2 e^{-0.5t}$$

The initial conditions are now employed to determine the constants K_1 and K_2. Since $v(t) = v_C(t)$,

$$v_C(0) = v(0) = 4 = K_1 + K_2$$

The second equation needed to determine K_1 and K_2 is normally obtained from the expression

$$\frac{dv(t)}{dt} = -2K_1 e^{-2t} - 0.5K_2 e^{-0.5t}$$

However, the second initial condition is not $dv(0)/dt$. If this were the case, we would simply evaluate the equation at $t = 0$. This would produce a second equation in the unknowns K_1 and K_2. We can, however, circumvent this problem by noting that the node equation for the circuit can be written as

$$C\frac{dv(t)}{dt} + \frac{v(t)}{R} + i_L(t) = 0$$

or

$$\frac{dv(t)}{dt} = \frac{-1}{RC}v(t) - \frac{i_L(t)}{C}$$

At $t = 0$,

$$\frac{dv(0)}{dt} = \frac{-1}{RC}v(0) - \frac{1}{C}i_L(0)$$
$$= -2.5(4) - 5(-1)$$
$$= -5$$

But since

$$\frac{dv(t)}{dt} = -2K_1e^{-2t} - 0.5K_2e^{-0.5t}$$

then when $t = 0$

$$-5 = -2K_1 - 0.5K_2$$

This equation, together with the equation

$$4 = K_1 + K_2$$

produces the constants $K_1 = 2$ and $K_2 = 2$. Therefore, the final equation for the voltage is

$$v(t) = 2e^{-2t} + 2e^{-0.5t}\,\text{V}$$

Note that the voltage equation satisfies the initial condition $v(0) = 4$ V. The response curve for this voltage $v(t)$ is shown in Fig. 7.4.

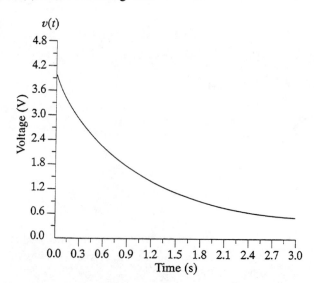

Figure 7.4 Overdamped response.

The inductor current is related to $v(t)$ by the equation

$$i_L(t) = \frac{1}{L} \int v(t)\, dt$$

Substituting our expression for $v(t)$ yields

$$i_L(t) = \frac{1}{5} \int \left[2e^{-2t} + 2e^{-0.5t} \right] dt$$

or

$$i_L(t) = -\frac{1}{5} e^{-2t} - \frac{4}{5} e^{-0.5t} \text{ A}$$

Note that in comparison with the RL and RC circuits we analyzed in Chapter 6, the response of this RLC circuit is controlled by two time constants. The first term has a time constant of $\frac{1}{2}$, and the second term has a time constant of 2.

❏

EXAMPLE 7.2

The series RLC circuit shown in Fig. 7.5 has the following parameters: $C = 0.04$ F, $L = 1$ H, $R = 6$ Ω, $i_L(0) = 4$ A, and $v_C(0) = -4$ V. The equation for the current in the circuit is given by the expression

$$\frac{d^2 i}{dt^2} + \frac{R}{L} \frac{di}{dt} + \frac{i}{LC} = 0$$

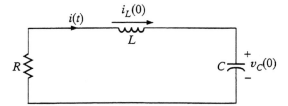

Figure 7.5 Series RLC circuit.

A comparison of this equation with Eqs. (7.3) and (7.4) illustrates that for a series RLC circuit the damping term is $R/2L$ and the resonant frequency is $1/\sqrt{LC}$. Substituting the circuit element values into the preceding equation yields

$$\frac{d^2 i}{dt^2} + 6\frac{di}{dt} + 25i = 0$$

Let us determine the expression for both the current and the capacitor voltage.

SOLUTION The characteristic equation is then

$$s_2 + 6s + 25 = 0$$

and the roots are

$$s_1 = -3 + j4$$
$$s_2 = -3 - j4$$

Since the roots are complex, the circuit is underdamped, and the expression for $i(t)$ is

$$i(t) = K_1 e^{-3t} \cos 4t + K_2 e^{-3t} \sin 4t$$

Using the initial conditions, we find that

$$i(0) = 4 = K_1$$

and

$$\frac{di}{dt} = -4K_1 e^{-3t} \sin 4t - 3K_1 e^{-3t} \cos 4t + 4K_2 e^{-3t} \cos 4t - 3K_2 e^{-3t} \sin 4t$$

and thus

$$\frac{di(0)}{dt} = -3K_1 + 4K_2$$

Although we do not know $di(0)/dt$, we can find it via KVL. From the circuit we note that

$$Ri(0) + L\frac{di(0)}{dt} + v_C(0) = 0$$

or

$$\frac{di(0)}{dt} = -\frac{R}{L} i(0) - \frac{v_C(0)}{L}$$

$$= -\frac{6}{1}(4) + \frac{4}{1}$$

$$= -20$$

Therefore,

$$-3K_1 + 4K_2 = -20$$

and since $K_1 = 4$, $K_2 = -2$, the expression then for $i(t)$ is

$$i(t) = 4e^{-3t} \cos 4t - 2e^{-3t} \sin 4t \ \text{A}$$

Note that this expression satisfies the initial condition $i(0) = 4$. The voltage across the capacitor could be determined via KVL using this current:

$$Ri(t) + L\frac{di(t)}{dt} + v_C(t) = 0$$

or

$$v_C(t) = -Ri(t) - L\frac{di(t)}{dt}$$

Substituting the preceding expression for $i(t)$ into this equation yields

$$v_C(t) = -4e^{-3t} \cos 4t + 22e^{-3t} \sin 4t \ \text{V}$$

Note that this expression satisfies the initial condition $v_C(0) = -4 \ V$. A plot of the function $v_C(t)$ is shown in Fig. 7.6.

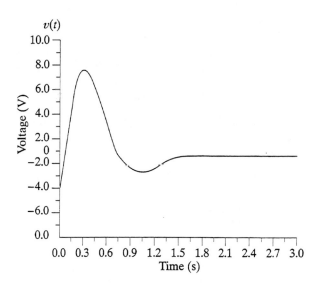

Figure 7.6 Underdamped response.

❑

EXAMPLE 7.3

Let us examine the circuit in Fig. 7.7, which is slightly more complicated than the two we have considered earlier. The two equations that describe the network are

$$L\frac{di(t)}{dt} + R_1 i(t) + v(t) = 0$$

$$i(t) = C\frac{dv(t)}{dt} + \frac{v(t)}{R_2}$$

Substituting the second equation into the first yields

$$\frac{d^2 v}{dt^2} + \left(\frac{1}{R_2 C} + \frac{R_1}{L}\right)\frac{dv}{dt} + \frac{R_1 + R_2}{R_2 LC} v = 0$$

If the circuit parameters and initial conditions are

$$R_1 = 10\,\Omega \qquad C = \tfrac{1}{8}\text{F} \qquad v_C(0) = 1\,\text{V}$$
$$R_2 = 8\,\Omega \qquad L = 2\,\text{H} \qquad i_L(0) = \tfrac{1}{2}\text{A}$$

Figure 7.7 Series-parallel *RLC* circuit.

the differential equation becomes

$$\frac{d^2v}{dt^2} + 6\frac{dv}{dt} + 9v = 0$$

We wish to find expressions for the current $i(t)$ and the voltage $v(t)$.

SOLUTION The characteristic equation is then

$$s^2 + 6s + 9 = 0$$

and hence the roots are

$$s_1 = -3$$
$$s_2 = -3$$

Since the roots are real and equal, the circuit is critically damped. The term $v(t)$ is then given by the expression

$$v(t) = K_1 e^{-3t} + K_2 te^{-3t}$$

Since $v(t) = v_C(t)$,

$$v(0) = v_C(0) = 1 = K_1$$

In addition,

$$\frac{dv(t)}{dt} = -3K_1 e^{-3t} + K_2 e^{-3t} - 3K_2 te^{-3t}$$

However,

$$\frac{dv(t)}{dt} = \frac{i(t)}{C} - \frac{v(t)}{R_2 C}$$

Setting these two expressions equal to one another and evaluating the resultant equation at $t = 0$ yields

$$\frac{1/2}{1/8} - \frac{1}{1} = -3K_1 + K_2$$

$$3 = -3K_1 + K_2$$

Since $K_1 = 1, K_2 = 6$ and the expression for $v(t)$ is

$$v(t) = e^{-3t} + 6te^{-3t} \, \text{V}$$

Note that the expression satisfies the initial condition $v(0) = 1$.

The current $i(t)$ can be determined from the nodal analysis equation at $v(t)$.

$$i(t) = C\frac{dv(t)}{dt} + \frac{v(t)}{R_2}$$

Substituting $v(t)$ from the preceding equation, we find

$$i(t) = \frac{1}{8}\left[-3e^{-3t} + 6e^{-3t} - 18te^{-3t}\right] + \frac{1}{8}\left[e^{-3t} + 6te^{-3t}\right]$$

or

$$i(t) = \tfrac{1}{2}e^{-3t} - \tfrac{3}{2}te^{-3t} \text{ A}$$

If this expression for the current is employed in the circuit equation,

$$v(t) = -L\frac{di(t)}{dt} - R_1 i(t)$$

we obtain

$$v(t) = e^{-3t} + 6te^{-3t} \text{ V}$$

which is identical to the expression derived earlier. This function is shown in Fig. 7.8.

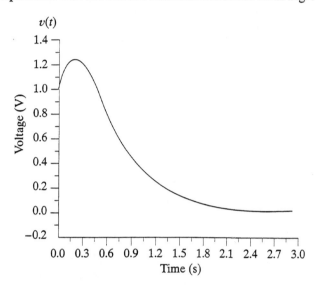

Figure 7.8 Critically damped response.

EXTENSION EXERCISES

E7.3 The switch in the network in Fig. E7.3 opens at $t = 0$. Find $i(t)$ for $t > 0$.

Figure E7.3

ANSWER: $i(t) = -2e^{-t/2} + 4e^{-t}$ A.

E7.4   The switch in the network in Fig. E7.4 moves from position 1 to position 2 at $t = 0$. Find $v_o(t)$ for $t > 0$.

Figure E7.4

ANSWER: $v_o(t) = 2(e^{-t} - 3e^{-3t})$ V.

EXAMPLE 7.4

Consider the circuit shown in Fig. 7.9. This circuit is the same as that analyzed in Example 7.2, except that a constant forcing function is present. The circuit parameters are the same as those used in Example 7.2:

$$C = 0.04 \text{ F} \qquad i_L(0) = 4 \text{ A}$$
$$L = 1 \text{ H} \qquad v_C(0) = -4 \text{ V}$$
$$R = 6 \text{ }\Omega$$

We want to find an expression for $v_C(t)$ for $t > 0$.

SOLUTION   From our earlier mathematical development we know that the general solution of this problem will consist of a particular solution plus a complementary solution. From Example 7.2 we know that the complementary solution is of the form $K_3 e^{-3t} \cos 4t + K_4 e^{-3t} \sin 4t$. The particular solution is a constant, since the input is a constant and therefore the general solution is

$$v_C(t) = K_3 e^{-3t} \cos 4t + K_4 e^{-3t} \sin 4t + K_5$$

Figure 7.9 Series *RLC* circuit with a step function input.

An examination of the circuit shows that in the steady state the final value of $v_C(t)$ is 12 V, since in the steady-state condition, the inductor is a short circuit and the capacitor is an open circuit. Thus, $K_5 = 12$. The steady-state value could also be immediately calculated from the differential equation. The form of the general solution is then

$$v_C(t) = K_3 e^{-3t} \cos 4t + K_4 e^{-3t} \sin 4t + 12$$

The initial conditions can now be used to evaluate the constants K_3 and K_4.

$$v_C(0) = -4 = K_3 + 12$$
$$-16 = K_3$$

Since the derivative of a constant is zero, the results of Example 7.2 show that

$$\frac{dv_C(0)}{dt} = \frac{i(0)}{C} = 100 = -3K_3 + 4K_4$$

and since $K_3 = -16$, $K_4 = 13$. Therefore, the general solution for $v_C(t)$ is

$$v_C(t) = 12 - 16e^{-3t} \cos 4t + 13e^{-3t} \sin 4t \text{ V}$$

Note that this equation satisfies the initial condition $v_C(0) = -4$ and the final condition $v_C(\infty) = 12$ V.   ❑

EXAMPLE 7.5

Let us examine the circuit shown in Fig. 7.10. A close examination of this circuit will indicate that it is identical to that shown in Example 7.3 except that a constant forcing function is present. We assume the circuit is in steady state at $t = 0-$. The equations that describe the circuit for $t > 0$ are

$$L \frac{di(t)}{dt} + R_1 i(t) + v(t) = 24$$

$$i(t) = C \frac{dv(t)}{dt} + \frac{v(t)}{R_2}$$

Combining these equations, we obtain

$$\frac{d^2 v(t)}{dt^2} + \left(\frac{1}{R_2 C} + \frac{R_1}{L} \right) \frac{dv(t)}{dt} + \frac{R_1 + R_2}{R_2 LC} v(t) = \frac{24}{LC}$$

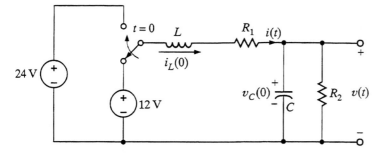

Figure 7.10 Series-parallel *RLC* circuit with a constant forcing function.

If the circuit parameters are $R_1 = 10\ \Omega$, $R_2 = 2\ \Omega$, $L = 2\ \text{H}$, and $C = \frac{1}{4}\text{F}$, the differential equation for the output voltage reduces to

$$\frac{d^2 v(t)}{dt^2} + 7\frac{dv(t)}{dt} + 12v(t) = 48$$

Let us determine the output voltage $v(t)$.

SOLUTION   The characteristic equation is

$$s^2 + 7s + 12 = 0$$

and hence the roots are

$$s_1 = -3$$

$$s_2 = -4$$

The circuit response is overdamped, and therefore the general solution is of the form

$$v(t) = K_1 e^{-3t} + K_2 e^{-4t} + K_3$$

The steady-state value of the voltage, K_3, can be computed from Fig. 7.11a. Note that

$$v(\infty) = 4\,\text{V} = K_3$$

(a) $t = \infty$
(b) $t = 0-$

(c) $t = 0+$

Figure 7.11 Equivalent circuits at $t = \infty$, $t = 0-$, and $t = 0+$ for the circuit in Fig. 7.10.

The initial conditions can be calculated from Figs. 7.11b and c, which are valid at $t = 0-$ and $t = 0+$, respectively. Note that $v(0+) = 2$ V and, hence, from the response equation

$$v(0+) = 2 \text{ V} = K_1 + K_2 + 4$$
$$-2 = K_1 + K_2$$

Figure 7.11c illustrates that $i(0+) = 1$. From the response equation we see that

$$\frac{dv(0)}{dt} = -3K_1 - 4K_2$$

and since

$$\frac{dv(0)}{dt} = \frac{i(0)}{C} - \frac{v(0)}{R_2 C}$$
$$= 4 - 4$$
$$= 0$$

then

$$0 = -3K_1 - 4K_2$$

Solving the two equations for K_1 and K_2 yields $K_1 = -8$ and $K_2 = 6$. Therefore, the general solution for the voltage response is

$$v(t) = 4 - 8e^{-3t} + 6e^{-4t} \text{ V}$$

Note that this equation satisfies both the initial and final values of $v(t)$.   ❑

EXTENSION EXERCISE

E7.5   The switch in the network in Fig. E7.5 moves from position 1 to position 2 at $t = 0$. Compute $i_o(t)$ for $t > 0$ and use this current to determine $v_o(t)$ for $t > 0$.

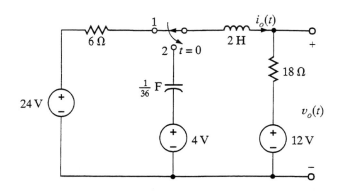

Figure E7.5

ANSWER: $i_o(t) = -\frac{11}{6}e^{-3t} + \frac{14}{6}e^{-6t}$ A, $v_o(t) = 12 + 18i_o(t)$ V.

The examples clearly indicate that the general solution of even very simple RLC circuits can be quite complicated. Although we can continue our current mathematical analysis and present techniques for solving more general networks, better methods exist. For example, in Chapter 14 the Laplace transform technique will be applied to analyze RLC circuits with a variety of forcing functions, and the reader will find that the transform technique is not only easier to use but provides some additional insight in the analysis.

7.4 TRANSIENT PSPICE ANALYSIS USING SCHEMATIC CAPTURE

Introduction

In transient analysis, we determine voltages and currents as functions of time. Typically, the time dependence is demonstrated by plotting the waveforms using time as the independent variable. PSPICE can perform this kind of analysis, called a TRANSIENT simulation, in which all voltages and currents are determined over a specified time duration. To facilitate plotting, PSPICE uses what is known as the PROBE utility, which will be described later. As an introduction to transient analysis, let us simulate the circuit in Fig. 7.12, plot the voltage $v_C(t)$ and the current $i(t)$, and extract the time constant. While we will introduce some new PSPICE topics in this section, *Schematics* fundamentals such as getting parts and wiring and editing part names and values have already been covered in Chapter 4. Also, uppercase text refers to PSPICE utilities and dialog boxes while boldface denotes keyboard or mouse inputs.

Figure 7.12 A circuit used for *Transient* simulation.

The Switch Parts

The inductor and capacitor parts are called L and C, respectively, and are in the ANALOG library. The switch, called SW_TCLOSE, is in the EVAL library. There is also a SW_TOPEN part that models an opening switch. After placing and wiring the switch along with the other parts, the *Schematics* circuit appears as that shown in Fig. 7.13.

To edit the switch's attributes, double click on the switch symbol, and the ATTRIBUTES box shown in Fig. 7.14 will appear. Deselecting the **Include Non-changeable Attributes** and **Include System-defined Attributes** fields limits the attribute list to those we can edit and is highly recommended.

The attribute **tClose** is the time at which the switch begins to close and **ttran** is the time required to complete the closure. Switch attributes **Rclosed** and **Ropen** are the switch's

Figure 7.13 The *Schematics* circuit.

Figure 7.14 The switch's ATTRIBUTES box.

resistance in the closed and open positions, respectively. During simulations, the resistance of the switch changes linearly from **Ropen** at $t = $ **tClose** to **Rclosed** at $t = $ **tClose** $+$ **ttran**.

When using the SW_TCLOSE and SW_TOPEN parts to simulate ideal switches, care should be taken to ensure that the values for **ttran**, **Rclosed**, and **Ropen** are appropriate for valid simulation results. In our present example, we see that the switch and R_1 are in series; thus, their resistances add. Using the default values listed in Figure 7.14, we find that when the switch is closed, the switch resistance, **Rclosed**, is 0.01 Ω, 100,000 times smaller than that of the resistor. The resulting series equivalent resistance is essentially that of the resistor. Alternatively, when the switch is open, the switch resistance is 1 MΩ, 1000 times larger than that of the resistor. Now the equivalent resistance is much larger than that of the resistor. Both are desirable scenarios.

To determine a reasonable value for **ttran**, we first estimate the duration of the transient response. The component values yield a time constant of 1 ms, and thus all voltages and currents will reach steady state in about 5 ms. For accurate simulations, **ttran** should be much less than 5 ms. Therefore, the default value of 1 μs is viable in this case.

The Importance of Pin Numbers

As mentioned in Chapter 4, each component within the various Parts libraries has two or more terminals. Within PSPICE, these terminals are called pins and are numbered sequentially starting with pin 1, as shown in Fig. 7.15 for several two-terminal parts. The significance of the pin numbers is their effect on currents plotted using the PROBE utility. PROBE always plots the current entering pin 1 and exiting pin 2. So if the current through an element is to be plotted, the part should be oriented in the *Schematics* diagram such that the defined current direction enters the part at pin 1. This can be done by using the ROTATE command in the EDIT menu. ROTATE causes the part to spin 90° counterclockwise. In our example, we will plot the current $i(t)$ by plotting the current through the capacitor, I(C1). Therefore, when the *Schematics* circuit in Fig. 7.13 was created, the capacitor was rotated 270°. As a result, pin 1 is at the top of the diagram and the assigned current direction in Fig. 7.12 matches the direction presumed by PROBE. If a component's current direction in PROBE is opposite the desired direction, simply go to the *Schematics* circuit, rotate the part in question 180°, and resimulate.

Figure 7.15 Pin numbers for common PSPICE parts.

Setting Initial Conditions

To set the initial condition of the capacitor voltage, double click on the capacitor symbol in Fig. 7.13 to open its ATTRIBUTE box, as shown in Fig. 7.16. Click on the **IC** field and set the value to the desired voltage, 0 V in this example. Setting the initial condition on an inductor current is done in a similar fashion. Be forewarned that the initial condition for a capacitor voltage is positive at pin 1 versus pin 2. Similarly, the initial condition for an inductor's current will flow into pin 1 and out of pin 2.

Setting Up a Transient Analysis

The simulation duration is selected using **SETUP** from the **ANALYSIS** menu. When the SETUP window shown in Fig. 7.17 appears, double click on the text **TRANSIENT** and the TRANSIENT window in Fig. 7.18 will appear. The simulation period described by **Final time** is selected as 6 ms. All simulations start at $t = 0$. The **No-Print Delay** field sets the time the simulation runs before data collection begins. **Print Step** is the interval used for printing data to the output file. **Print Step** has no effect on the data used to create PROBE plots. The **Detailed Bias Pt.** option is useful when simulating circuits containing transistors and diodes, and thus will not be used here. When **Skip initial transient solution** is en-

Figure 7.16 Setting the capacitor initial condition.

Figure 7.17 The ANALYSIS SETUP window.

abled, all capacitors and inductors that do not have specific initial condition values in their ATTRIBUTES boxes will use zero initial conditions.

Sometimes plots created in PROBE are not smooth. This is caused by an insufficient number of data points. More data points can be requested by inputting a **Step Ceiling** value. A reasonable first guess would be a hundredth of the **Final Time**. If the resulting PROBE plots are still unsatisfactory, reduce the **Step Ceiling** further. As soon as the TRANSIENT window is complete, simulate the circuit by selecting Simulate and the ANALYSIS menu.

Plotting in PROBE

When the PSPICE simulation is finished, the PROBE window shown in Fig. 7.19 will open. If not, select **Run Probe** from the **Analysis** menu. In order to plot the voltage, $v_C(t)$, select **ADD** from the **TRACE** menu. The ADD TRACES window is shown in Fig. 7.20. Notice that the options **Alias Names** and **Subcircuit Nodes** have been deselected, which

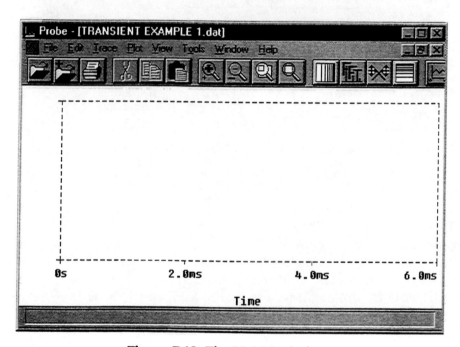

Figure 7.18 The TRANSIENT window.

Figure 7.19 The PROBE window.

greatly simplifies the ADD TRACES window. The capacitor voltage is obtained by clicking on V(Vc) in the left column. The PROBE window should look like that shown in Fig. 7.21.

Before adding the current $i(t)$ to the plot, we note that the dc source is 10 V and the resistance is 1 kΩ, which results in a loop current of a few milliamps. Since the capacitor

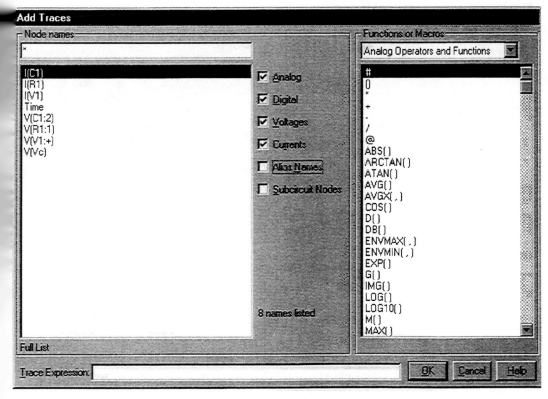

Figure 7.20 The ADD TRACES window.

Figure 7.21 The capacitor voltage.

voltage span is much greater, we will plot the current on a second y-axis. From the **PLOT** menu, select **ADD Y AXIS**. To add the current to the plot, select **ADD** from the **TRACE** menu, then select I(C1). Figure 7.22 shows the PROBE plot for $v_C(t)$ and $i(t)$.

Figure 7.22 The capacitor voltage and the clockwise loop current.

Finding the Time Constant

Given that the final value of $v_C(t)$ is 10 V, we can write

$$v_C(t) = 10[1 - e^{-t/\tau}] \text{ V}$$

When $t = \tau$,

$$v_C(\tau) = 10(1 - e^{-1}) \text{ V} = 6.32 \text{ V}$$

In order to determine the time at which the capacitor voltage is 6.32 V, we activate the cursors by selecting **CURSORS/DISPLAY** in the **TOOLS** menu. There are two cursors that can be used to extract x-y data from the plots. Use the ← and → arrow keys to move the first cursor. By holding the SHIFT key down, the arrow keys move the second cursor. Moving the first cursor along the voltage plot, as shown in Fig. 7.23, we find that 6.32 V occurs at a time of 1 ms. Therefore, the time constant is 1 ms—exactly the RC product.

Saving and Printing PROBE Plots

Saving plots within PROBE requires the use of the **DISPLAY CONTROL** command in the **TOOLS** menu. Using the SAVE/RESTORE DISPLAY window shown in Fig. 7.24, we simply name the plot and click on **SAVE**. Fig. 7.24 shows that one plot has already been saved; TranExample. This procedure saves the plot attributes such as axes settings additional text, and cursor settings in a file with a .prb extension. Additionally, the .prb file contains a reference to the appropriate data file. The data file, which has a .dat extension,

Figure 7.23 PROBE plot for $v_C(t)$ and time constant extraction.

contains the actual simulation results. Therefore, in using **DISPLAY CONTROL** to save a plot, we do not save the actual data on the plot, only the plot settings and the .dat file's name. To access an old PROBE plot, enter PROBE and select **DISPLAY CONTROL**. In the **DISPLAY CONTROL** window, select the file of interest and click on **RESTORE**. Using the **SAVE AS** option in Fig. 7.24 allows you to save the .prb file to any directory on any disk, hard or floppy.

To copy the PROBE plot to other programs such as word processors, use the **COPY TO CLIPBOARD** command in the **TOOLS** menu. To print a PROBE plot, select **PRINT**

Figure 7.24 The SAVE/RESTORE DISPLAY window used in PROBE to save plots.

from the **FILE** menu and the PRINT window in Fig. 7.25 will open. The options in this window are self-explanatory.

Figure 7.25 The PRINT window in PROBE.

Other PROBE Features

There are several features within PROBE for plot manipulation and data extraction. Within the **PLOT** menu reside commands for editing the plot itself. These include altering the axes, adding axes, and adding additional plots to the page. Also in the **TOOLS** menu is the **LABEL** command, which allows one to add explanatory text, lines, and shapes to the plot.

EXAMPLE 7.6

Using the PSPICE *Schematics* editor, draw the circuit in Fig. 7.26, and use the PROBE utility to find the time at which the capacitor and inductor current are equal.

Figure 7.26 Circuit used in Example 7.6.

SOLUTION   Figure 7.27 shows the *Schematics* circuit, while the simulation results are shown in Fig. 7.28. Based on the PROBE plot, the currents are equal at 556 ns.

Figure 7.27

Figure 7.28 Simulation results for $i_C(t)$ and $i_L(t)$.

EXAMPLE 7.7

Using the PSPICE *Schematics* editor, draw the circuit in Fig. 7.29, and use the PROBE utility to plot $v_C(t)$ and $v_L(t)$. At what time after switching does $v_C(t)$ reach half its maximum value?

Figure 7.29

SOLUTION   Figure 7.30 shows the *Schematics* circuit, while simulation results are shown in Fig. 7.31. From the plot, we see that the maximum value of $v_C(t)$ is 40 V, which occurs at $t \leq 1$ ms. The capacitor voltage has decreased to 20 V at 4.236 ms, which is 3.236 ms after switching.

Figure 7.30

7.5 APPLICATIONS

Although there is a wide variety of practical applications for *RLC* circuits, we concentrate our discussion here on a particular class of circuits in which power is supplied to a network by a discharged capacitor.

EXAMPLE 7.8

Let us describe the use of a discharged capacitor in what is known as a *slapper* detonator.

SOLUTION   A device known as a capacitive discharge unit (CDU) is designed to produce very high power (e.g., megawatts in a very short time such as nanoseconds). One practical application of the CDU is its use as what is called a slapper detona-

Figure 7.31 Simulation results for $v_C(t)$ and $v_L(t)$.

tor. The slapper is shown in Fig. 7.32a and consists of a polished insulating substance that is selectively coated with copper in the form of a *bowtie*. Over the center of the bowtie is placed a polyimide material called a *flyer*. The device is positioned close to an explosive pellet, as shown in Fig. 7.32a. When a large current is discharged through the bowtie, the foil heats so quickly that it explodes and the pressure produced by the explosion propels the thin piece of polyimide upward at velocities of 3 to 7 km/s. When the flyer impacts the explosive pellet, the kinetic energy of the flyer is imparted to the pellet and denotation results.

The CDU is modeled as shown in Fig. 7.32b. The resistor represents the central region of the bowtie, the inductor represents the inductance in the circuit, the switch is typically a spark-gap, and the charged capacitor is the energy source. Typical values for the components are shown in the figure. A plot of the current as a function of time following switch closure is shown in Fig. 7.33. An approximate value of the average power delivered by the device can be obtained by dividing the total energy stored in the capacitor by the total time of the discharge. The total energy stored in the capacitor is approximately 1 J and, as shown in Fig. 7.33, this energy is discharged in about 1 μs and 1 J/1μs is a megawatt.   ❑

EXAMPLE 7.9

An experimental schematic for a railgun is shown in Fig. 7.34. With switch sw-2 open, switch sw-1 is closed and the power supply charges the capacitor bank to 10 kV.

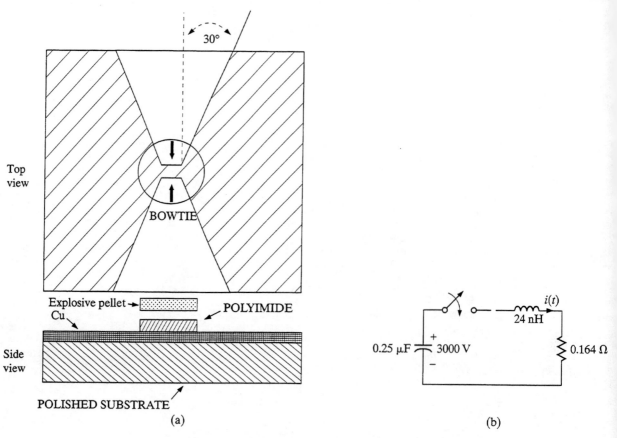

(a)

(b)

Figure 7.32 Slapper detonator and its equivalent circuit.

Figure 7.33 Detonator current as a function of time.

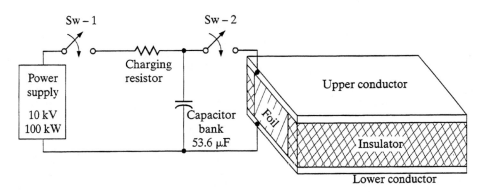

Figure 7.34 Experimental schematic for a railgun.

Then switch sw-1 is opened. The railgun is fired by closing switch sw-2. When the capacitor discharges, the current causes the foil at the end of the gun to explode, creating a hot plasma that accelerates down the tube. The voltage drop in vaporizing the foil is negligible and, therefore, more than 95% of the energy remains available for accelerating the plasma. The current flow establishes a magnetic field, and the force on the plasma caused by the magnetic field, which is proportional to the square of the current at any instant of time, accelerates the plasma. The higher the initial voltage, the more the acceleration.

The circuit diagram for the discharge circuit is shown in Fig. 7.35. The resistance of the bus (a heavy conductor) includes the resistance of the switch. The resistance of the foil and resultant plasma is negligible and, therefore, the current flowing between the upper and lower conductors is dependent upon the remaining circuit components in the closed path, as specified in Fig. 7.34.

The differential equation for the natural response of the current is

$$\frac{d^2 i(t)}{dt^2} + \frac{R_{bus}}{L_{bus}}\frac{di(t)}{dt} + \frac{i(t)}{L_{bus}C} = 0$$

Let us use the characteristic equation to describe the current waveform.

Figure 7.35 Railgun discharge circuit.

SOLUTION   Using the circuit values, the characteristic equation is

$$s^2 + 37.5 \times 10^4 s + 58.3 \times 10^{10} = 0$$

and the roots of the equation are

$$s_1, s_2 = (-18.75 \pm j74) \times 10^4$$

and hence the network is underdamped.

The roots of the characteristic equation illustrate that the damped resonant frequency is

$$\omega_d = 740 \text{ krad/s}$$

Therefore,

$$f_d = 118 \text{ kHz}$$

and the period of the waveform is

$$T = \frac{1}{f_d} = 8.5 \text{ }\mu\text{s}$$

An actual plot of the current is shown in Fig. 7.36, and this plot verifies that the period of the damped response is indeed 8.5 μs.

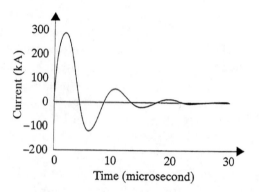

Figure 7.36 Load current with a capacitor bank charged to 10 kV.

Two additional applications for the CDU are the automotive airbag and the exploding bolts used by NASA to provide for reliable stage separation in a multistage rocket.

In the airbag application, a thin wire is welded between two pins. The wire is surrounded by a pyrotechnic mix. Upon impact, a current is passed through the wire via the pins. When the wire heats sufficiently to ignite the mix, a large volume of cool gas is produced, and the bag expands.

NASA employs a number of bolts, which are electrically insulated from the rocket, to provide the mechanical support necessary to hold two stages together. Each bolt is the resistive load in a CDU. At the time of separation, a common fire-control signal is sent to the switches in the CDUs and the energy in the capacitor is discharged through the bolt, causing it to vaporize. All bolts explode simultaneously and stage separation occurs. Clearly all bolts must release simultaneously to avoid a catastrophic accident.

7.6 CIRCUIT DESIGN

The following examples provide an introduction to some typical design strategies for second-order circuits. In these examples the problem involves the selection of the proper circuit parameters to achieve a specified transient response.

EXAMPLE 7.10

The network in Fig. 7.37 models an automobile ignition system. The voltage source represents the standard 12-V battery. The inductor is the ignition coil, which is magnetically coupled to the starter (not shown). The inductor's internal resistance is modeled by the resistor and the switch is the keyed ignition switch. Initially the switch connects the ignition circuitry to the battery, and thus the capacitor is charged to 12 V. To start the motor, we close the switch, thereby discharging the capacitor through the inductor. Assuming that optimum starter operation requires an overdamped response for $i_L(t)$ that reaches at least 1 A within 100 ms after switching, and remains above 1 A for between 1 and 1.5 s, let us find a value for the capacitor that will produce such a current waveform. In addition, let us plot the response including the time interval just prior to moving the switch and verify our design.

Figure 7.37 Circuit model for ignition system.

SOLUTION   Before the switch is moved at $t = 0$, the capacitor looks like an open circuit and the inductor acts like a short circuit. Thus,

$$i_L(0^-) = i_L(0^+) = 0 \text{ A} \quad \text{and} \quad v_C(0^-) = v_C(0^+) = 12 \text{ V}$$

After switching, the circuit is a series RLC unforced network described by the characteristic equation

$$s^2 + \frac{R}{L}s + \frac{1}{LC} = 0$$

with roots at $s = -s_1$ and $-s_2$. The characteristic equation is of the form

$$(s + s_1)(s + s_2) = s^2 + (s_1 + s_2)s + s_1 s_2 = 0$$

Comparing the two expressions, we see that

$$\frac{R}{L} = s_1 + s_2 = 20$$

and

$$\frac{1}{LC} = s_1 s_2$$

Since the network must be overdamped, the inductor current is of the form

$$i_L(t) = K_1 e^{-s_1 t} + K_2 e^{-s_2 t}$$

Just after switching,

$$i_L(0^+) = K_1 + K_2 = 0$$

or

$$K_2 = -K_1$$

Also, at $t = 0^+$, the inductor voltage equals the capacitor voltage because $i_L = 0$ and therefore $i_L R = 0$. Thus, we can write

$$v_L(0^+) = L \frac{di_L(0^+)}{dt} \Rightarrow -s_1 K_1 + s_2 K_1 = \frac{12}{L}$$

or

$$K_1 = \frac{60}{s_2 - s_1}$$

At this point, let us arbitrarily choose $s_1 = 3$ and $s_2 = 17$, which satisfies the condition $s_1 + s_2 = 20$, and furthermore,

$$K_1 = \frac{60}{s_2 - s_1} = \frac{60}{14} = 4.29$$

$$C = \frac{1}{L s_1 s_2} = \frac{1}{(0.2)(3)(17)} = 98 \text{ mF}$$

Hence, $i_L(t)$ is

$$i_L(t) = 4.29 \left[e^{-3t} - e^{-17t} \right] \text{ A}$$

Figure 7.38a shows a plot of $i_L(t)$. At 100 ms the current has increased to 2.39 A, which meets the initial magnitude specifications. However, one second later at $t = 1.1$ s, $i_L(t)$ has fallen to only 0.16 A—well below the magnitude-over-time requirement. Simply put, the current falls too quickly. In order to make an informed estimate for s_1 and s_2, let us investigate the effect the roots exhibit on the current waveform when $s_2 > s_1$.

Since $s_2 > s_1$, the exponential associated with s_2 will decay to zero faster than that associated with s_1. This causes $i_L(t)$ to rise—the larger the value of s_2, the faster the rise. After $5(1/s_2)$ seconds have elapsed, the exponential associated with s_2 is approximately zero and $i_L(t)$ decreases exponentially with a time constant of $\tau = 1/s_1$. Thus, to slow the fall of $i_L(t)$ we should reduce s_1. Hence, let us choose $s_1 = 1$. Since $s_1 + s_2$ must equal 20, $s_2 = 19$. Under these conditions

$$C = \frac{1}{L s_1 s_2} = \frac{1}{(0.2)(1)(19)} = 263 \text{ mF}$$

and

$$K_1 = \frac{60}{s_2 - s_1} = \frac{60}{18} = 3.33$$

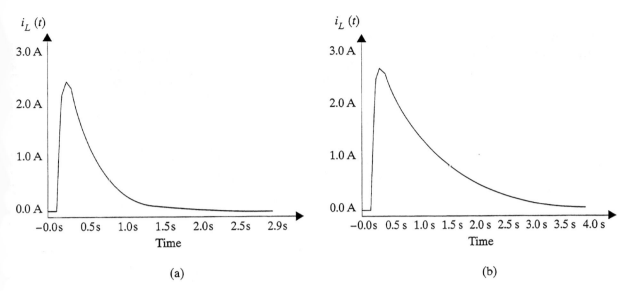

Figure 7.38 Ignition current as a function of time.

Thus, the current is

$$i_L(t) = 3.33\left[e^{-t} - e^{-19t}\right] \text{A}$$

which is shown in Fig. 7.38b. At 100 ms the current is 2.52 A. Also, at $t = 1.1$ s, the current is 1.11 A—above the 1-A requirement. Therefore, the choice of $C = 263$ mF meets all starter specifications. ❏

EXAMPLE 7.11

A defibrillator is a device that is used to stop heart fibrillations—erratic uncoordinated quivering of the heart muscle fibers—by delivering an electric shock to the heart. The Lown defibrillator was developed by Dr. Bernard Lown in 1962. Its key feature, shown in Fig. 7.39a, is its voltage waveform. A simplified circuit diagram that is capable of producing the Lown waveform is shown in Fig. 7.39b. Let us find the necessary values for the inductor and capacitor.

SOLUTION   Since the Lown waveform is oscillatory in nature, we know that the circuit is underdamped ($\zeta < 1$) and the voltage applied to the patient is of the form

$$v_o(t) = K_1 e^{-\zeta \omega_o t} \sin[\omega t]$$

where

$$\zeta \omega_o = \frac{R}{2L}$$

$$\omega = \omega_o \sqrt{1 - \zeta^2}$$

and

$$\omega_o = \frac{1}{\sqrt{LC}}$$

Figure 7.39 Lown defribillator waveform and simplified circuit. Reprinted with permission from John Wiley & Sons Inc., *Introduction to Biomedical Equipment Technology.*

for the series *RLC* circuit. From Fig. 7.39a, we see that the period of the sine function is

$$T = 10 \text{ ms}$$

Thus, we have one expression involving ω_o and ζ

$$\omega = \frac{2\pi}{T} = \omega_o \sqrt{1 - \zeta^2} = 200\pi$$

A second expression can be obtained by solving for the ratio of $v_o(t)$ at $t = T/4$ to that at $t = 3T/4$. At these two instants of time the sine function is equal to +1 and −1, respectively. Using values from Fig. 7.39a, we can write

$$\frac{v_o(T/4)}{-v_o(3T/4)} = \frac{K_1 e^{-\zeta\omega_o(T/4)}}{K_1 e^{-\zeta\omega_o(3T/4)}} = e^{\zeta\omega_o(T/2)} \approx \frac{3000}{250} = 12$$

or

$$\zeta\omega_o = 497.0$$

Given $R = 50\ \Omega$, the necessary inductor value is

$$L = 50.3 \text{ mH}$$

Using our expression for ω,

$$\omega^2 = (200\pi)^2 = \omega_o^2 - (\zeta\omega_o)^2$$

or

$$(200\pi)^2 = \frac{1}{LC} - (497.0)^2$$

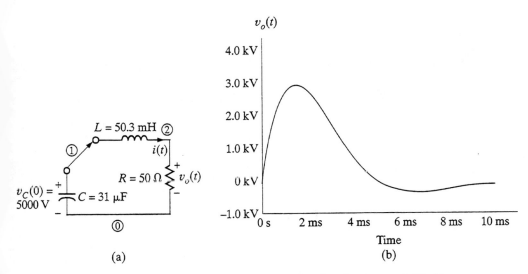

Figure 7.40 PSPICE circuit and output plot for the Lown defibrillator.

Solving for the capacitor value, we find

$$C = 31.0 \ \mu F$$

Let us verify our design using PSPICE. The PSPICE circuit is shown in Fig. 7.40a. The output voltage plot shown in Fig. 7.40b matches the Lown waveform in Fig. 7.39a; thus, we can consider the design to be a success.

It is important to note that while this solution is a viable one, it is not the only one. Like many design problems, there are often various ways to satisfy the design specifications. ❑

7.7 SUMMARY

- The voltage or current in an RLC transient circuit can be described by a constant coefficient differential equation of the form

$$\frac{d^2x(t)}{dt^2} + 2\zeta\omega_o\frac{dx(t)}{dt} + \omega_o^2 x(t) = f(t)$$

where $f(t)$ is the network forcing function.

- The characteristic equation for a second-order circuit is $s^2 + 2\zeta\omega_o s + \omega_o^2 = 0$, where ζ is the damping ratio of ω_o is the undamped resonant frequency.

- If the two roots of the characteristic equation are

    real and unequal, then $\zeta > 1$ and the network response is overdamped
    real and equal, then $\zeta = 1$ and the network response is critically damped
    complex conjugates, then $\zeta < 1$ and the network response is underdamped

- The three types of damping together with the corresponding network response are as follows:

  1. Overdamped: $x(t) = K_1 e^{-(\zeta\omega_o - \omega_o\sqrt{\zeta^2-1})t} + K_2 e^{-(\zeta\omega_o + \omega_o\sqrt{\zeta^2-1})t}$

  2. Critically damped: $x(t) = B_1 e^{-\zeta\omega_o t} + B_2 t e^{-\zeta\omega_o t}$

  3. Underdamped: $x(t) = e^{-\sigma t}(A_1 \cos \omega_d t + A_2 \sin \omega_d t)$, where $\sigma = \zeta\omega_o$ and $\omega_d = \omega_o\sqrt{1 - \zeta^2}$

- Two initial conditions are required to derive the two unknown coefficients in the network response equations.

PROBLEMS

Section 7.2

7.1 The differential equation that describes the current $i_o(t)$ in a network is

See Drill 7.2 and select the type of response from the damping ratio.

$$\frac{d^2 i_o(t)}{dt^2} + 6\frac{di_o(t)}{dt} + 4i_o(t) = 0$$

Find (a) the characteristic equation of the network, (b) the network's natural frequencies, and (c) the expression for $i_o(t)$.

7.2 The terminal current in a network is described by the equation

Similar to P7.1

$$\frac{d^2 i_o(t)}{dt^2} + 8\frac{di_o(t)}{dt} + 16i_o(t) = 0$$

Find (a) the characteristic equation of the network, (b) the network's natural frequencies, and (c) the equation for $i_o(t)$.

7.3 The voltage $v_1(t)$ in a network is defined by the equation

Similar to P7.1

$$\frac{d^2 v_1(t)}{dt^2} + 4\frac{dv_1(t)}{dt} + 5v_1(t) = 0$$

Find (a) the characteristic equation of the network, (b) the circuit's natural frequencies, and (c) the expression for $v_1(t)$.

7.4 The output voltage of a circuit is described by the differential equation

$$\frac{d^2 v_o(t)}{dt^2} + 8\frac{dv_o(t)}{dt} + 10v_o(t) = 0$$

Similar to P7.1

Find (a) the characteristic equation of the circuit, (b) the network's natural frequencies, and (c) the equation for $v_o(t)$.

7.5 The parameters for a parallel RLC circuit are $R = 1\,\Omega, L = \frac{1}{2}\,H$, and $C = \frac{1}{2}F$. Determine the type of damping exhibited by the circuit.

Derive the characteristic equation and compare it to Eq. (7.4).

7.6 A series RLC circuit contains a resistor $R = 2\,\Omega$ and a capacitor $C = \frac{1}{2}F$. Select the value of the inductor so that the circuit is critically damped.

Similar to P7.5

7.7 A parallel RLC circuit contains a resistor $R = 1\,\Omega$ and an inductor $L = 2\,H$. Select the value of the capacitor so that the circuit is critically damped.

Similar to P7.6

Section 7.3

7.8 For the underdamped circuit shown in Fig. P7.8, determine the voltage $v(t)$ if the initial conditions on the storage elements are $i_L(0) = 1\,A$ and $v_C(0) = 12\,V$.

Figure P7.8

Use the problem-solving strategy for second-order transient circuits.

7.9 Given the circuit and the initial conditions of Problem 7.8, determine the current through the inductor.

The $i_L(t)$ can be obtained from $v(t)$.

7.10 In the critically damped circuit shown in Fig. P7.10, the initial conditions on the storage elements are $i_L(0) = 2\,A$ and $v_c(0) = 5\,V$. Determine the voltage $v(t)$.

Figure P7.10

Similar to P7.8

Smilar to P7.9

7.11 Given the circuit and the initial conditions from Problem 7.10, determine the current $i_L(t)$ that is flowing through the inductor.

Smilar to P7.8

7.12 Find $v_o(t)$ for $t > 0$ in the circuit in Fig. P7.12 and plot the response including the time interval just prior to moving the switch.

Figure P7.12

Similar to P7.8

7.13 Find $v_o(t)$ for $t > 0$ in the circuit in Fig. P7.13 and plot the response including the time interval just prior to moving the switch.

Figure P7.13

Similar to P7.8

7.14 Find $v_C(t)$ for $t > 0$ in the circuit in Fig. P7.14 if $v_C(0) = 0$.

Figure P7.14

7.15 Find $v_c(t)$ for $t > 0$ in the circuit in Fig. P7.15.

Figure P7.15

Similar to P7.8

7.16 Find $i_L(t)$ for $t > 0$ in the circuit in Fig. P7.15.

Similar to P7.8

7.17 Given the circuit in Fig. P7.17, find the equation for $i(t), t > 0$.

Figure P7.17

Similar to P7.8

7.18 Find $v_o(t)$ for $t > 0$ in the circuit in Fig. P7.18 and plot the response including the time interval just prior to closing the switch.

Figure P7.18

Similar to P7.8

7.19 In the circuit shown in Fig. P7.19, find $v(t), t > 0$.

Similar to P7.8

Figure P7.19

Section 7.4

7.20 Using the PSPICE *Schematics* editor, draw the circuit in Figure P7.20, and use the PROBE utility to plot $v_C(t)$ and determine the time constants for $0 < t < 1$ ms and 1 ms $< t < \infty$. Also, find the maximum voltage on the capacitor.

Figure P7.20

7.21 Using the PSPICE *Schematics* editor, draw the circuit in Figure P7.21, and use the PROBE utility to find the maximum values of $v_L(t), i_C(t)$, and $i(t)$.

Figure P7.21

Section 7.6

7.22 Design a series RCL circuit with $R \geqq 1\ k\Omega$ that has the characteristic equation

$$s^2 + 4 \times 10^7 s + 4 \times 10^{14} = 0$$

Similar to Example 7.10

7.23 Design a parallel RLC circuit with $R \geqq 1\ k\Omega$ that has the characteristic equation

$$s^2 + 4 \times 10^7 s + 3 \times 10^{14} = 0$$

Similar to P7.2

Advanced Problems

7.24 Find $i_o(t)$ for $t > 0$ in the circuit in Fig. P7.24 and plot the response including the time interval just prior to closing the switch.

Figure P7.24

7.25 Find $v_o(t)$ for $t > 0$ in the circuit in Fig. P7.25 and plot the response including the time interval just prior to $t = 0$.

Figure P7.25

7.26 Find $i_o(t)$ for $t > 0$ in the circuit in Fig. P7.26 and plot the response including the time interval just prior to opening the switch.

Figure P7.26

7.27 Find $v_o(t)$ for $t > 0$ in the circuit in Fig. P7.27 and plot the response including the time interval just prior to closing the switch.

Figure P7.27

7.28 Find $v_o(t)$ for $t > 0$ in the circuit in Fig. P7.28 and plot the response including the time interval just prior to closing the switch.

Figure P7.28

7.29 In the circuit shown in Fig. P7.29, switch action occurs at $t = 0$. Determine the voltage $v_o(t), t > 0$.

Figure P7.29

7.30 Find $v_o(t)$ for $t > 0$ in the circuit in Fig. P7.30 and plot the response including the time interval just prior to moving the switch.

Figure P7.30

7.31 Find $v_o(t)$ for $t > 0$ in the circuit in Fig. P7.31 and plot the response including the time interval just prior to moving the switch.

Figure P7.31

7.32 Find $v_o(t)$ for $t > 0$ in the circuit in Fig. P7.32 and plot the response including the time interval just prior to moving the switch.

Figure P7.32

7.33 For the network in Fig. P7.33, choose C and L such that for $t > 0$, the characteristic equation is

$$s^2 + 7 \times 10^5 s + 10^{10} = 0$$

Figure P7.33

AC Steady-State Analysis

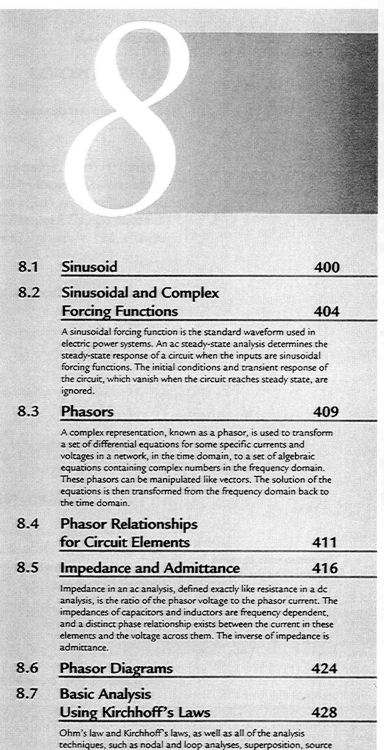

*I*n the preceding chapters we have considered in some detail both the natural and forced response of a network. We found that the natural response was a characteristic of the network and was independent of the forcing function. The forced response, however, depends directly on the type of forcing function, which until now has generally been a constant. At this point we will diverge from this tack to consider an extremely important excitation: the *sinusoidal forcing function*. Nature is replete with examples of sinusoidal phenomena, and although this is important to us as we examine many different types of physical systems, one reason that we can appreciate at this point for studying this forcing function is that it is the dominant waveform in the electric power industry. The signal present at the ac outlets in our home, office, laboratory, and so on is sinusoidal. In addition, we will demonstrate in Chapter 15 that via Fourier analysis we can represent any periodic electrical signal by a sum of sinusoids.

In this chapter we concentrate on the steady-state forced response of networks with sinusoidal driving functions. We will ignore the initial conditions and the transient or natural response, which will eventually vanish for the type of circuits with which we will be dealing. We refer to this as an *ac steady-state analysis*.

Our approach will be to begin by first studying the characteristics of a sinusoidal function as a prelude to using it as a forcing function for a circuit. We will mathematically relate this sinusoidal forcing function to a complex forcing function, which will lead us to define a phasor. By employing phasors we effectively transform a set of differential equations with sinusoidal forcing functions in the time domain into a set of algebraic equations containing complex numbers in the frequency domain. We will show that in this frequency domain Kirchhoff's laws are valid and, thus, all the analysis techniques that we have learned for dc analysis are applicable in ac steady-state analysis. Finally, we demonstrate the power of PSPICE in the solution of ac steady-state circuits.

8.1 SINUSOIDS

Let us begin our discussion of sinusoidal functions by considering the sine wave

$$x(\omega t) = X_M \sin \omega t \qquad 8.1$$

where $x(t)$ could represent either $v(t)$ or $i(t)$. X_M is the *amplitude* or *maximum value*, ω is the *radian* or *angular frequency*, and ωt is the *argument* of the sine function. A plot of the function in Eq. (8.1) as a function of its argument is shown in Fig. 8.1a. Obviously, the function repeats itself every 2π radians. This condition is described mathematically as $x(\omega t + 2\pi) = x(\omega t)$ or in general for period T as

$$x\big[\omega(t + T)\big] = x(\omega t) \qquad 8.2$$

meaning that the function has the same value at time $t + T$ as it does at time t.

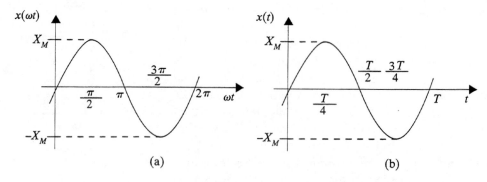

Figure 8.1 Plots of a sine wave as a function of both ωt and t.

The waveform can also be plotted as a function of time, as shown in Fig. 8.1b. Note that this function goes through one period every T seconds, or in other words, in 1 second it goes through $1/T$ periods or cycles. The number of cycles per second, called hertz, is the frequency f, where

The relationship between frequency and period.

$$f = \frac{1}{T} \qquad 8.3$$

Now since $\omega T = 2\pi$, as shown in Fig. 8.1a, we find that

$$\omega = \frac{2\pi}{T} = 2\pi f \qquad 8.4$$

The relationship among frequency, period, and radian frequency.

which is, of course, the general relationship among period in seconds, frequency in hertz, and radian frequency.

Now that we have discussed some of the basic properties of a sine wave, let us consider the following general expression for a sinusoidal function:

$$x(t) = X_M \sin(\omega t + \theta) \qquad 8.5$$

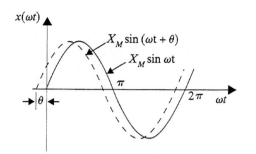

Figure 8.2 Graphical illustration of $X_M \sin(\omega t + \theta)$ leading $X_M \sin \omega t$ by θ radians.

In this case ($\omega t + \theta$) is the argument of the sine function, and θ is called the *phase angle*. A plot of this function is shown in Fig. 8.2, together with the original function in Eq. (8.1) for comparison. Because of the presence of the phase angle, any point on the waveform $X_M \sin(\omega t + \theta)$ occurs θ radians earlier in time than the corresponding point on the waveform $X_M \sin \omega t$. Therefore, we say that $X_M \sin \omega t$ *lags* $X_M \sin(\omega t + \theta)$ by θ radians. In the more general situation, if

$$x_1(t) = X_{M_1} \sin(\omega t + \theta)$$

and

$$x_2(t) = X_{M_2} \sin(\omega t + \phi)$$

then $x_1(t)$ leads $x_2(t)$ by $\theta - \phi$ radians and $x_2(t)$ lags $x_1(t)$ by $\theta - \phi$ radians. If $\theta = \phi$, the waveforms are identical and the functions are said to be *in phase*. If $\theta \neq \phi$, the functions are *out of phase*.

The phase angle is normally expressed in degrees rather than radians, and therefore we will simply state at this point that we will use the two forms interchangeably; that is,

$$x(t) = X_M \sin\left(\omega t + \frac{\pi}{2}\right) = X_M \sin(\omega t + 90°) \qquad \textbf{8.6}$$

Rigorously speaking, since ωt is in radians, the phase angle should be also. However, it is common practice and convenient to use degrees for phase, and therefore, that will be our practice in this text.

In addition, it should be noted that adding to the argument integer multiples of either 2π radians or 360° does not change the original function. This can easily be shown mathematically but is visibly evident when examining the waveform, as shown in Fig. 8.2.

Although our discussion has centered the sine function, we could just as easily have used the cosine function, since the two waveforms differ only by a phase angle; that is,

$$\cos \omega t = \sin\left(\omega t + \frac{\pi}{2}\right) \qquad \textbf{8.7}$$

$$\sin \omega t = \cos\left(\omega t - \frac{\pi}{2}\right) \qquad \textbf{8.8}$$

It should be noted that when comparing one sinusoidal function with another *of the same frequency* to determine the phase difference, it is necessary to express both functions as either sines or cosines with positive amplitudes. Once in this format, the phase angle between the functions can be computed as outlined previously. Two other trigonometric identities that normally prove useful in phase angle determination are

$$-\cos(\omega t) = \cos(\omega t \pm 180°) \qquad \textbf{8.9}$$

$$-\sin(\omega t) = \sin(\omega t \pm 180°) \qquad \textbf{8.10}$$

Finally, the angle-sum and angle-difference relationships for sines and cosines may be useful in the manipulation of sinusoidal functions. These relations are

$$\sin(\alpha + \beta) = \sin\alpha\cos\beta + \cos\alpha\sin\beta$$
$$\cos(\alpha + \beta) = \cos\alpha\cos\beta - \sin\alpha\sin\beta$$
$$\sin(\alpha - \beta) = \sin\alpha\cos\beta - \cos\alpha\sin\beta \qquad \textbf{8.11}$$
$$\cos(\alpha - \beta) = \cos\alpha\cos\beta + \sin\alpha\sin\beta$$

Our interest in these relations will revolve about the cases where $\alpha = \omega t$ and $\beta = \theta$. Since θ is a constant, the preceding equations indicate that either a sinusoidal or cosinusoidal function with a phase angle can be written as the sum of a sine function and a cosine function with the same arguments. For example, using Eq. (8.11), Eq. (8.5) can be written as

$$x(t) = X_M \sin(\omega t + \theta)$$
$$= X_M(\sin\omega t\cos\theta + \cos\omega t\sin\theta)$$
$$= A\sin\omega t + B\cos\omega t$$

where

$$A = X_M \cos\theta$$
$$B = X_M \sin\theta$$

or

$$x(t) = \sqrt{A^2 + B^2}\,\sin\left(\omega t + \tan^{-1}(B/A)\right)$$

EXAMPLE 8.1

We wish to plot the waveforms for the following functions: (a) $v(t) = 1\cos(\omega t + 45°)$, (b) $v(t) = 1\cos(\omega t + 225°)$, and (c) $v(t) = 1\cos(\omega t - 315°)$.

SOLUTION   Figure 8.3a shows a plot of the function $v(t) = 1\cos\omega t$. Figure 8.3b is a plot of the function $v(t) = 1\cos(\omega t + 45°)$. Figure 8.3c is a plot of the function $v(t) = 1\cos(\omega t + 225°)$. Note that since

$$v(t) = 1\cos(\omega t + 225°) = 1\cos(\omega t + 45° + 180°)$$

this waveform is 180° out of phase with the waveform in Fig. 8.3b; that is, $\cos(\omega t + 225°) = -\cos(\omega t + 45°)$, and Fig. 8.3c is the negative of Fig. 8.3b. Finally, since the function

$$v(t) = 1\cos(\omega t - 315°) = 1\cos(\omega t - 315° + 360°) = 1\cos(\omega t + 45°)$$

this function is identical to that shown in Fig. 8.3b.   ❑

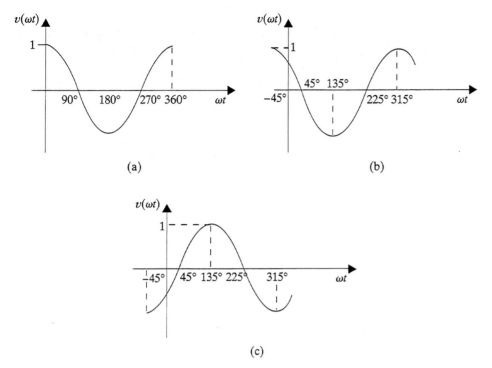

Figure 8.3 Cosine waveforms with various phase angles.

EXAMPLE 8.2

Determine the frequency and the phase angle between the two voltages $v_1(t) = 12 \sin(1000t + 60°)$ V and $v_2(t) = -6 \cos(1000t + 30°)$ V.

SOLUTION   The frequency in hertz (Hz) is given by the expression

$$f = \frac{\omega}{2\pi} = \frac{1000}{2\pi} = 159.2 \text{ Hz}$$

Using Eq. (8.9), $v_2(t)$ can be written as

$$v_2(t) = -6\cos(\omega t + 30°) = 6\cos(\omega t + 210°) \text{ V}$$

Then employing Eq. (8.7), we obtain

$$6\sin(\omega t + 300°) \text{ V} = 6\sin(\omega t - 60°) \text{ V}$$

Now that both voltages of the same frequency are expressed as sine waves with positive amplitudes, the phase angle between $v_1(t)$ and $v_2(t)$ is $60° - (-60) = 120°$; that is, $v_1(t)$ leads $v_2(t)$ by 120° or $v_2(t)$ lags $v_1(t)$ by 120°.   ❏

DRILL

D8.1   The local power company supplies voltage to the home at a radian frequency of $\omega = 377$ r/s. What is the frequency in Hertz, or equivalent, cycles per second.

ANSWER: 60 Hz.

EXTENSION EXERCISES

E8.1   Given the voltage $v(t) = 120 \cos(314t + \pi/4)$V, determine the frequency of the voltage in hertz and the phase angle in degrees.

ANSWER: $f = 50$ Hz, $\theta = 45°$.

E8.2   Three branch currents in a network are known to be

$$i_1(t) = 2 \sin(377t + 45°)\text{A}$$

$$i_2(t) = 0.5 \cos(377t + 10°)\text{A}$$

$$i_3(t) = -0.25 \sin(377t + 60°)\text{A}$$

Determine the phase angles by which $i_1(t)$ leads $i_2(t)$ and $i_1(t)$ leads $i_3(t)$.

ANSWER: i_1 leads i_2 by $-55°$, i_1 leads i_3 by $165°$.

8.2 SINUSOIDAL AND COMPLEX FORCING FUNCTIONS

In the preceding chapters we applied a constant forcing function to a network and found that the steady-state response was also constant.

In a similar manner, if we apply a sinusoidal forcing function to a linear network, the steady-state voltages and currents in the network will also be sinusoidal. This should also be clear from the KVL and KCL equations. For example, if one branch voltage is a sinusoid of some frequency, the other branch voltages must be sinusoids of the same frequency if KVL is to apply around any closed path. This means, of course, that the forced solutions of the differential equations that describe a network with a sinusoidal forcing function are sinusoidal functions of time. For example, if we assume that our input function is a voltage $v(t)$ and our output response is a current $i(t)$ as shown in Fig. 8.4, then if $v(t) = A \sin(\omega t + \theta)$, $i(t)$ will be of the form $i(t) = B \sin(\omega t + \phi)$. The critical point here is that we know the form of the output response, and therefore the solution involves simply determining the values of the two parameters B and ϕ.

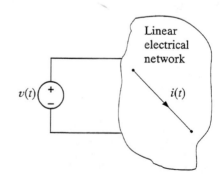

Figure 8.4 Current response to an applied voltage in an electrical network.

EXAMPLE 8.3

Consider the circuit in Fig. 8.5. Let us derive the expression for the current.

$$v(t) = V_M \cos \omega t$$

Figure 8.5 A simple *RL* circuit.

$i(t)$

SOLUTION   The KVL equation for this circuit is

$$L \frac{di(t)}{dt} + Ri(t) = V_M \cos \omega t$$

Since the input forcing function is $X_M \cos \omega t$, we assume that the forced response component of the current $i(t)$ is of the form

$$i(t) = A \cos(\omega t + \phi)$$

which can be written using Eq. (8.11) as

$$i(t) = A \cos \phi \cos \omega t - A \sin \phi \sin \omega t$$
$$= A_1 \cos \omega t + A_2 \sin \omega t$$

Note that this is, as we observed in Chapter 6, of the form of the forcing function $\cos \omega t$ and its derivative $\sin \omega t$. Substituting this form for $i(t)$ into the preceding differential equation yields

$$L \frac{d}{dt} (A_1 \cos \omega t + A_2 \sin \omega t) + R(A_1 \cos \omega t + A_2 \sin \omega t) = V_M \cos \omega t$$

Evaluating the indicated derivative produces

$$-A_1 \omega L \sin \omega t + A_2 \omega L \cos \omega t + RA_1 \cos \omega t + RA_2 \cos \omega t = V_M \cos \omega t$$

By equating coefficients of the sine and cosine functions, we obtain

$$-A_1 \omega L + A_2 R = 0$$
$$A_1 R + A_2 \omega L = V_M$$

that is, two simultaneous equations in the unknowns A_1 and A_2. Solving these two equations for A_1 and A_2 yields

$$A_1 = \frac{RV_M}{R^2 + \omega^2 L^2}$$

$$A_2 = \frac{\omega L V_M}{R^2 + \omega^2 L^2}$$

Therefore,

$$i(t) = \frac{RV_M}{R^2 + \omega^2 L^2} \cos \omega t + \frac{\omega L V_M}{R^2 + \omega^2 L^2} \sin \omega t$$

which, using the last identity in Eq. (8.11), can be written as

$$i(t) = A \cos(\omega t + \phi)$$

where A and ϕ are determined as follows:

$$A \cos \phi = \frac{RV_M}{R^2 + \omega^2 L^2}$$

$$A \sin \phi = \frac{-\omega L V_M}{R^2 + \omega^2 L^2}$$

Hence,

$$\tan \phi = \frac{A \sin \phi}{A \cos \phi} = -\frac{\omega L}{R}$$

and therefore,

$$\phi = -\tan^{-1} \frac{\omega L}{R}$$

and since

$$(A \cos \phi)^2 + (A \sin \phi)^2 = A^2(\cos^2 \phi + \sin^2 \phi) = A^2$$

$$A^2 = \frac{R^2 V_M^2}{\left(R^2 + \omega^2 L^2\right)^2} + \frac{(\omega L)^2 V_M^2}{\left(R^2 + \omega^2 L^2\right)^2} = \frac{V_M^2}{R^2 + \omega^2 L^2}$$

$$A = \frac{V_M}{\sqrt{R^2 + \omega^2 L^2}}$$

Hence, the final expression for $i(t)$ is

$$i(t) = \frac{V_M}{\sqrt{R^2 + \omega^2 L^2}} \cos\left(\omega t - \tan^{-1} \frac{\omega L}{R}\right)$$

The preceding analysis indicates that ϕ is zero if $L = 0$ and hence $i(t)$ is in phase with $v(t)$. If $R = 0$, $\phi = 90°$ and the current lags the voltage by 90°. If L and R are both present, the current lags the voltage by some angle between 0° and 90°. ❏

This example illustrates an important point—solving even a simple one-loop circuit containing one resistor and one inductor is very complicated when compared to the solution of a single-loop circuit containing only two resistors. Imagine for a moment how laborious it would be to solve a more complicated circuit using the procedure we have employed in Example 8.3. To circumvent this approach, we will establish a correspondence between sinusoidal time functions and complex numbers. We will then show that this relationship leads to a set of algebraic equations for currents and voltages in a network (e.g., loop currents or node voltages) in which the coefficients of the variables are complex numbers. Hence, once again we will find that determining the currents or voltages in a circuit can be accomplished by solving a set of algebraic equations; however, in this case, their solution is complicated by the fact that variables in the equations have complex, rather than real, coefficients.

The vehicle we will employ to establish a relationship between time-varying sinusoidal functions and complex numbers is Euler's equation, which for our purposes is written as

$$e^{j\omega t} = \cos \omega t + j \sin \omega t \qquad \textbf{8.12}$$

This complex function has a real part and an imaginary part:

$$\mathrm{Re}(e^{j\omega t}) = \cos \omega t$$
$$\mathrm{Im}(e^{j\omega t}) = \sin \omega t \qquad \textbf{8.13}$$

where $\mathrm{Re}(\cdot)$ and $\mathrm{Im}(\cdot)$ represent the real part and the imaginary part, respectively, of the function in the parentheses.

Now suppose that we select as our forcing function in Fig. 8.4 the nonrealizable voltage

$$v(t) = V_M e^{j\omega t} \qquad \textbf{8.14}$$

which because of Euler's identity can be written as

$$v(t) = V_M \cos \omega t + j V_M \sin \omega t \qquad \textbf{8.15}$$

The real and imaginary parts of this function are each realizable. We think of this complex forcing function as two forcing functions, a real one and an imaginary one, and as a consequence of linearity, the principle of superposition applies and thus the current response can be written as

$$i(t) = I_M \cos(\omega t + \phi) + j I_M \sin(\omega t + \phi) \qquad \textbf{8.16}$$

where $I_M \cos(\omega t + \phi)$ is the response due to $V_M \cos \omega t$ and $j I_M \sin(\omega t + \phi)$ is the response due to $j V_M \sin \omega t$. This expression for the current containing both a real and an imaginary term can be written via Euler's equation as

$$i(t) = I_M e^{j(\omega t + \phi)} \qquad \textbf{8.17}$$

Because of the preceding relationships we find that rather than applying the forcing function $V_M \cos \omega t$ and calculating the response $I_M \cos(\omega t + \phi)$, we can apply the complex forcing function $V_M e^{j\omega t}$ and calculate the response $I_M e^{j(\omega t + \phi)}$, the real part of which is the desired response $I_M \cos(\omega t + \phi)$. Although this procedure may initially appear to be more complicated, it is not. It is via this technique that we will convert the differential equation to an algebraic equation that is much easier to solve.

EXAMPLE 8.4

Once again, let us determine the current in the RL circuit examined in Example 8.3. However, rather than applying $V_M \cos \omega t$ we will apply $V_M e^{j\omega t}$.

SOLUTION   The forced response will be of the form

$$i(t) = I_M e^{j(\omega t + \phi)}$$

where only I_M and ϕ are unknown. Substituting $v(t)$ and $i(t)$ into the differential equation for the circuit, we obtain

$$RI_M e^{j(\omega t + \phi)} + L \frac{d}{dt} \left(I_M e^{j(\omega t + \phi)} \right) = V_M e^{j\omega t}$$

Taking the indicated derivative, we obtain

$$RI_M e^{j(\omega t + \phi)} + j\omega L I_M e^{j(\omega t + \phi)} = V_M e^{j\omega t}$$

Dividing each term of the equation by the common factor $e^{j\omega t}$ yields

$$RI_M e^{j\phi} + j\omega L I_M e^{j\phi} = V_M$$

which is an algebraic equation with complex coefficients. This equation can be written as

$$I_M e^{j\phi} = \frac{V_M}{R + j\omega L}$$

Converting the right-hand side of the equation to exponential or polar form produces the equation

$$I_M e^{j\phi} = \frac{V_M}{\sqrt{R^2 + \omega^2 L^2}} e^{j[-\tan^{-1}(\omega L / R)]}$$

Summary of complex number relationships:

$$x + jy = re^{j\theta}$$
$$r = \sqrt{x^2 + y^2}$$
$$\theta = \tan^{-1} \frac{y}{x}$$
$$x = r \cos \theta$$
$$y = r \sin \theta$$
$$\frac{1}{e^{j\theta}} = e^{-j\theta}$$

(A quick refresher on complex numbers is given in Appendix C for readers who need to sharpen their skills in this area.) The preceding form clearly indicates that the magnitude and phase of the resulting current are

$$I_M = \frac{V_M}{\sqrt{R^2 + \omega^2 L^2}}$$

and

$$\phi = -\tan^{-1} \frac{\omega L}{R}$$

However, since our actual forcing function was $V_M \cos \omega t$ rather than $V_M e^{j\omega t}$, our actual response is the real part of the complex response:

$$i(t) = I_M \cos(\omega t + \phi)$$
$$= \frac{V_M}{\sqrt{R^2 + \omega^2 L^2}} \cos\left(\omega t - \tan^{-1} \frac{\omega L}{R} \right)$$

Note that this is identical to the response obtained in the previous example by solving the differential equation for the current $i(t)$. ❏

8.3 PHASORS

Once again let us assume that the forcing function for a linear network is of the form

$$v(t) = V_M e^{j\omega t}$$

Then every steady-state voltage or current in the network will have the same form and the same frequency ω; for example, a current $i(t)$ will be of the form $i(t) = I_M e^{j(\omega t + \phi)}$.

As we proceed in our subsequent circuit analyses, we will simply note the frequency and then drop the factor $e^{j\omega t}$ since it is common to every term in the describing equations. Dropping the term $e^{j\omega t}$ indicates that every voltage or current can be fully described by a magnitude and phase. For example, a voltage $v(t)$ can be written in exponential form as

$$v(t) = V_M \cos(\omega t + \theta) = \text{Re}\left[V_M e^{j(\omega t + \theta)}\right] \qquad \textbf{8.18}$$

or as a complex number

$$v(t) = \text{Re}\left(V_M \underline{/\theta}\, e^{j\omega t}\right) \qquad \textbf{8.19}$$

Since we are working with a complex forcing function, the real part of which is the desired answer, and each term in the equation will contain $e^{j\omega t}$, we can drop $\text{Re}(\cdot)$ and $e^{j\omega t}$ and work only with the complex number $V_M \underline{/\theta}$. This complex representation is commonly called a *phasor*. As a distinguishing feature, phasors will be written in boldface type. In a completely identical manner a voltage $v(t) = V_M \cos(\omega t + \theta) = \text{Re}\left[V_M e^{j(\omega t + \theta)}\right]$ and a current $i(t) = I_M \cos(\omega t + \phi) = \text{Re}\left[I_M e^{j(\omega t + \phi)}\right]$ are written in phasor notation as $\mathbf{V} = V_M \underline{/\theta}$ and $\mathbf{I} = V_M \underline{/\phi}$, respectively.

If $v(t) = V_m \cos(\omega t + \theta)$ and $i(t) = I_m \cos(\omega t + \phi)$, then in phasor notation

$$\mathbf{V} = V_m \underline{/\theta} \quad \text{and}$$

$$\mathbf{I} = I_m \underline{/\phi}$$

EXAMPLE 8.5

Again, we consider the *RL* circuit in Example 8.3. Let us use phasors to determine the expression for the current.

SOLUTION   The differential equation is

$$L \frac{di(t)}{dt} + Ri(t) = V_M \cos \omega t$$

The forcing function can be replaced by a complex forcing function that is written as $\mathbf{V}e^{j\omega t}$ with phasor $\mathbf{V} = V_M \underline{/0°}$. Similarly, the forced response component of the current $i(t)$ can be replaced by a complex function that is written as $\mathbf{I}e^{j\omega t}$ with phasor $\mathbf{I} = I_M \underline{/\phi}$. From our previous discussions we recall that the solution of the differential equation is the real part of this current.

Using the complex forcing function, we find that the differential equation becomes

$$L \frac{d}{dt}\left(\mathbf{I}e^{j\omega t}\right) + R\mathbf{I}e^{j\omega t} = \mathbf{V}e^{j\omega t}$$

$$j\omega L\mathbf{I}e^{j\omega t} + R\mathbf{I}e^{j\omega t} = \mathbf{V}e^{j\omega t}$$

Note that $e^{j\omega t}$ is a common factor and, as we have already indicated, can be eliminated, leaving the phasors; that is,

The differential equation is reduced to a phasor equation.

$$j\omega L\mathbf{I} + R\mathbf{I} = \mathbf{V}$$

Therefore,

$$\mathbf{I} = \frac{\mathbf{V}}{R + j\omega L} = I_M \underline{/\phi} = \frac{V_M}{\sqrt{R^2 + \omega^2 L^2}} \underline{\bigg/ -\tan^{-1}\frac{\omega L}{R}}$$

Thus,

$$i(t) = \frac{V_M}{\sqrt{R^2 + \omega^2 L^2}} \cos\left(\omega t - \tan^{-1}\frac{\omega L}{R}\right)$$

which once again is the function we obtained earlier.   ❏

Phasor analysis

1 Using phasors, transform a set of differential equations in the time domain into a set of algebraic equations in the frequency domain.

2 Solve the algebric equations for the unknown phasors.

3 Transform the now-known phasors back to the time domain.

We define relations between phasors after the $e^{j\omega t}$ term has been eliminated as "phasor, or frequency domain, analysis." Thus we have transformed a set of differential equations with sinusoidal forcing functions in the time domain into a set of algebraic equations containing complex numbers in the frequency domain. In effect, we are now faced with solving a set of algebraic equations for the unknown phasors. The phasors are then simply transformed back to the time domain to yield the solution of the original set of differential equations. In addition, we note that the solution of sinusoidal steady-state circuits would be relatively simple if we could write the phasor equation directly from the circuit description. In Section 8.4 we will lay the groundwork for doing just that.

Note that in our discussions we have tacitly assumed that sinusoidal functions would be represented as phasors with a phase angle based on a cosine function. Therefore, if sine functions are used, we will simply employ the relationship in Eq. (8.7) to obtain the proper phase angle.

In summary, while $v(t)$ represents a voltage in the time domain, the phasor $\mathbf{V}$ represents the voltage in the frequency domain. The phasor contains only magnitude and phase information, and the frequency is implicit in this representation. The transformation from the time domain to the frequency domain, as well as the reverse transformation, is shown in Table 8.1. Recall that the phase angle is based on a cosine function and, therefore, if a sine function is involved, a 90° shift factor must be employed, as shown in the table.

Table 8.1 Phasor Representation

Time Domain	Frequency Domain
$A \cos(\omega t \pm \theta)$	$A \underline{/\pm\theta}$
$A \sin(\omega t \pm \theta)$	$A \underline{/\pm\theta - 90°}$

The following examples illustrate the use of the phasor transformation.

It is important to note, however, that if a network contains only sine sources, there is no need to perform the 90° shift. We simply perform the normal phasor analysis and then the *imaginary* part of the time-varying complex solution is the desired response. Simply put, cosine sources generate a cosine response and sine sources generate a sine response.

EXAMPLE 8.6

Convert the time functions $v(t) = 24 \cos(337t - 45°)$ V and $i(t) = 12 \sin(337t + 120°)$ A to phasors.

SOLUTION Using the phasor transformation shown previously, we have

$$\mathbf{V} = 24\ \underline{/-45°}\ \text{V}$$
$$\mathbf{I} = 12\ \underline{/120° - 90°} = 12\ \underline{/30°}\ \text{A} \qquad \square$$

EXAMPLE 8.7

Convert the phasors $\mathbf{V} = 16\ \underline{/20°}$ V and $\mathbf{I} = 10\ \underline{/-75°}$ A from the frequency domain to the time domain if the frequency is 1 kHz.

SOLUTION Employing the reverse transformation for phasors, we find that

$$v(t) = 16 \cos(2000\pi t + 20°)\text{V}$$
$$i(t) = 10 \cos(2000\pi t - 75°)\text{A} \qquad \square$$

EXTENSION EXERCISES

E8.3 Convert the following voltage functions to phasors.

$$v_1(t) = 12 \cos(377t - 425°)\text{V}$$
$$v_2(t) = 18 \sin(2513t + 4.2°)\text{V}$$

ANSWER: $\mathbf{V}_1 = 12\ \underline{/-425°}$ V, $\mathbf{V}_2 = 18\ \underline{/-85.8°}$ V.

E8.4 Convert the following phasors to the time domain if the frequency is 400 Hz.

$$\mathbf{V}_1 = 10\ \underline{/20°}\text{V}$$
$$\mathbf{V}_2 = 12\ \underline{/-60°}\text{V}$$

ANSWER: $v_1(t) = 10 \cos(800\pi t + 20°)$ V, $v_2(t)) = 12 \cos(800\pi t - 60°)$ V.

8.4 PHASOR RELATIONSHIPS FOR CIRCUIT ELEMENTS

As we proceed in our development of the techniques required to analyze circuits in the sinusoidal steady state, we are now in a position to establish the phasor relationships between voltage and current for the three passive elements R, L, and C.

In the case of a resistor as shown in Fig. 8.6a, the voltage–current relationship is known to be

$$v(t) = Ri(t) \qquad \qquad \textbf{8.20}$$

Current and voltage are in phase.

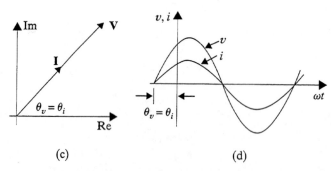

Figure 8.6 Voltage–current relationships for a resistor.

Applying the complex voltage $V_M e^{j(\omega t + \theta_v)}$ results in the complex current $I_M e^{j(\omega t + \theta_i)}$, and therefore Eq. (8.20) becomes

$$V_M e^{j(\omega t + \theta_v)} = R I_M e^{j(\omega t + \theta_i)}$$

which reduces to

$$V_M e^{j\theta_v} = R I_M e^{j\theta_i} \qquad \textbf{8.21}$$

Equation (8.21) can be written in phasor form as

$$\mathbf{V} = R\mathbf{I} \qquad \textbf{8.22}$$

where

$$\mathbf{V} = V_M e^{j\theta_v} = V_M \ \underline{/\theta_v} \quad \text{and} \quad \mathbf{I} = I_M e^{j\theta_i} = I_M \ \underline{/\theta_i}$$

From Eq. (8.21) we see that $\theta_v = \theta_i$ and thus the current and voltage for this circuit are *in phase*.

Historically, complex numbers have been represented as points on a graph in which the x-axis represents the real axis and the y-axis the imaginary axis. The line segment connecting the origin with the point provides a convenient representation of the magnitude and angle when the complex number is written in a polar form. A review of Appendix C will indicate how these complex numbers or line segments can be added, subtracted, and so on. Since phasors are complex numbers, it is convenient to represent the phasor voltage and current graphically as line segments. A plot of the line segments representing the phasors is called a *phasor diagram*. This pictorial representation of phasors provides immediate information on the relative magnitude of one phasor with another, the angle between two phasors, and the relative position of one phasor with respect to another (i.e., leading or lagging). A phasor diagram and the sinusoidal waveforms for the resistor are shown in Figs. 8.6c and d, respectively. A phasor diagram will be drawn for each of the other circuit elements in the remainder of this section.

EXAMPLE 8.8

If the voltage $v(t) = 24 \cos(377t + 75°)$ V is applied to a 6-Ω resistor as shown in Fig. 8.6a, we wish to determine the resultant current.

SOLUTION   Since the phasor voltage is

$$\mathbf{V} = 24 \,\underline{/75°}\, \text{V}$$

the phasor current from (Eq. (8.22) is

$$\mathbf{I} = \frac{24 \,\underline{/75°}}{6} = 4 \,\underline{/75°}\, \text{A}$$

which in the time domain is

$$i(t) = 4 \cos(377t + 75°) \text{ A} \qquad \square$$

EXTENSION EXERCISE

E8.5 The current in a 4-Ω resistor is known to be $\mathbf{I} = 12 \,\underline{/60°}$ A. Express the voltage across the resistor as a time function if the frequency of the current is 4 kHz.

ANSWER: $v(t) = 48 \cos(8000\pi t + 60°)$ V.

The voltage–current relationship for an inductor, as shown in Fig. 8.7a, is

$$v(t) = L\frac{di(t)}{dt} \qquad \textbf{8.23}$$

Substituting the complex voltage and current into this equation yields

$$V_M e^{j(\omega t + \theta_v)} = L\frac{d}{dt} I_M e^{j(\omega t + \theta_i)}$$

which reduces to

$$V_M e^{j\theta_v} = j\omega L I_M e^{j\theta_i} \qquad \textbf{8.24}$$

Equation (8.24) in phasor notation is

$$\mathbf{V} = j\omega L \mathbf{I} \qquad \textbf{8.25}$$

Note that the differential equation in the time domain (8.23) has been converted to an algebraic equation with complex coefficients in the frequency domain. This relationship is

The derivative process yields a frequency-dependent impedance.

Figure 8.7 Voltage–current relationships for an inductor.

The voltage leads the current or the current lags the voltage.

shown in Fig. 8.7b. Since the imaginary operator $j = 1e^{j90°} = 1 \underline{/90°} = \sqrt{-1}$, Eq. (8.24) can be written as

$$V_M e^{j\theta_v} = \omega L I_M e^{j(\theta_i + 90°)} \qquad \textbf{8.26}$$

Therefore, the voltage and current are *90° out of phase* and in particular the voltage leads the current by 90° or the current lags the voltage by 90°. The phasor diagram and the sinusoidal waveforms for the inductor circuit are shown in Figs. 8.7c and d, respectively.

EXAMPLE 8.9

The voltage $v(t) = 12 \cos(377t + 20°)$ V is applied to a 20-mH inductor as shown in Fig. 8.7a. Find the resultant current.

SOLUTION The phasor current is

Applying $\mathbf{V} = j\omega L\mathbf{I}$

$$\frac{x_1 \underline{/\theta_1}}{x_2 \underline{/\theta_2}} = \frac{x_1}{x_2} \underline{/\theta_1 - \theta_2}$$

$$\mathbf{I} = \frac{\mathbf{V}}{j\omega L} = \frac{12 \underline{/20°}}{\omega L \underline{/90°}}$$

$$= \frac{12 \underline{/20°}}{(377)(20 \times 10^{-3}) \underline{/90°}}$$

$$= 1.59 \underline{/-70°} \text{ A}$$

or

$$i(t) = 1.59 \cos(377t - 70°) \text{ A} \qquad \square$$

EXTENSION EXERCISE

E8.6 The current in a 0.05-H inductor is $\mathbf{I} = 4 \underline{/-30°}$ A. If the frequency of the current is 60 Hz, determine the voltage across the inductor.

ANSWER: $v_L(t) = 75.4 \cos(377t + 60°)$ V.

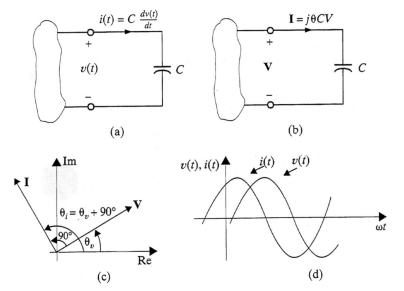

Figure 8.8 Voltage–current relationships for a capacitor.

The voltage–current relationship for our last passive element, the capacitor, as shown in Fig. 8.8a, is

$$i(t) = C \frac{dv(t)}{dt} \qquad\qquad \textbf{8.27}$$

Once again employing the complex voltage and current, we obtain

$$I_M e^{j(\omega t + \theta_i)} = C \frac{d}{dt} V_M e^{j(\omega t + \theta_v)}$$

which reduces to

$$I_M e^{j\theta_i} = j\omega C V_M e^{j\theta_v} \qquad\qquad \textbf{8.28}$$

In phasor notation this equation becomes

> The derivative process yields a frequency-dependent impedance.

$$\mathbf{I} = j\omega C \mathbf{V} \qquad\qquad \textbf{8.29}$$

Equation (8.27), a differential equation in the time domain, has been transformed into Eq. (8.29), an algebraic equation with complex coefficients in the frequency domain. The phasor relationship is shown in Fig. 8.8b. Substituting $j = 1e^{j90°}$ into Eq. (8.28) yields

$$I_M e^{j\theta_i} = \omega C V_M e^{j(\theta_v + 90°)} \qquad\qquad \textbf{8.30}$$

Note that the voltage and current are *90° out of phase*. Equation (8.30) states that the current leads the voltage by 90° or the voltage lags the current by 90°. The phasor diagram and the sinusoidal waveforms for the capacitor circuit are shown in Figs. 8.8c and d, respectively.

EXAMPLE 8.10

The voltage $v(t) = 100 \cos(314t + 15°)$ V is applied to a 100-μF capacitor as shown in Fig. 8.8a. Find the current.

Applying $\mathbf{I} = j\omega C\mathbf{V}$

SOLUTION   The resultant phasor current is

$$\begin{aligned}
\mathbf{I} &= j\omega C\left(100\ \underline{/15°}\right) \\
&= (314)\left(100 \times 10^{-6}\ \underline{/90°}\right)\left(100\ \underline{/15°}\right) \\
&= 3.14\ \underline{/105°}\ \text{A}
\end{aligned}$$

Therefore, the current written as a time function is

$$i(t) = 3.14 \cos(314t + 105°)\ \text{A} \qquad \square$$

EXTENSION EXERCISE

E8.7   The current in a 150-μF is $\mathbf{I} = 3.6\ \underline{/-145°}$ A. If the frequency of the current is 60 Hz, determine the voltage across the capacitor.

ANSWER: $\mathbf{V}_C = 63.66\ \underline{/-235°}$ V.

8.5 IMPEDANCE AND ADMITTANCE

We have examined each of the circuit elements in the frequency domain on an individual basis. We now wish to treat these passive circuit elements in a more general fashion. We now define the two-terminal input *impedance* $\mathbf{Z}$, also referred to as the driving point impedance, in exactly the same manner in which we defined resistance earlier. Later we will examine another type of impedance, called transfer impedance.

Impedance is defined as the ratio of the phasor voltage $\mathbf{V}$ to the phasor current $\mathbf{I}$:

$$\mathbf{Z} = \frac{\mathbf{V}}{\mathbf{I}} \qquad \textbf{8.31}$$

at the two terminals of the element related to one another by the passive sign convention, as illustrated in Fig 8.9. Since $\mathbf{V}$ and $\mathbf{I}$ are complex, the impedance $\mathbf{Z}$ is complex and

$$\mathbf{Z} = \frac{V_M\ \underline{/\theta_v}}{I_M\ \underline{/\theta_i}} = \frac{V_M}{I_M}\ \underline{/\theta_v - \theta_i} = Z\ \underline{/\theta_z} \qquad \textbf{8.32}$$

Since $\mathbf{Z}$ is the ratio of $\mathbf{V}$ to $\mathbf{I}$, the units of $\mathbf{Z}$ are ohms. Thus, impedance in an ac circuit is analogous to resistance in a dc circuit. In rectangular form, impedance is expressed as

$$\mathbf{Z}(\omega) = R(\omega) + jX(\omega) \qquad \textbf{8.33}$$

Figure 8.9   General impedance relationship.

where $R(\omega)$ is the real, or resistive, component and $X(\omega)$ is the imaginary, or reactive, component. In general, we simply refer to R as the resistance and X as the reactance. It is important to note that R and X are real functions of ω and therefore $\mathbf{Z}(\omega)$ is frequency dependent. Equation (8.33) clearly indicates that $\mathbf{Z}$ is a complex number; however, it is not a phasor, since phasors denote sinusoidal functions.

Equations (8.32) and (8.33) indicate that

$$Z \underline{/\theta_z} = R + jX \qquad\qquad \textbf{8.34}$$

Therefore,

$$Z = \sqrt{R^2 + X^2} \qquad\qquad \textbf{8.35}$$

$$\theta_z = \tan^{-1}\frac{X}{R}$$

where

$$R = Z\cos\theta_z$$

$$X = Z\sin\theta_z$$

For the individual passive elements the impedance is as shown in Table 8.2. However, just as it was advantageous to know how to determine the equivalent resistance in dc circuits, we want to learn how to determine the equivalent impedance in ac circuits..

Table 8.2 Passive element impedance

Passive element	Impedance
R	$\mathbf{Z} = R$
L	$\mathbf{Z} = j\omega L = jX_L = \omega L \underline{/90°}, X_L = \omega L$
C	$\mathbf{Z} = \dfrac{1}{j\omega C} = jX_C = -\dfrac{1}{\omega C}\underline{/90°}, X_C = -\dfrac{1}{\omega C}$

KCL and KVL are both valid in the frequency domain and we can use this fact, as was done in Chapter 2 for resistors, to show that impedances can be combined using the same rules that we established for resistor combinations. That is, if $\mathbf{Z}_1, \mathbf{Z}_2, \mathbf{Z}_3, \ldots, \mathbf{Z}_n$ are connected in series, the equivalent impedance $\mathbf{Z}_s$ is

$$\mathbf{Z}_s = \mathbf{Z}_1 + \mathbf{Z}_2 + \mathbf{Z}_3 + \cdots + \mathbf{Z}_n \qquad\qquad \textbf{8.36}$$

and if $\mathbf{Z}_1, \mathbf{Z}_2, \mathbf{Z}_3, \ldots, \mathbf{Z}_n$ are connected in parallel, the equivalent impedance is given by

$$\frac{1}{\mathbf{Z}_p} = \frac{1}{\mathbf{Z}_1} + \frac{1}{\mathbf{Z}_2} + \frac{1}{\mathbf{Z}_3} + \cdots + \frac{1}{\mathbf{Z}_n} \qquad\qquad \textbf{8.37}$$

EXAMPLE 8.11

Technique

1 Express $v(t)$ as a phasor and determine the impedance of each passive element.

2 Combine impedances and solve for the phasor $\mathbf{I}$.

3 Convert the phasor $\mathbf{I}$ to $i(t)$.

Determine the equivalent impedance of the network shown in Fig. 8.10 if the frequency is $f = 60$ Hz. Then compute the current $i(t)$ if the voltage source is $v(t) = 50 \cos(\omega t + 30°)$ V. Finally, calculate the equivalent impedance if the frequency is $f = 400$ Hz.

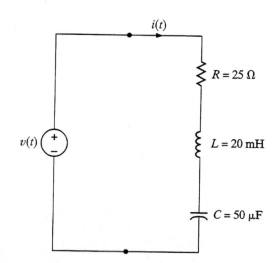

Figure 8.10 Series ac circuit.

SOLUTION The impedances of the individual elements at 60 Hz are

$$\mathbf{Z}_R = 25\Omega$$

$$\mathbf{Z}_L = j\omega L = j(2\pi \times 60)(20 \times 10^{-3}) = j7.54 \ \Omega$$

$$\mathbf{Z}_C = \frac{-j}{\omega C} = \frac{-j}{(2\pi \times 60)(50 \times 10^{-6})} = -j53.05 \ \Omega$$

Since the elements are in series,

$$\mathbf{Z} = \mathbf{Z}_R + \mathbf{Z}_L + \mathbf{Z}_C$$

$$= 25 - j45.51 \ \Omega$$

The current in the circuit is given by

$$\mathbf{I} = \frac{\mathbf{V}}{\mathbf{Z}} = \frac{50 \ \underline{/30°}}{25 - j45.51} = \frac{50 \ \underline{/30°}}{51.93 \ \underline{/-61.22°}} = 0.96 \ \underline{/91.22°} \ \text{A}$$

or in the time domain, $i(t) = 0.96 \cos(377t + 91.22°)$ A.

If the frequency is 400 Hz, the impedance of each element is

$$\mathbf{Z}_R = 25 \ \Omega$$

$$\mathbf{Z}_L = j\omega L = j50.27 \ \Omega$$

$$\mathbf{Z}_C = \frac{-j}{\omega C} = -j7.96 \ \Omega$$

The total impedance is then

$$\mathbf{Z} = 25 + j42.31 = 49.14 \ \underline{/59.42°} \ \Omega$$

It is important to note that at the frequency $f = 60$ Hz, the reactance of the circuit is capacitive; that is, if the impedance is written as $R + jX$, $X < 0$; however, at $f = 400$ Hz the reactance is inductive since $X > 0$.   ❏

EXTENSION EXERCISE

E8.8   Find the current $i(t)$ in the network in Fig. E8.8.

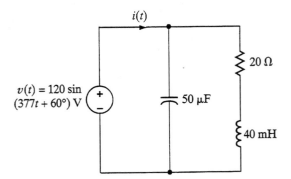

Figure E8.8

ANSWER: $i(t) = 3.88 \cos(377t - 39.2°)$ A.

Another quantity that is very useful in the analysis of ac circuits is the two-terminal input *admittance*, which is the reciprocal of impedance; that is,

$$\mathbf{Y} = \frac{1}{\mathbf{Z}} = \frac{\mathbf{I}}{\mathbf{V}} \qquad \text{8.38}$$

The units of $\mathbf{Y}$ are siemens, and this quantity is analogous to conductance in resistive dc circuits. Since $\mathbf{Z}$ is a complex number, $\mathbf{Y}$ is also a complex number.

$$\mathbf{Y} = Y_M \underline{/\theta_y} \qquad \text{8.39}$$

which is written in rectangular form as

$$\mathbf{Y} = G + jB \qquad \text{8.40}$$

where G and B are called *conductance* and *susceptance*, respectively. Because of the relationship between $\mathbf{Y}$ and $\mathbf{Z}$, we can express the components of one quantity as a function of the components of the other. From the expression

$$G + jB = \frac{1}{R + jX} \qquad \text{8.41}$$

we can show that

$$G = \frac{R}{R^2 + X^2}, \qquad B = \frac{-X}{R^2 + X^2} \qquad \text{8.42}$$

Technique for taking the reciprocal:

$$\frac{1}{R + jX} = \frac{R - jX}{(R + jX)(R - jX)}$$

$$= \frac{R - jX}{R^2 + X^2}$$

D8.2 Find $\mathbf{Y}_p$

ANSWER: $\mathbf{Y}_p = 0.1 + j1$ S.

D8.3 Find $\mathbf{Y}_s$

ANSWER:

$\mathbf{Y}_s = 0.05 - j0.05$ S.

and in a similar manner, we can show that

$$R = \frac{G}{G^2 + B^2}, \qquad X = \frac{-B}{G^2 + B^2} \qquad \text{8.43}$$

It is very important to note that in general R and G are *not* reciprocals of one another. The same is true for X and B. The purely resistive case is an exception. In the purely reactive case the quantities are negative reciprocals of one another.

The admittance of the individual passive elements is

$$\mathbf{Y}_R = \frac{1}{R} = G$$

$$\mathbf{Y}_L = \frac{1}{j\omega L} = -\frac{1}{\omega L} \underline{/90^\circ} \qquad \text{8.44}$$

$$\mathbf{Y}_C = j\omega C = \omega C \underline{/90^\circ}$$

Once again, since KCL and KVL are valid in the frequency domain, we can show, using the same approach outlined in Chapter 2 for conductance in resistive circuits, that the rules for combining admittances are the same as those for combining conductances; that is, if $\mathbf{Y}_1, \mathbf{Y}_2, \mathbf{Y}_3, \ldots, \mathbf{Y}_n$ are connected in parallel, the equivalent admittance is

$$\mathbf{Y}_p = \mathbf{Y}_1 + \mathbf{Y}_2 + \cdots + \mathbf{Y}_n \qquad \text{8.45}$$

and if $\mathbf{Y}_1, \mathbf{Y}_2, \ldots, \mathbf{Y}_n$ are connected in series, the equivalent admittance is

$$\frac{1}{\mathbf{Y}_S} = \frac{1}{\mathbf{Y}_1} + \frac{1}{\mathbf{Y}_2} + \cdots + \frac{1}{\mathbf{Y}_n} \qquad \text{8.46}$$

EXAMPLE 8.12

Calculate the equivalent admittance $\mathbf{Y}_p$ for the network in Fig. 8.11 and use it to determine the current $\mathbf{I}$ if $\mathbf{V}_S = 60 \underline{/45^\circ}$ V.

Figure 8.11 An Example parallel circuit.

SOLUTION From Fig. 8.11 we note that

$$\mathbf{Y}_R = \frac{1}{\mathbf{Z}_R} = \frac{1}{2}\,\text{S}$$

$$\mathbf{Y}_L = \frac{1}{\mathbf{Z}_L} = \frac{-j}{4}\,\text{S}$$

Therefore,

$$\mathbf{Y}_p = \frac{1}{2} - j\frac{1}{4}\,\text{S}$$

Admittances add in parallel.

and hence,

$$\mathbf{I} = \mathbf{Y}_p\mathbf{V}_S$$
$$= \left(\frac{1}{2} - j\frac{1}{4}\right)(60\,\underline{/45°})$$
$$= 33.5\,\underline{/18.43°}\,\text{A} \qquad \square$$

EXTENSION EXERCISE

E8.9 Find the current **I** in the network in Fig. E8.9.

$$V = 10\,\underline{/20°}\,\text{V} \qquad \mathbf{Z}_R = 2\,\Omega \qquad \mathbf{Z}_L = j2\,\Omega \qquad \mathbf{Z}_C = -j1\,\Omega \qquad \mathbf{Z}_R = 4\,\Omega$$

Figure E8.9

ANSWER: $\mathbf{I} = 9\,\underline{/53.7°}\,\text{A}.$

As a prelude to our analysis of more general ac circuits, let us examine the techniques for computing the impedance or admittance of circuits in which numerous passive elements are interconnected. The following example illustrates that our technique is analogous to our earlier computations of equivalent resistance.

EXAMPLE 8.13

Consider the network shown in Fig. 8.12a. The impedance of each element is given in the figure. We wish to calculate the equivalent impedance of the network $\mathbf{Z}_{eq}$ at terminals A-B.

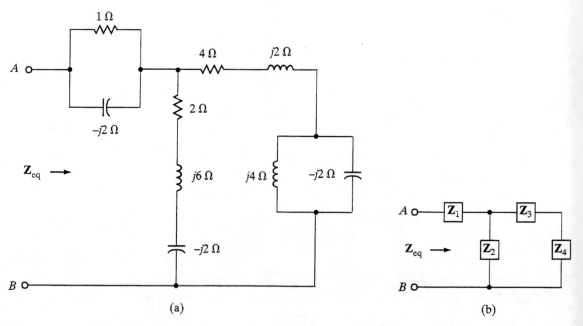

Figure 8.12 Example circuit for determining equivalent impedance in two steps.

Technique:

1 Add the admittances of elements in parallel.

2 Add the impedances of elements in series.

3 Convert back and forth between admittance and impedance in order to combine neighboring elements.

SOLUTION   The equivalent impedance $\mathbf{Z}_{eq}$ could be calculated in a variety of ways; we could use only impedances, or only admittances, or a combination of the two. We will use the latter. We begin by noting that the circuit in Fig. 8.12a can be represented by the circuit in Fig. 8.12b.

Note that

$$\mathbf{Y}_4 = \mathbf{Y}_L + \mathbf{Y}_C$$
$$= \frac{1}{j4} + \frac{1}{-j2}$$
$$= j\frac{1}{4}\ \text{S}$$

Therefore,

$$\mathbf{Z}_4 = -j4\ \Omega$$

Now

$$\mathbf{Z}_{34} = \mathbf{Z}_3 + \mathbf{Z}_4$$
$$= (4 + j2) + (-j4)$$
$$= 4 - j2\ \Omega$$

and hence,

$$\mathbf{Y}_{34} = \frac{1}{\mathbf{Z}_{34}}$$

$$= \frac{1}{4 - j2}$$

$$= 0.20 + j0.10 \text{ S}$$

Since

$$\mathbf{Z}_2 = 2 + j6 - j2$$

$$= 2 + j4 \ \Omega$$

then

$$\mathbf{Y}_2 = \frac{1}{2 + j4}$$

$$= 0.10 - j0.20 \text{ S}$$

$$\mathbf{Y}_{234} = \mathbf{Y}_2 + \mathbf{Y}_{34}$$

$$= 0.30 - j0.10 \text{ S}$$

The reader should note carefully our approach—we are adding impedances in series and adding admittances in parallel.

From $\mathbf{Y}_{234}$ we can compute $\mathbf{Z}_{234}$ as

$$\mathbf{Z}_{234} = \frac{1}{\mathbf{Y}_{234}}$$

$$= \frac{1}{0.30 - j0.10}$$

$$= 3 + j1 \ \Omega$$

Now

$$\mathbf{Y}_1 = \mathbf{Y}_R + \mathbf{Y}_C$$

$$= \frac{1}{1} + \frac{1}{-j2}$$

$$= 1 + j\frac{1}{2} \text{ S}$$

and then

$$\mathbf{Z}_1 = \frac{1}{1 + j\frac{1}{2}}$$

$$= 8.0 - j0.4 \ \Omega$$

Therefore,

$$\mathbf{Z}_{eq} = \mathbf{Z}_1 + \mathbf{Z}_{234}$$
$$= 0.8 - j0.4 + 3 + j1$$
$$= 3.8 + j0.6 \ \Omega$$

EXTENSION EXERCISE

E8.10 Compute the impedance $\mathbf{Z}_T$ in the network in Fig. E8.10.

Figure E8.10

ANSWER: $\mathbf{Z}_T = 3.38 + j1.08 \ \Omega$.

8.6 PHASOR DIAGRAMS

Impedance and admittance are functions of frequency, and therefore their values change as the frequency changes. These changes in $\mathbf{Z}$ and $\mathbf{Y}$ have a resultant effect on the current–voltage relationships in a network. This impact of changes in frequency on circuit parameters can be easily seen via a phasor diagram. The following examples will serve to illustrate these points.

EXAMPLE 8.14

Let us sketch the phasor diagram for the network shown in Fig. 8.13.

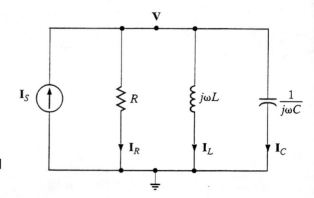

Figure 8.13 Example parallel circuit.

SOLUTION The pertinent variables are labeled on the figure. For convenience in forming a phasor diagram, we select **V** as a reference phasor and arbitrarily assign it a 0° phase angle. We will, therefore, measure all currents with respect to this phasor. We suffer no loss of generality by assigning **V** a 0° phase angle, since if it is actually, for example, 30°, we will simply rotate the entire phasor diagram by 30° because all the currents are measured with respect to this phasor.

At the upper node in the circuit KCL is

$$\mathbf{I}_S = \mathbf{I}_R + \mathbf{I}_L + \mathbf{I}_C = \frac{\mathbf{V}}{R} + \frac{\mathbf{V}}{j\omega L} + \frac{\mathbf{V}}{1/j\omega C}$$

Since $\mathbf{V} = V_M \underline{/0°}$, then

$$\mathbf{I}_S = \frac{V_M \underline{/0°}}{R} + \frac{V_M \underline{/-90°}}{\omega L} + V_M \omega C \underline{/90°}$$

The phasor diagram that illustrates the phase relationship between $\mathbf{V}, \mathbf{I}_R, \mathbf{I}_L$, and $\mathbf{I}_C$ is shown in Fig. 8.14a. For small values of ω such that the magnitude of $\mathbf{I}_L$ is greater than that of $\mathbf{I}_C$, the phasor diagram for the currents is shown in Fig. 8.14b. In the case of large values of ω, that is, those for which $\mathbf{I}_C$ is greater than $\mathbf{I}_L$, the phasor diagram for the currents is shown in Fig. 8.14c. Note that as ω increases, the phasor $\mathbf{I}_S$ moves from $\mathbf{I}_{S_1}$ to $\mathbf{I}_{S_n}$ along a locus of points specified by the dashed line shown in Fig. 8.14d.

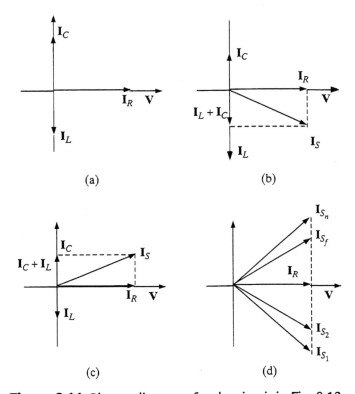

From a graphical standpoint, phasors can be manipulated like vectors.

Figure 8.14 Phasor diagrams for the circuit in Fig. 8.13.

DRILL

D8.4 Find the frequency at which $v(t)$ and $i(t)$ are in phase.

ANSWER:
$f = 5.03 \times 10^3$ Hz.

Note that $\mathbf{I}_S$ is in phase with $\mathbf{V}$ when $\mathbf{I}_C = \mathbf{I}_L$ or, in other words, when $\omega L = 1/\omega C$. Hence, the node voltage $\mathbf{V}$ is in phase with the current source $\mathbf{I}_S$ when

$$\omega = \frac{1}{\sqrt{LC}}$$

This can also be seen from the KCL equation

$$\mathbf{I} = \left[\frac{1}{R} + j\left(\omega C - \frac{1}{\omega L}\right)\right]\mathbf{V}$$

EXAMPLE 8.15

Let us determine the phasor diagram for the series circuit shown in Fig. 8.15a.

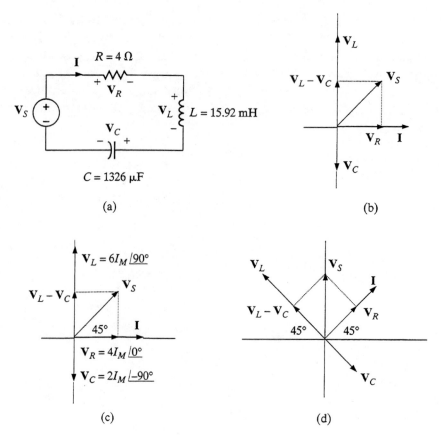

Figure 8.15 Series circuit and certain specific phasor diagrams.

SOLUTION KVL for this circuit is of the form

$$\mathbf{V}_S = \mathbf{V}_R + \mathbf{V}_L + \mathbf{V}_C$$

$$= \mathbf{I}R + \omega L\mathbf{I}\,\underline{/90°} + \frac{\mathbf{I}}{\omega C}\,\underline{/-90°}$$

If we select $\mathbf{I}$ as a reference phasor so that $\mathbf{I} = I_M \underline{/0°}$, then if $\omega L I_M > I_M/\omega C$, the phasor diagram will be of the form shown in Fig. 8.15b. Specifically, if $\omega = 377$ rad/s (i.e., $f = 60$ Hz), then $\omega L = 6$ and $1/\omega C = 2$. Under these conditions the phasor diagram is as shown in Fig. 8.15c. If, however, we select $\mathbf{V}_S$ as reference with, for example,

$$v_S(t) = 12\sqrt{2}\cos(377t + 90°) \text{ V}$$

then

$$\mathbf{I} = \frac{\mathbf{V}}{\mathbf{Z}} = \frac{12\sqrt{2}\underline{/90°}}{4 + j6 - j2}$$

$$= \frac{12\sqrt{2}\underline{/90°}}{4\sqrt{2}\underline{/45°}}$$

$$= 3\underline{/45°} \text{ A}$$

and the entire phasor diagram, as shown in Figs. 8.15b and c, is rotated 45°, as shown in Fig. 8.15d. ❑

EXTENSION EXERCISE

E8.11 Draw a phasor diagram illustrating all currents and voltages for the network in Fig. E8.11.

Figure E8.11

ANSWER:

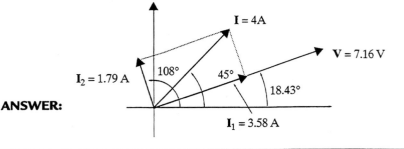

8.7 BASIC ANALYSIS USING KIRCHOFF'S LAWS

We have shown that Kirchhoff's laws apply in the frequency domain, and therefore they can be used to compute steady-state voltages and currents in ac circuits. This approach involves expressing these voltages and currents as phasors, and once this is done, the ac steady-state analysis employing phasor equations is performed in an identical fashion to that used in the dc analysis of resistive circuits. Complex number algebra is the tool that is used for the mathematical manipulation of the phasor equations, which, of course, have complex coefficients. We will begin by illustrating that the techniques we have applied in the solution of dc resistive circuits are valid in ac circuit analysis also—the only difference being that in steady-state ac circuit analysis the algebraic phasor equations have complex coefficients.

Problem-solving Strategies for AC Steady-state Analysis

- For relatively simple circuits (e.g., those with a single source), use
  - Ohm's law for ac analysis, i.e., $\mathbf{V} = \mathbf{IZ}$
  - The rules for combining $\mathbf{Z}_s$ and $\mathbf{Y}_s$
  - KCL and KVL
  - Current and voltage division
- For more complicated circuits with multiple sources, use
  - Nodal analysis
  - Loop of mesh analysis
  - Superposition
  - Source exchange
  - Thévenin's and Norton's theorems

Technique

1. Compute $\mathbf{I}_1$.

2. Determine $\mathbf{V}_1 =$
   $\mathbf{V}_S - \mathbf{I}_1\mathbf{Z}_1$.

Then $\mathbf{I}_2 = \dfrac{\mathbf{V}_1}{\mathbf{Z}_2}$ and

$$\mathbf{I}_3 = \frac{\mathbf{V}_1}{\mathbf{Z}_3}$$

Current and voltage division are also applicable.

At this point, it is important for the reader to understand that in our manipulation of algebraic phasor equations with complex coefficients we will, for the sake of simplicity, normally carry only two digits to the right of the decimal point. In doing so, we will introduce round-off errors in our calculations. Nowhere are these errors more evident than when two or more approaches are used to solve the same problem, as is done in the following example.

EXAMPLE 8.16

We wish to calculate all the voltages and currents in the circuit shown in Fig. 8.16a.

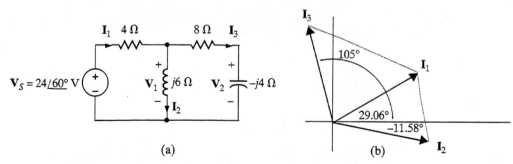

Figure 8.16 (a) Example ac circuit, (b) phasor diagram for the currents.

SOLUTION   Our approach will be as follows. We will calculate the total impedance seen by the source $\mathbf{V}_S$. Then we will use this to determine $\mathbf{I}_1$. Knowing $\mathbf{I}_1$, we can compute $\mathbf{V}_1$ using KVL. Knowing $\mathbf{V}_1$, we can compute $\mathbf{I}_2$ and $\mathbf{I}_3$, and so on.

The total impedance seen by the source $\mathbf{V}_S$ is

$$\begin{aligned}
\mathbf{Z}_{eq} &= 4 + \frac{(j6)(8-j4)}{j6+8-j4} \\
&= 4 + \frac{24+j48}{8+j2} \\
&= 4 + 4.24 + j4.94 \\
&= 9.61 \,\underline{/30.94°}\ \Omega
\end{aligned}$$

Then

$$\begin{aligned}
\mathbf{I}_1 = \frac{\mathbf{V}_S}{\mathbf{Z}_{eq}} &= \frac{24\,\underline{/60°}}{9.61\,\underline{/30.94°}} \\
&= 2.5\,\underline{/29.06°}\ \text{A}
\end{aligned}$$

$\mathbf{V}_1$ can be determined using KVL:

$$\begin{aligned}
\mathbf{V}_1 &= \mathbf{V}_S - 4\mathbf{I}_1 \\
&= 24\,\underline{/60°} - 10\,\underline{/29.06°} \\
&= 3.26 + j15.93 = 16.26\,\underline{/78.43°}\ \text{V}
\end{aligned}$$

Note that $\mathbf{V}_1$ could also be computed via voltage division:

$$\mathbf{V}_1 = \frac{\mathbf{V}_S \dfrac{(j6)(8-j4)}{j6+8-j4}}{4 + \dfrac{(j6)(8-j4)}{j6+8-4}}\ \text{V}$$

which from our previous calculation is

$$\begin{aligned}
\mathbf{V}_1 &= \frac{\left(24\,\underline{/60°}\right)\left(6.51\,\underline{/49.36°}\right)}{9.61\,\underline{/30.94°}} \\
&= 16.26\,\underline{/78.42°}\ \text{V}
\end{aligned}$$

Knowing $\mathbf{V}_1$, we can calculate both $\mathbf{I}_2$ and $\mathbf{I}_3$:

$$\begin{aligned}
\mathbf{I}_2 = \frac{\mathbf{V}_1}{j6} &= \frac{16.26\,\underline{/78.43°}}{6\,\underline{/90°}} \\
&= 2.71\,\underline{/-11.58°}\ \text{A}
\end{aligned}$$

and

$$\begin{aligned}
\mathbf{I}_3 &= \frac{\mathbf{V}_1}{8-j4} \\
&= 1.82\,\underline{/105°}\ \text{A}
\end{aligned}$$

Note that I_2 and I_3 could have been calculated by current division. For exampl
I_2 could be determined by

$$
\begin{aligned}
I_2 &= \frac{I_1(8 - j4)}{8 - j4 + j6} \\
&= \frac{(2.5\ \underline{/29.06°})(8.94\ \underline{/-26.57°})}{8 + j2} \\
&= 2.71\ \underline{/-11.55°}\ \text{A}
\end{aligned}
$$

Finally, V_2 can be computed as

$$
\begin{aligned}
V_2 &= I_3(-j4) \\
&= 7.28\ \underline{/15°}\ \text{V}
\end{aligned}
$$

This value could also have been computed by voltage division. The phasor diagram
for the currents I_1, I_2, and I_3 is shown in Fig. 8.16b and is an illustration of KCL.

Finally, the reader is encouraged to work the problem in reverse; that is, given
V_2, find V_S. Note that if V_2 is known, I_3 can be computed immediately using the capacitor impedance. Then $V_2 + I_3(8)$ yields V_1. Knowing V_1 we can find I_2. Then
$I_2 + I_3 = I_1$, and so on. Note that this analysis, which is the subject of Extension Exercise E8.12, involves simply a repeated application of Ohm's law, KCL, and KVL.☐

EXTENSION EXERCISE

E8.12 In the network in Fig. E8.12, V_o is known to be $8\ \underline{/45°}$ V. Compute V_S.

Figure E8.12

ANSWER: $V_S = 17.89\ \underline{/-18.43°}$ V.

8.8 ANALYSIS TECHNIQUES

In this section we revisit the circuit analysis methods that were successfully applied earlier to dc circuits and illustrate their applicability to ac steady-state analysis. The vehicle
we employ to present these techniques is examples in which all the theorems, together with
nodal analysis and loop analysis, are used to obtain a solution.

EXAMPLE 8.17

Let us determine the current $\mathbf{I}_o$ in the network in Fig. 8.17a using nodal analysis, loop analysis, superposition, source exchange, Thévenin's theorem, and Norton's theorem.

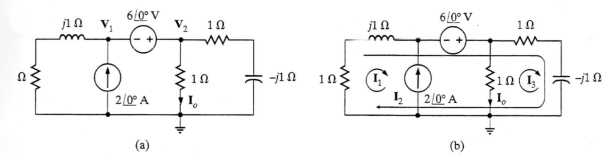

(a)   **(b)**

Figure 8.17 Circuits used in Example 8.17 for node and loop analysis.

SOLUTION

(1) Nodal Analysis   We begin with a nodal analysis of the network. The KCL equation for the supernode that includes the voltage source is

> Summing the current, leaving the supernode. Outbound currents have a positive sign.

$$\frac{\mathbf{V}_1}{1 + j} - 2 \underline{/0°} + \frac{\mathbf{V}_2}{1} + \frac{\mathbf{V}_2}{1 - j} = 0$$

and the associated KVL constraint equation is

$$\mathbf{V}_1 + 6 \underline{/0°} = \mathbf{V}_2$$

Solving for $\mathbf{V}_1$ in the second equation and using this value in the first equation yields

$$\frac{\mathbf{V}_2 - 6 \underline{/0°}}{1 + j} - 2 \underline{/0°} + \mathbf{V}_2 + \frac{\mathbf{V}_2}{1 - j} = 0$$

or

$$\mathbf{V}_2\left[\frac{1}{1 + j} + 1 + \frac{1}{1 - j}\right] = \frac{6 + 2 + 2j}{1 + j}$$

Solving for $\mathbf{V}_2$, we obtain

$$\mathbf{V}_2 = \left(\frac{4 + j}{1 + j}\right)\mathbf{V}$$

D8.5 Use nodal analysis to find I_o.

ANSWER: $I_o = 1.581 \underline{/-18.4°}$ A.

Therefore,

$$I_o = \frac{4 + j}{1 + j} = \left(\frac{5}{2} - \frac{3}{2}j\right) A$$

(2) Loop Analysis The network in Fig. 8.17b is used to perform a loop analysis. Note that one loop current is selected that passes through the independent current source. The three loop equations are

Just as in a dc analysis, the loop equations assume a decrease in potential level is $+$ and an increase is $-$.

$$I_1 = -2 \underline{/0°}$$
$$1(I_1 + I_2) + j1(I_1 + I_2) - 6 \underline{/0°} + 1(I_2 + I_3) - j1(I_2 + I_3) = 0$$
$$1I_3 + 1(I_2 + I_3) - j1(I_2 + I_3) = 0$$

Combining the first two equations yields

$$I_2(2) + I_3(1 - j) = 8 + 2j$$

The third loop equation can be simplified to the form

$$I_2(1 - j) + I_3(2 - j) = 0$$

Solving this last equation for I_2 and substituting the value into the previous equation yields

$$I_3\left[\frac{-4 + 2j}{1 - j} + 1 - j\right] = 8 + 2j$$

or

$$I_3 = \frac{-10 + 6j}{4}$$

D8.6 Find I_o using the loop currents shown in the circuit.

ANSWER: $I_o = I_3 = 1.581 \underline{/-18.40}$ A.

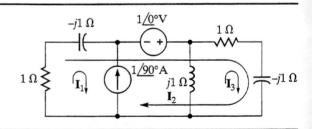

and finally

$$\mathbf{I}_o = -\mathbf{I}_3 = \left(\frac{5}{2} - \frac{3}{2} j \right) \mathrm{A}$$

(3) Superposition   In using superposition, we apply one independent source at a time. The network in which the current source acts alone is shown in Fig. 8.18a. By combining the two parallel impedances on each end of the network, we obtain the circuit in Fig. 8.18b, where

$$\mathbf{Z}' = \frac{(1 + j)(1 - j)}{(1 + j) + (1 - j)} = 1\,\Omega$$

Therefore, using current division

$$\mathbf{I}'_o = 1\ \underline{/0°}\ \mathrm{A}$$

The circuit in which the voltage source acts alone is shown in Fig. 8.18c. The voltage $\mathbf{V}''_1$ obtained using voltage division is

$$\mathbf{V}''_1 = \frac{(6\ \underline{/0°})\left[\dfrac{1(1 - j)}{1 + 1 - j} \right]}{1 + j + \left[\dfrac{1(1 - j)}{1 + 1 - j} \right]}$$

$$= \frac{6(1 - j)}{4}\ \mathrm{V}$$

and hence,

$$\mathbf{I}''_0 = \frac{6}{4}(1 - j)\mathrm{A}$$

> In applying superposition in this case, each source is applied independently and the results are added to obtain the solution.

DRILL

D8.7   Find $\mathbf{I}_o$ in Drill 8.6 using superposition.

Figure 8.18 Circuits used in Example 8.17 for a superposition analysis.

Then

$$\mathbf{I}_o = \mathbf{I}_o' + \mathbf{I}_o'' = 1 + \frac{6}{4}(1 - j) = \left(\frac{5}{2} - \frac{3}{2}j\right)A$$

In source exchange, a voltage source in series with an impedance can be exchanged for a current source in parallel with the impedance, and vice versa. Repeated application systematically reduces the number of circuit elements.

(4) Source exchange As a first step in the source exchange approach, we exchang the current source and parallel impedance for a voltage source in series with the im pedance, as shown in Fig. 8.19a.

(a)

(b)   (c)

Figure 8.19 Circuits used in Example 8.17 for a source exchange analysis.

Adding the two voltage sources and transforming them and the series impedance into a current source in parallel with that impedance is shown in Fig. 8.19b. Combining the two impedances that are in parallel with the 1-Ω resistor produces the network in Fig. 8.19c, where

DRILL

D8.8 Find $\mathbf{V}_o$ using source exchange.

ANSWER: $\mathbf{V}_o = 1.486 \,\underline{/110.7°}$ V.

$$\mathbf{Z} = \frac{(1 + j)(1 - j)}{1 + j + 1 - j} = 1\,\Omega$$

Therefore, using current division,

$$\mathbf{I}_o = \left(\frac{8 + 2j}{1 + j}\right)\left(\frac{1}{1 + j}\right) = \frac{4 + j}{1 + j}$$

$$= \left(\frac{5}{2} - \frac{3}{2}j\right)\mathbf{A}$$

(5) Thévenin Analysis   In applying Thévenin's theorem to the circuit in Fig. 8.20a, we first find the open-circuit voltage, $\mathbf{V}_{oc}$, as shown in Fig. 8.20a. In order to simplify the analysis, we perform a source exchange on the left end of the network, which results in the circuit in Fig. 8.20b. Now using voltage division

$$\mathbf{V}_{oc} = [6 + 2(1 + j)]\left[\frac{1 - j}{1 - j + 1 + j}\right]$$

or

$$\mathbf{V}_{oc} = (5 - 3j)\mathbf{V}$$

The Thévenin equivalent impedance, $\mathbf{Z}_{Th}$, obtained at the open-circuit terminals when the current source is replaced with an open circuit and the voltage source is replaced with a short circuit is shown in Fig. 8.20c and calculated to be

$$\mathbf{Z}_{Th} = \frac{(1 + j)(1 - j)}{1 + j + 1 - j} = 1\,\Omega$$

In this Thévenin analysis

1 Remove the 1-Ω load and find the voltage across the open terminals, $\mathbf{V}_{oc}$.

2 Determine the impedance $\mathbf{Z}_{Th}$ at the open terminals with all sources made zero.

3 Construct the following circuit and determine $\mathbf{I}_o$.

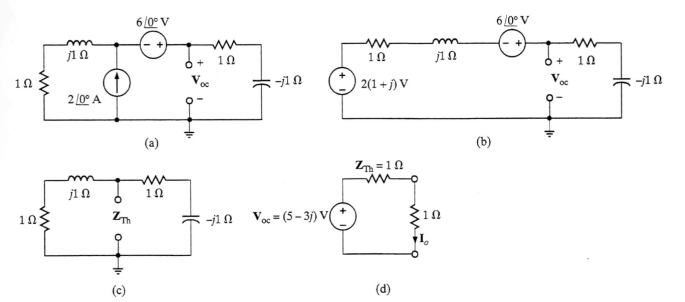

Figure 8.20 Circuits used in Example 8.17 for a Thévenin analysis.

D8.9 Use Thévenin's theorem to find $\mathbf{V}_o$ in the following circuit.

ANSWER:
1. Removing just the resistor produces an unsolvable (impossible) circuit.

2. Removing just the capaitor produces
$\mathbf{V}_{oc} = 1.5 + 0.5j$ V, $\mathbf{Z}_{Th} = 0.5 + 0.5j$ Ω.

3. Removing both resistor and capacitor produces $\mathbf{V}_{oc} = 1 + 2j$ V, $\mathbf{Z}_{Th} = j$ Ω. Either approach 2 or 3 produces $\mathbf{V}_o = 2 - j$ V.

In this Norton analysis

1 Remove the 1-Ω load and find the current $\mathbf{I}_{sc}$ through the short-circuited terminals.

2 Determine the impedance $\mathbf{Z}_{Th}$ at the open load terminals with all sources made zero.

3 Construct the following circuit and determine $\mathbf{I}_o$.

Connecting the Thévenin equivalent circuit to the 1-Ω resistor containing $\mathbf{I}_o$ in the original network yields the circuit in Fig. 8.20d. The current $\mathbf{I}_o$ is then

$$\mathbf{I}_o = \left(\frac{5}{2} - \frac{3}{2}j\right)\text{A}$$

(6) Norton Analysis Finally, in applying Norton's theorem to the circuit in Fig. 8.17a, we calculate the short-circuit current, $\mathbf{I}_{sc}$, using the network in Fig. 8.21a. Note that because of the short circuit, the voltage source is directly across the impedance in the left-most branch. Therefore,

$$\mathbf{I}_1 = \frac{6\,\underline{/0°}}{1 + j}$$

Then using KCL

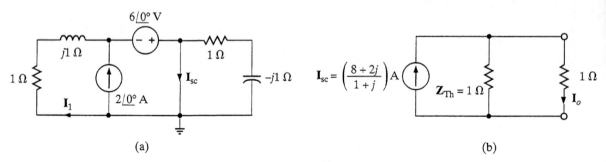

(a)   (b)

Figure 8.21 Circuits used in Example 8.17 for a Norton analysis.

$$\mathbf{I}_{sc} = \mathbf{I}_1 + 2 \underline{/0^\circ} = 2 + \frac{6}{1 + j}$$

$$= \left(\frac{8 + 2j}{1 + j}\right)\mathrm{A}$$

DRILL

D8.10   Use Norton's theorem to find $\mathbf{V}_o$ in Drill 8.9.

The Thévenin equivalent impedance, $\mathbf{Z}_{Th}$, is known to be 1 Ω and, therefore, connecting the Norton equivalent to the 1-Ω resistor containing $\mathbf{I}_o$ yields the network in Fig. 8.21b. Using current division, we find that

$$\mathbf{I}_o = \frac{1}{2}\left(\frac{8 + 2j}{1 + j}\right)$$

$$= \left(\frac{5}{2} - \frac{3}{2}j\right)\mathrm{A} \qquad \square$$

Let us now consider an example containing a dependent source.

EXAMPLE 8.18

Let us determine the voltage $\mathbf{V}_o$ in the circuit in Fig. 8.22a. In this example we will use node equations, loop equations, Thévenin's theorem, and Norton's theorem. We will omit the techniques of superposition and source transformation. Why?

How does the presence of a dependent source affect superposition and source exchange?

SOLUTION

(1) Nodal Analysis   In order to perform a nodal analysis, we label the node voltages and identify the supernode as shown in Fig. 8.22b. The constraint equation for the supernode is

$$\mathbf{V}_3 + 12 \underline{/0^\circ} = \mathbf{V}_1$$

and the KCL equations for the nodes of the network are

$$\frac{\mathbf{V}_1 - \mathbf{V}_2}{-j1} + \frac{\mathbf{V}_3 - \mathbf{V}_2}{1} - 4 \underline{/0^\circ} + \frac{\mathbf{V}_3 - \mathbf{V}_o}{1} + \frac{\mathbf{V}_3}{j1} = 0$$

$$\frac{\mathbf{V}_2 - \mathbf{V}_1}{-j1} + \frac{\mathbf{V}_2 - \mathbf{V}_3}{1} - 2\left(\frac{\mathbf{V}_3 - \mathbf{V}_o}{1}\right) = 0$$

$$4 \underline{/0^\circ} + \frac{\mathbf{V}_o - \mathbf{V}_3}{1} + \frac{\mathbf{V}_o}{1} = 0$$

At this point we can solve the foregoing equations using a matrix analysis or, for example, substitute the first and last equations into the remaining two equations, which yields

$$3\mathbf{V}_o - (1 + j)\mathbf{V}_2 = -(4 + j12)$$
$$-(4 + j2)\mathbf{V}_o + (1 + j)\mathbf{V}_2 = 12 + j16$$

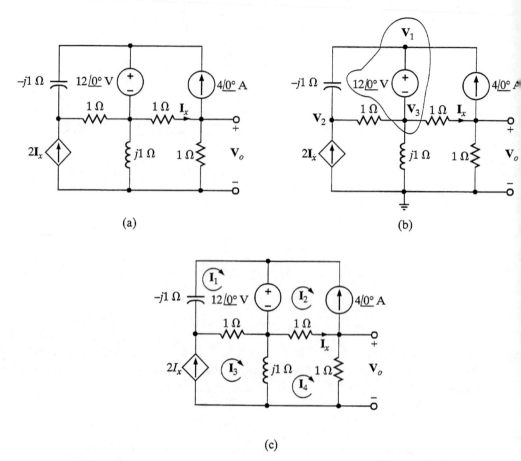

(a)   (b)

(c)

Figure 8.22 Circuits used in Example 8.18 for nodal and loop analysis.

Solving these equations for $\mathbf{V}_o$ yields

$$\mathbf{V}_o = \frac{-(8 + j4)}{1 + j2}$$
$$= -4 \underline{/143.13°} \text{ V}$$

(2) Loop Analysis   The mesh currents for the network are defined in Fig. 8.22c. The constraint equations for the circuit are

$$\mathbf{I}_2 = -4 \underline{/0°}$$
$$\mathbf{I}_x = \mathbf{I}_4 - \mathbf{I}_2 = \mathbf{I}_4 + 4 \underline{/0°}$$
$$\mathbf{I}_3 = 2\mathbf{I}_x = 2\mathbf{I}_4 + 8 \underline{/0°}$$

The KVL equations for mesh 1 and mesh 4 are

$$-j1\mathbf{I}_1 + 1(\mathbf{I}_1 - \mathbf{I}_3) = -12 \underline{/0°}$$
$$j1(\mathbf{I}_4 - \mathbf{I}_3) + 1(\mathbf{I}_4 - \mathbf{I}_2) + 1\mathbf{I}_4 = 0$$

Note that if the constraint equations are substituted into the second KVL equation, the only unknown in the equation is $\mathbf{I}_4$. This substitution yields

$$\mathbf{I}_4 = -4 \ \underline{/143.13°} \ \text{A}$$

and hence,

$$\mathbf{V}_o = -4 \ \underline{/143.13°} \ \text{V}$$

(3) Thévenin's Theorem   In applying Thévenin's theorem, we will find the open-circuit voltage and then determine the Thévenin equivalent impedance using a test source at the open-circuit terminals. We could determine the Thévenin equivalent impedance by calculating the short-circuit current; however, we will determine this current when we apply Norton's theorem.

The open-circuit voltage is determined from the network in Fig. 8.23a. Note that $\mathbf{I}'_x = 4 \ \underline{/0°} \ \text{A}$ and since $2\mathbf{I}'_x$ flows through the inductor, the open-circuit voltage $\mathbf{V}_{oc}$ is

$$\mathbf{V}_{oc} = -1(4 \ \underline{/0°}) + j1(2\mathbf{I}'_x)$$
$$= -4 + j8 \ \text{V}$$

To determine the Thévenin equivalent impedance, we turn off the independent sources, apply a test voltage source to the output terminals, and compute the current leaving the test source. As shown in Fig. 8.23b, since $\mathbf{I}''_x$ flows in the test source, KCL requires that the current in the inductor be $\mathbf{I}''_x$ also. KVL around the mesh containing the test source indicates that

$$j1\mathbf{I}''_x - 1\mathbf{I}''_x - \mathbf{V}_{test} = 0$$

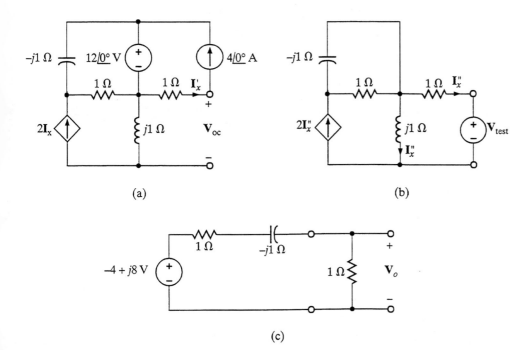

(a)

(b)

(c)

Figure 8.23 Circuits used in Example 8.18 when applying Thévenin's theorem.

Therefore,

$$\mathbf{I}_x'' = \frac{-\mathbf{V}_{\text{test}}}{1 - j}$$

Then

$$\mathbf{Z}_{\text{Th}} = \frac{\mathbf{V}_{\text{test}}}{-\mathbf{I}_x''}$$

$$= 1 - j \, \Omega$$

If the Thévenin equivalent network is now connected to the load, as shown in Fig. 8.23c, the output voltage $\mathbf{V}_o$ is found to be

$$\mathbf{V}_o = \frac{-4 + 8j}{2 - j1} \quad (1)$$

$$= -4 \,\underline{/143.13°} \, \text{V}$$

(4) Norton's Theorem   In using Norton's theorem, we will find the short-circuit current from the network in Fig. 8.24a. Once again, using the supernode, the constraint and KCL equations are

$$\mathbf{V}_3 + 12 \,\underline{/0°} = \mathbf{V}_1$$

$$\frac{\mathbf{V}_2 - \mathbf{V}_1}{-j1} + \frac{\mathbf{V}_2 - \mathbf{V}_3}{1} - 2\mathbf{I}_x''' = 0$$

$$\frac{\mathbf{V}_1 - \mathbf{V}_2}{-j1} + \frac{\mathbf{V}_3 - \mathbf{V}_2}{1} - 4 \,\underline{/0°} + \frac{\mathbf{V}_3}{j1} + \mathbf{I}_x''' = 0$$

$$\mathbf{I}_x''' = \frac{\mathbf{V}_3}{1}$$

Substituting the first and last equations into the remaining equations yields

$$(1 + j)\mathbf{V}_2 - (3 + j)\mathbf{I}_x''' = j12$$
$$-(1 + j)\mathbf{V}_2 + (2)\mathbf{I}_x''' = 4 - j12$$

Solving these equations for $\mathbf{I}_x'$ yields

$$\mathbf{I}_x''' = \frac{-4}{1 + j} \, \mathbf{A}$$

The KCL equation at the right-most node in the network in Fig. 8.24a is

$$\mathbf{I}_x''' = 4 \,\underline{/0°} + \mathbf{I}_{sc}$$

Solving for $\mathbf{I}_{sc}$, we obtain

$$\mathbf{I}_{sc} = \frac{-(8 + j4)}{1 + j} \, \mathbf{A}$$

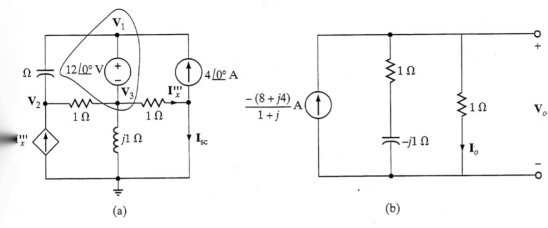

Figure 8.24 Circuits used in Example 8.18 when applying Norton's theorem.

The Thévenin equivalent impedance was found earlier to be

$$\mathbf{Z}_{Th} = 1 - j\,\Omega$$

Using the Norton equivalent network, the original network is reduced to that shown in Fig. 8.24b. The voltage $\mathbf{V}_o$ is then

$$\mathbf{V}_o = \frac{-(8 + j4)}{1 + j}\left[\frac{(1)(1 - j)}{1 + 1 - j}\right]$$

$$= -4\left[\frac{3 - j}{3 + j}\right]$$

$$= -4\,\underline{/-36.87°}\ \text{V} \qquad\square$$

EXTENSION EXERCISES

E8.13 Use nodal analysis to find $\mathbf{V}_o$ in the network in Fig. E8.13.

Figure E8.13

ANSWER: $\mathbf{V}_o = 2.12\,\underline{/75°}\ \text{V}.$

E8.14 Use mesh equations to find $\mathbf{V}_o$ in the network in Fig. E8.14.

Figure E8.14

ANSWER: $\mathbf{V}_o = 10.88 \angle 36° \text{ V}.$

E8.15 Using superposition, find $\mathbf{V}_o$ in the network in Fig. E8.15.

Figure E8.15

ANSWER: $\mathbf{V}_o = 12 \angle 90° \text{ V}.$

E8.16 Find $\mathbf{V}_o$ in the network in Fig. E8.15 using source transformation.

ANSWER: $\mathbf{V}_o = 12 \angle 90° \text{ V}.$

E8.17 Find $\mathbf{V}_o$ in the network in Fig. E8.14 using Thévenin's theorem.

ANSWER: $\mathbf{V}_o = 10.88 \angle 36.3° \text{ V}.$

E8.18 Use Norton's theorem to find $\mathbf{V}_o$ in the network in Fig. E8.15.

ANSWER: $\mathbf{V}_o = 12 \angle 90° \text{ V}.$

.9 AC PSPICE ANALYSIS USING SCHEMATIC CAPTURE

ntroduction

this chapter we have found that an ac steady-state analysis is facilitated by the use of
nasors. PSPICE can perform ac steady-state simulations, outputting magnitude and phase
ata for any voltage or current phasors of interest. Additionally, PSPICE can perform an
C SWEEP in which the frequency of the sinusoidal sources is varied over a user de-
ned range. In this case, the simulation results are the magnitude and phase of every node
oltage and branch current as a function of frequency.

We will introduce five new *Schematics*/PSPICE topics in this section: defining AC
ources, simulating at a single frequency, simulating over a frequency range, using the
ROBE feature to create plots and finally saving and printing these plots. *Schematics*
undamentals such as getting parts, wiring and changing part names and values have al-
eady been discussed in Chapter 4. As in Chapter 4, we will use the following font con-
entions. Uppercase text refers to programs and utilities within PSPICE such as the
C SWEEP feature and PROBE graphing utility. All boldface text, whether upper or
owercase, denotes keyboard or mouse entries. For example, when placing a resistor into
a circuit schematic, one must specify the resistor **VALUE** using the keyboard. The case
of the boldface text matches that used in PSPICE.

Defining AC Sources

Figure 8.25 shows the circuit we will simulate at a frequency of 60 Hz. We will continue
to follow the flowchart shown in Figure 4.21 in performing this simulation. Inductor and
capacitor parts are in the ANALOG library and are called L and C, respectively. The AC
source, VAC, is in the SOURCE library. Figure 8.26 shows the resulting *Schematics* diagram after wiring and editing the part's names and values.

Figure 8.25 Circuit for ac
simulation.

Figure 8.26 The *Schematics*
diagram for the circuit in
Figure 8.25.

To set up the AC source for simulation, double-click on the source symbol to open
ATTRIBUTES box which is shown, after editing, in Figure 8.27. As indicated earlier, v
deselect the fields **Include Non-changeable Attributes and Include System-defined A**
tributes. Each line in the ATTRIBUTES box is called an attribute of the ac source. Eac
attribute has a name and a value. The **DC** attribute is the dc value of the source for c
analyses. The **ACMAG** and **ACPHASE** attributes set the magnitude and phase of th
phasor representing $\mathbf{V}_{in}$ for ac analyses. Each of these attributes defaults to zero. Th
value of the **ACMAG** attribute was set to 4 V when we created the schematic in Fig. 8.2€
To set the **ACPHASE** attribute to 10°, click on the **ACPHASE** attribute line, enter 10 i
the **Value** field, press **Save Attr** and **OK**. When the ATTRIBUTE box looks like tha
shown in Fig. 8.27, the source is ready for simulation.

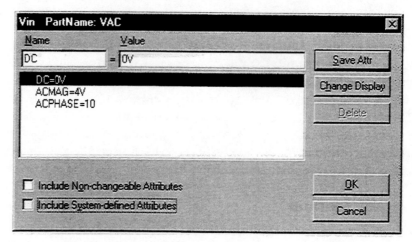

Figure 8.27 Setting the ac source phase angle.

Single Frequency AC Simulations

Next, we must specify the frequency for simulation. This is done by selecting **Setup** from
the **Analysis** menu. The SETUP box in Fig. 8.28 should appear. If we double-click on the
text **AC Sweep**, the AC SWEEP AND NOISE ANALYSIS window in Fig. 8.29 will open.
All of the fields in Fig. 8.29 have been set for our 60 Hz simulation.

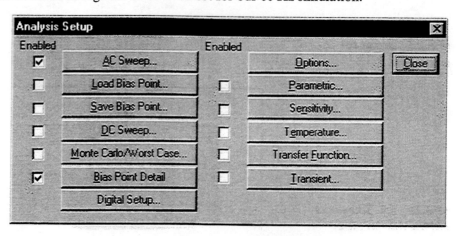

Figure 8.28 The ANALYSIS SETUP window.

Figure 8.29 Setting the frequency range for a single frequency simulation.

Since the simulation will be performed at only one frequency, 60 Hz, graphing the simulation results is not an attractive option. Instead, we will write the magnitude and phase of the phasors V_{OUT} and I to the output file using the VPRINT1 and IPRINT parts, from the SPECIAL library, which have been added to the circuit diagram, as shown in Fig. 8.30. The VPRINT1 part acts as a voltmeter, measuring the voltage at any single node with respect to the ground node. There is also a VPRINT2 part that measures the voltage between any two nonreference nodes. Similarly, the IPRINT part acts as an ammeter and must be placed in series with the branch current of interest. By convention, current in the IPRINT part is assumed to exit from the negatively marked terminal. To find the clockwise loop current, as defined in Fig. 8.25, the IPRINT part has been flipped. The FLIP command is in the EDIT menu.

After placing the VPRINT1 part, double click on it to open its ATTRIBUTES box, shown in Fig. 8.31. The VPRINT1 part can be configured to meter the node voltage in any kind of simulation; dc, ac, or transient. Since an ac analysis was specified in the SETUP window in Fig. 8.28, the values of the **AC**, **MAG**, and **PHASE** attributes are set to **Y**, where "Y" stands for YES. This process is repeated for the IPRINT part. When we return to *Schematics*, the simulation is ready to run.

Figure 8.30 A *Schematics* diagram ready for single frequency ac simulation.

When an AC SWEEP is performed, PSPICE, unless instructed otherwise, will attempt to plot the results using the PROBE plotting program. To turn off this feature, select **Probe Setup** in the **Analysis** menu. When the PROBE SETUP window shown in Fig. 8.32 appears, select **Do Not Auto-Run Probe** and **OK**.

Figure 8.31 Setting the VPRINT1 measurements for ac magnitude and phase.

Figure 8.32 The PROBE
SETUP window.

The circuit is simulated by selecting **Simulate** from the **Analysis** menu. Since the results are in the output file, select **Examine Output** from the **Analysis** menu to view the data. At the bottom of the file, we find the results as seen in Fig. 8.33: $\mathbf{V}_{OUT} = 2.651\ \underline{/-38.54°}$ V and $\mathbf{I} = 2.998\ \underline{/51.46°}$ mA.

Variable Frequency AC Simulations

In order to sweep the frequency over a range, 1 Hz to 10 MHz, for example, return to the AC SWEEP AND NOISE ANALYSIS box shown in Fig. 8.29. Change the fields to that shown in Fig. 8.34. Since the frequency range is so large, we have chosen a log axis for frequency with 50 data points in each decade. We can now plot the data using the PROBE utility. This procedure requires two steps. First, we remove the VPRINT1 and IPRINT

Figure 8.33 Magnitude and phase data for V_{OUT} and **I** are at the bottom of the output file.

Figure 8.34 Setting the frequency range for a swept frequency simulation.

parts in Fig. 8.30. Second, we return to the PROBE SETUP window shown in Fig. 8.32, and select **Automatically Run Probe After Simulation** and **OK**.

Creating Plots in PROBE

When the PSPICE simulation is finished, the PROBE window shown in Fig. 8.35 will open. Note that the frequency is displayed on a log axis, as requested. In order to plot the magnitude and phase of V_{OUT}, select **Add** from the **Trace** menu. The ADD TRACES window shown in Fig. 8.36 will appear. To display the magnitude of V_{OUT}, we select **V(V_{OUT})** from the left column. When a voltage or current is selected, the magnitude of the phasor will be plotted. Now the PROBE window should look like that shown in Fig. 8.37.

Before adding the phase to the plot, we note that V_{OUT} spans a small range (i.e., 0 to 4 V). Since the phase change could span a much greater range, we will plot the phase on

Figure 8.35 The PROBE window.

Figure 8.36 The Add Traces window.

Figure 8.37 The magnitude of $\mathbf{V}_{OUT}$.

a second *y*-axis. From the **Plot** menu, select **Add Y Axis**. To add the phase to the plot, select **Add** from the **Trace** menu. On the right side of the ADD TRACES window in Fig. 8.36, scroll down to the entry, **P()**. Click on that, and then click on **V(Vout)** in the left column. The TRACE EXPRESSION line at the bottom of the window will contain the expression P(V(Vout))—the phase of Vout. Figure 8.38 shows the PROBE plot for both magnitude and phase of $\mathbf{V}_{OUT}$.

Figure 8.38 The magnitude and phase of $\mathbf{V}_{OUT}$.

Figure 8.39 The magnitude and phase of the current, I.

In order to plot the current, I, on a new plot, we select **New** from the **Window** menu. Then, add the traces for the magnitude and phase of the current through R1 (PSPICE calls it I(R1)) using the process described above for plotting the magnitude and phase of Vout. The results are shown in Fig. 8.39.

The procedures for saving and printing PROBE plots, as well as the techniques for plot manipulation and data extraction, are described in Chapter 6.

EXAMPLE 8.19

Using the PSPICE *Schematics* editor, draw the circuit in Figure 8.40, and use the PROBE utility to create plots for the magnitude and phase of $\mathbf{V_{OUT}}$ and $\mathbf{I}$. At what frequency is $|\mathbf{V_{OUT}}| = 1$ V? What are the phases of $\mathbf{V_{OUT}}$ and $\mathbf{I}$ and the magnitude of $\mathbf{I}$ at that frequency?

Figure 8.40

Circuit: $v_{IN}(t) = 4\cos(\omega t + 10°)$ V source, R = 1 kΩ, L = 2 mH, C = 3 μF, output $v_{OUT}(t)$

SOLUTION The *Schematics* diagram for the simulation is shown in Figure 8.41 where the ac source voltage and phase have been set to 4 V and 10° respectively. An AC SWEEP analysis was run over the frequency range 1 Hz to 10 MHz at 50 data points per decade. Simulation results for $\mathbf{V_{OUT}}$ magnitude and phase are shows in Figure 8.42. Using the CURSOR tool in PROBE, $\mathbf{V_{OUT}}$ was found to be $1\ \underline{/-65.6°}$ V

AC Analysis Example

Figure 8.41

at f = 205.65 Hz. A second plot, shown in Figure 8.43, was created for the current **I** by plotting I(R1). Using the CURSOR, the current at 205.65 Hz was found to be 3.88 $\underline{/24.3°}$ mA.

When plotting currents, remember that current direction is determined by the netlist as shown below.

```
*   Schematics Netlist   *
R_R1          $N_0002 $N_0001  1k
L_L1          $N_0001 Vout   2mH
V_Vin         SN_0002 0 DC OV AC AV 10
C_C1          0 Vout   3uF
```

Note that R_R1 is connected between nodes 2 and 1 (node 0 is always ground). So, PROBE thinks the current through R_1 flows left-to-right, the same direction we want to plot. If the netlist current direction were opposite our assumption, we would have to plot negative I(RI) with PROBE.   ❏

Figure 8.42 Simulation results for V_{OUT}.

Figure 8.43 Simulation results for **I**.

EXAMPLE 8.20

Using the PSPICE Schematics editor, draw the circuit in Figure 8.44, and use the PROBE utility to create plots for the magnitude and phase of V_{OUT} and **I**. At what frequency does maximum $|I|$ occur? What are the phasors V_{OUT} and **I** at that frequency?

Figure 8.44

SOLUTION The Schematics diagram for the simulation is shown in Figure 8.45 where an AC SWEEP has been setup for the frequency range 10 Hz to 10 MHz at 100 data points per decade. The phase angle of the ac current source part, IAC, is zero degrees and is not changeable in PSPICE. Therefore, we will perform the simulation with the phase of LAC at zero and add 20° to all simulation data. Plots for V_{OUT} and **I** magnitudes and phases are given in Figures 8.46 and 8.47 respectively. From Figure 8.47 we see that maximum inductor current magnitude occurs at my TI-86 says 31.35, using the fmax features kHz. The corresponding phasors are $I = 5.94 \underline{/16.09°}$ and $V_{OUT} = \underline{/-68.11°}$ V.

AC Example 2

Figure 8.45

Figure 8.46 Simulation results for V_{OUT}.

EXTENSION PROBLEM

E8.19 Using the PSPICE Schematics editor, draw the circuit in Figure E8.19, and use the VPRINT1 and IPRINT parts to determine the phasors V_{OUT} and I at a frequency of 60 Hz?

Figure 8.47 Simulation results for **I**.

Figure E8.19

ANSWER: $\mathbf{V}_{OUT} = 63.22\ \underline{/-83.24°}\ \text{V}$ and $\mathbf{I} = 70.64\ \underline{/-83.24°}\ \text{mA}$.

8.10 SUMMARY

The sinusoidal function definition   The sinusoidal function $x(t) = X_M \sin(\omega t + \theta)$ has an amplitude of X_M, a radian frequency of ω, a period of $2\pi/\omega$, and a phase angle of θ.

The phase lead and phase lag definitions   If $x_1(t) = X_{M_1} \sin(\omega t + \theta)$ and $x_2(t) = X_{M_2} \sin(\omega t + \phi)$, $x_1(t)$ leads $x_2(t)$ by $\theta - \phi$ radians and $x_2(t)$ lags $x_1(t)$ by $\theta - \phi$ radians.

The phasor definition   The sinusoidal voltage $v(t) = V_M \cos(\omega t + \theta)$ can be written in exponential form as $v(t) = \text{Re}\left[V_M e^{j(\omega t + \theta)}\right]$ and in phasor form as $\mathbf{V} = V_M\ \underline{/\theta}$.

he phase relationship in θ_v and θ_i for elements R, L and C   If θ_v and θ_i represent the phase angles of the voltage across and the current through a circuit element, then $\theta_i = \theta_v$ if the element is a resistor, θ_i lags θ_v by 90° if the element is an inductor, θ_i leads θ_v by 90° if the element is a capacitor.

he impedances of R, L and C   Impedance, **Z**, is defined as the ratio of the phasor voltage, **V**, to the phasor current, **I**, where $\mathbf{Z} = R$ for a resistor, $\mathbf{Z} = j\omega L$ for an inductor, and $\mathbf{Z} = 1/j\omega C$ for a capacitor.

he phasor diagrams   Phasor diagrams can be used to display the magnitude and phase relationships of various voltages and currents in a network.

requency domain analysis:

1. Represent all voltages, $v_i(t)$, and all currents, $i_j(t)$, as phasors and represent all passive elements by their impedance or admittance.
2. Solve for the unknown phasors in the frequency (ω) domain.
3. Transform the now-known phasors back to the time domain.

Solution techniques for ac steady-state problems:

Ohm's law
KCL and KVL
PSPICE
Nodal and loop analysis
Superposition and source exchange
Thévenin's and Norton's theorems

PROBLEMS

Section 8.1

8.1   Given $v(t) = 10 \cos(377t - 60°)$ V, determine the period and the frequency in hertz.

$$f = \frac{\omega}{2\pi}$$

$$T = \frac{1}{f}$$

8.2   Determine the relative phase relationship of the two waves.

$$v_1(t) = 10 \cos(377t - 30°) \text{ V}$$
$$v_2(t) = 10 \cos(377t + 90°) \text{ V}$$

Use the phase lead or lag relationship.

8.3   Given the following voltage and current

$$i(t) = 5 \sin(377t - 20°) \text{ V}$$
$$v(t) = 10 \cos(377t + 30°) \text{ V}$$

determine the phase relationship between $i(t)$ and $v(t)$.

Convert $i(t)$ to a cosine function and find the phase relationship.

8.4 Determine the phase angles by which $v_1(t)$ leads $i_1(t)$ and $v_1(t)$ leads $i_2(t)$, when

Similar to P8.3.

$$v_1(t) = 4 \sin(377t + 25°)$$
$$i_1(t) = 0.05 \cos(377t - 20°)$$
$$i_2(t) = -0.1 \sin(377t + 45°)$$

Sections 8.2, 8.3, 8.4, 8.5

$$i(t) = \frac{v(t)}{R}$$
$$\mathbf{I} = \mathbf{V}/R$$
if $v(t) = V_m \cos(\omega t + \theta)$,
$$\mathbf{V} = V_m e^{j\phi}$$

8.5 Calculate the current in the resistor in Fig. P8.5 if the voltage input is

(a) $v_1(t) = 10 \cos(377t + 80°)$ V
(b) $v_2(t) = 5 \sin(377t + 30°)$ V

Give the answers in both the time and frequency domains.

Figure P8.5

$$v(t) = L\frac{di(t)}{dt}$$
$$i(t) = \frac{1}{L}\int v(t)\,dt, \mathbf{I} = \frac{\mathbf{V}}{j\omega L}$$

8.6 Calculate the current in the inductor shown in Fig. P8.6 if the voltage input is

(a) $v_1(t) = 10 \cos(377t + 45°)$ V
(b) $v_2(t) = 5 \sin(377t - 90°)$ V

Give the answers in both the time and frequency domains.

Figure P8.6

$$i(t) = C\frac{dv(t)}{dt}$$
$$\mathbf{I} = j\omega C\mathbf{V}$$

8.7 Calculate the current in the capacitor shown in Fig. P8.7 if the voltage input is

(a) $v_1(t) = 10 \cos(377t - 30°)$ V
(b) $v_2(t) = 5 \sin(377t + 60°)$ V

Give the answers in both the time and frequency domains.

Figure P8.7

Impedances add in series.

8.8 Find the frequency domain impedance, $\mathbf{Z}$, for the network in Fig. P8.8.

Figure P8.8

9 Find the frequency domain impedance, **Z**, for the circuit in Fig. P8.9.

Figure P8.9

Admittances add in parallel.

.10 Find the frequency domain impedance, **Z**, in the network in Fig. P8.10.

Figure P8.10

A series parallel combination.

8.11 Find the impedance, **Z**, shown in Fig. P8.11 at a frequency of 60 Hz.

Figure P8.11

Use $j\omega L$ and $1/j\omega C$ for L and C, respectively. Then combine the impedances using the series and parallel relationships.

8.12 Find the impedance, **Z**, shown in Fig. P8.11 at a frequency of 400 Hz.

8.13 Find the frequency domain impedance, **Z**, shown in Fig. P8.13.

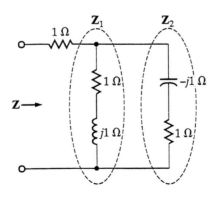

Figure P8.13

Find $\mathbf{Z}_1$ and $\mathbf{Z}_2$, and then combine them since they are in parallel.

8.14 Find the frequency domain impedance, **Z**, shown in Fig. P8.14.

There are two sets of parallel combinations, and the remaining impedances are in series with them.

Figure P8.14

8.15 Find the frequency domain impedance, **Z**, shown in P8.15.

Simliar to P8.14.

Figure P8.15

8.16 Draw the frequency domain circuit and calculate $i(t)$ for the circuit shown in Fig. P8.16 if $v_s(t) = 2 \cos(377t)$ V.

Replace L by $j\omega L$ and $v_s(t)$ by
$$\mathbf{V} = 2 \underline{/0°} \text{ V}, \mathbf{I} = \frac{V}{R + j\omega L}.$$

Figure P8.16

8.17 Draw the frequency domain circuit and calculate $v(t)$ for the circuit shown in Fig. P8.17 if $i_s(t) = 10 \cos(377t + 30°)$ A.

Replace C by $j\omega C$ and $i_s(t)$ by
$$\mathbf{I}_s = 10 \underline{/30°} \text{ A},$$
$$\mathbf{V} = \mathbf{I}_s \left(R \middle/\!\!\middle/ \frac{1}{j\omega C} \right).$$

Figure P8.17

18 Draw the frequency domain circuit and calculate $i(t)$ for the circuit shown in Fig. P8.18 if $v_s(t) = 10\cos(377t + 30°)$ V.

Figure P8.18

$$I = \dfrac{V}{2 + j\omega L + \dfrac{1}{j\omega C}}$$

.19 Draw the frequency domain circuit and calculate $v(t)$ for the circuit shown in Fig. P8.19 if $i_s(t) = 20\cos(377t + 120°)$ A.

Figure P8.19

Similar to P8.17.

8.20 Draw the frequency domain circuit and calculate $i(t)$ for the circuit shown in Fig. P8.20 if $v_s(t) = 10\cos(1000t + 30°)$ V.

Figure P8.20

Find **Z** and $I = V_s/Z$.

8.21 Draw the frequency domain circuit and calculate $v(t)$ for the circuit shown in Fig. P8.21 if $i_s(t) = 2\cos(1000t + 120°)$ A.

Figure P8.21

Find **Z** and $V = I_s Z$.

Section 8.6

Convert $v_R(t)$, $v_L(t)$ and $v_S(t)$ to phasors. Then use vector techniques to show that the sum of $v_R(t)$ and $v_L(t)$ is $v_S(t)$.

8.22 The voltages $v_R(t)$ and $v_L(t)$ in the circuit shown in Fig. P8.16, can be drawn as phasors in a phasor diagram. Use a phasor diagram to show that $v_R(t) + v_L(t) = v_s(t)$.

Similar to P8.22.

8.23 The currents $i_R(t)$ and $i_C(t)$ in the circuit shown in Fig. P8.17 can be drawn as phasors in a phasor diagram. Use the diagram to shown that $i_R(t) + v_C(t) = i_S(t)$.

Similar to P8.22.

8.24 The voltages $v_R(t)$, $v_L(t)$ and $v_C(t)$ in the circuit shown in Fig. P8.18, can be drawn as phasors in a phasor diagram. Show that $v_R(t) + v_L(t) + v_C(t) = v_S(t)$.

Similar to P8.22.

8.25 The currents $i_R(t)$, $i_L(t)$, and $i_C(t)$ in the circuit shown in Fig. P8.19, can be drawn as phasors in a phasor diagram. Show that $i_R(t) + i_L(t) + i_C(t) = i_S(t)$.

Similar to P8.22.

8.26 The currents $i_L(t)$ and $i_C(t)$ of the inductor and capacitor in the circuit shown in Fig. P8.20 can be drawn as phasors in a phasor diagram. Show that $i_L(t) + i_C(t) = i(t)$.

8.27 Find the value of C in the circuit shown in Fig. P8.27 so that **Z** is purely resistive at the frequency of 60 Hz.

When $\omega L = \dfrac{1}{\omega C}$, **Z** = 1 Ω.

Figure P8.27

8.28 Find the frequency at which the circuit shown in Fig. P8.28 is purely resistive.

Similar to P8.27.

Figure P8.28

9 The impedance of the circuit in Fig. P8.29 is real at $f = 60$ Hz. What is the value of L?

Figure P8.29

Similar to P8.27.

30 In the circuit shown in Fig. P8.30, determine the frequency at which $i(t)$ is in phase with $v_s(t)$.

Figure P8.30

The quantities are in phase when the imaginary part of $\mathbf{Z}$ is zero.

31 In the circuit shown in Fig. P8.31, determine the value of the inductance such that $v(t)$ is in phase with $i_s(t)$.

Figure P8.31

Similar to P8.30.

3.32 In the circuit shown in Fig. P8.32, determine the frequency at which $i(t)$ is in phase with $v_s(t)$.

Figure P8.32

Similar to P8.30.

Section 8.7

8.33 Find the current **I** shown in Fig. P8.33.

Use the current division rule.

Figure P8.33

8.34 Find the voltage **V** shown in Fig. P8.34.

Use the voltage division rule.

Figure P8.34

8.35 Find the voltage **V** shown in Fig. P8.35.

Find **Z**. $\mathbf{V} = (\mathbf{Z})1 \angle 30°$.

Figure P8.35

8.36 Find the frequency domain voltage, $\mathbf{V}_o$, as shown in Fig. P8.36.

Find $\mathbf{I}_o$ using the current division rule. Then $\mathbf{V}_o = \mathbf{I}_o(-j12)$V.

Figure P8.36

8.37 Find the voltage $\mathbf{V}_o$ as shown in Fig. P8.37.

Use the voltage division rule.

Figure P8.37

8 Find the voltage, V_o, shown in Fig. P8.38.

Figure P8.38

Find **Z**. Then use the voltage division rule.

---

39 Find the frequency domain current, I_o, shown in Fig. P8.39.

Figure P8.39

Similar to P8.38.

8.40 Given the network in Fig. P8.40, determine the value of V_o if $V_s = 24\ \underline{/0^\circ}$ V.

Figure P8.40

Find **Z**. Find V_1 using the voltage division rule. Find V_o using the voltage division rule again.

8.41 Determine I_o in the network shown in Fig. P8.41 if $V_s = 12\ \underline{/0^\circ}$ V.

Figure P8.41

Similar to P8.40.

8.42 Determine $\mathbf{I}_o$ in the network in Fig. P8.42 if $\mathbf{I}_s = 12\ \underline{/0°}$ A.

Find **Z**. Use the current division rule to find $\mathbf{I}_1$. Then use the current division rule to find $\mathbf{I}_o$.

Figure P8.42

8.43 Find $\mathbf{V}_o$ in the network in Fig. P8.43.

Use the voltage division rule to find $\mathbf{V}_o$.

Figure P8.43

Section 8.8

8.44 Find $\mathbf{V}_o$ in the circuit in Fig. P8.44.

Use nodal analysis, superposition or source exchange.

Figure P8.44

8.45 Determine $\mathbf{V}_o$ in the circuit in Fig. P8.45.

Use source exchange or superposition.

Figure P8.45

46 Draw the frequency domain network and calculate $v_o(t)$ in the circuit shown in Fig. P8.46 if $v_S(t)$ is $10\cos(10^4 t + 30°)$ V. Also, use a phasor diagram to determine $v_L(t)$.

Figure P8.46

See Example 8.15

47 Find $\mathbf{V}_o$ in the network in Fig. P8.47.

Figure P8.47

Use source exchange or superposition.

.48 Find $\mathbf{V}_o$ in the network in Fig. P8.48.

Figure P8.48

Find $\mathbf{Z}$. Then $\mathbf{I}$. KVL will then yield $\mathbf{V}_o$.

8.49 Use nodal analysis to find $\mathbf{I}_o$ in the circuit in Fig. P8.49.

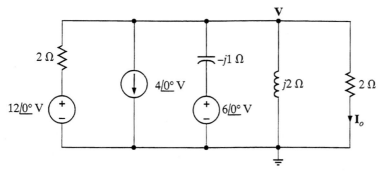

Figure P8.49

Assume that the node voltage of the top node is $\mathbf{V}$ and write the KCL equation for that node.

8.50 Using nodal analysis, find I_o in the circuit in Fig. P8.50.

Write the nodal equation for the two nodes with voltages V_1 and V_2.

Figure P8.50

8.51 Find I_o in the network in Fig. P8.51.

Use nodal, loop, or mesh analysis.

Figure P8.51

8.52 Find V_o in the network in Fig. P8.52 using nodal analysis.

Write the KCL equation for the node with voltage V. Then use voltage division to find V_o.

Figure P8.52

53 Find I_o in the circuit in Fig. P8.53 using nodal analysis.

Figure P8.53

Write the KCL equation for the nodes that are not connected to a voltage source.

.54 Use the supernode technique to find I_o in the circuit in Fig. P8.54.

Figure P8.54

Construct a supernode for the 12 $\underline{/0°}$ V source and write the node equation.

8.55 Find I_o in the network in Fig. P8.55.

Figure P8.55

Construct a supernode for the 12 $\underline{/0°}$ V source and write the node equations for the supernode and the remaining nonreference node.

8.56 Find I_o in the network in Fig. P8.56.

Figure P8.56

Similar to P8.55.

8.57 Find V_o in the network in Fig. P8.57.

Similar to P8.55.

Figure P8.57

8.58 Find V_o in the network in Fig. P8.58.

Use Thévenin's theorem and remove the $-j1$-Ω load.

Figure P8.58

8.59 Find V_o in the network in Fig. P8.59.

Similar to P8.58.

Figure P8.59

8.60 Use nodal analysis to find V_o in the circuit in Fig. P8.60.

Use the supernode technique.

Figure P8.60

51 Use nodal analysis to determine I_o in
the network in Fig. P8.61.

Figure P8.61

Only one nodal equation is
necessary.

.62 Find V_o in the network in Fig. P8.62.

Figure P8.62

Use voltage division.

8.63 Find the voltage across the inductor in the circuit shown in Fig. P8.63 using nodal
analysis.

Figure P8.63

V_2 is the only unknown
voltage.

8.64 The low-frequency equivalent circuit for a common-emitter transistor amplifier is
shown in Fig. P8.64. Compute the voltage gain V_o/V_S.

Figure P8.64

V can be obtained by voltage
division and then V_o by current
division.

8.65 Use nodal analysis to find $\mathbf{V}_o$ in the circuit in Fig. P8.65.

Use the supernode technique.

Figure P8.65

8.66 Use mesh analysis to find $\mathbf{V}_o$ in the circuit shown in Fig. P8.66.

$\mathbf{V}_o = 2\mathbf{I}_2$.

Figure P8.66

8.67 Use mesh analysis to find $\mathbf{V}_o$ in the circuit shown in Fig. P8.67.

$\mathbf{V}_o = 2(\mathbf{I}_1 - \mathbf{I}_2)$.

Figure P8.67

8.68 Find $\mathbf{V}_o$ in the circuit in Fig. P8.68 using mesh analysis.

The current sources make the mesh analysis straightforward.

Figure P8.68

69 Use mesh analysis to find V_o in the circuit in Fig. P8.69.

Figure P8.69

Similar to P8.68.

70 Using loop analysis, find I_o in the network in Fig. P8.70.

Figure P8.70

Use loop analysis or the supermesh technique to avoid a current source.

8.71 Using superposition, find V_o in the circuit in Fig. P8.71.

Figure P8.71

Use the 12 $\underline{/0°}$ V source and open the 2 $\underline{/0°}$ A source to find V_0'.
Use the 2 $\underline{/0°}$ A source and replace the voltage source with a short circuit to find V_o''. $V_o = V_o' + V_o''$.

8.72 Use superposition to determine V_o in the circuit in Fig. P8.72.

Figure P8.72

Use superposition and current division.

8.73 Find $\mathbf{V}_o$ in the network in Fig. P8.73 using superposition.

Similar to P8.72.

Figure P8.73

See Example 8.17.

8.74 Use source transformation to determine $\mathbf{I}_o$ in the network in Fig. P8.49.

Similar to P8.74.

8.75 Use source transformation to determine $\mathbf{I}_o$ in the network in Fig. P8.50.

8.76 Use source exchange to determine $\mathbf{V}_o$ in the network in Fig. P8.76.

Similar to P8.74.

Figure P8.76

See the steps in Section 8.7.

8.77 Solve Problem 8.45 using Thévenin's theorem.

Similar to P8.77.

8.78 Solve Problem 8.60 using Thévenin's theorem.

Similar to P8.77.

8.79 Solve Problem 8.62 using Thévenin's theorem.

8.80 Find $\mathbf{V}_o$ in the network in Fig. P8.80 using Thévenin's theorem.

See Example 8.18.

Figure P8.80

1 Solve Problem 8.72 using Norton's theorem.

See the steps in Section 8.7.

2 Solve Problem 8.73 using Norton's theorem.

Similar to P8.81.

83 Apply Thévenin's theorem to find $\mathbf{V}_o$ in the circuit in Fig. P8.83.

Figure P8.83

Use Thévenin's theorem twice to simplify the circuit.

8.84 Apply Norton's theorem to find $\mathbf{V}_o$ in the network in Fig. P8.84.

Figure P8.84

Similar to P8.81.

8.85 Find $\mathbf{V}_x$ in the circuit in Fig. P8.85 using Norton's theorem.

Figure P8.85

Similar to P8.81.

Advanced Problems

8.86 Find **Z** in the network in Fig. P8.86.

Figure P8.86

8.87 Find **Z** in the network in Fig. P8.87.

Figure P8.87

8.88 Find **Z** in the network in Fig. P8.88.

Figure P8.88

8.89 Determine the values of the two elements R_p and L_p or C_p in the network in Fig. P8.89 so that **Z** is $10 + j2 \ \Omega$ at a frequency of 60 Hz.

Figure P8.89

0 Determine the values of the two elements $(R, L,$ or $C)$ connected in parallel as shown in Fig. P8.90 so that $\mathbf{Z}$ is $5 - j5\,\Omega$ at 60 Hz.

Figure P8.90

.91 Draw the frequency domain network and calculate $v_o(t)$ in the circuit shown in Fig. P8.91 if $v_S(t)$ is $4\sin(500t + 45°)$ V and $i_S(t)$ is $1\cos(500t + 45°)$ A. Also, use a phasor diagram to determine $v_1(t)$.

Figure P8.91

8.92 Draw the frequency domain network and calculate $v_o(t)$ in the circuit shown in Fig. P8.92 if $i_1(t)$ is $200\cos(10^5 t + 60°)$ mA, $i_2(t)$ is $100\sin(10^5 t + 90°)$ mA, and $v_S(t) = 10\sin(10^5 t)$ V. Also, use a phasor diagram to determine $v_C(t)$.

Figure P8.92

8.93 Draw the frequency domain network and calculate $v_o(t)$ in the circuit shown in Fig. P8.93 if $v_S(t)$ is $10\cos(2\pi ft)$ V, where $f = 400$ Hz. Also, use a phasor diagram to determine $v(t)$.

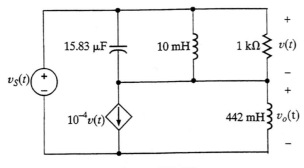

Figure P8.93

8.94 Find $v_o(t)$ in Problem 8.93 using Thévenin's theorem.

8.95 Find $\mathbf{V}_o$ using Norton's theorem for the circuit in Fig. P8.95.

Figure P8.95

8.96 Given the network in Fig. P8.96, find the Thévenin's equivalent of the network at the terminals A-B.

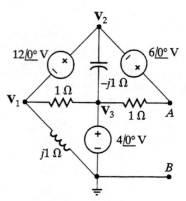

Figure P8.96

8.97 Use Thévenin's theorem to find $\mathbf{V}_o$ in the network in Fig. P8.97.

Figure P8.97

98 Use Norton's theorem to find $\mathbf{V}_o$ in the network in Fig. P8.98.

Figure P8.98

8.99 Using the PSPICE Schematics editor, draw the circuit in Figure P8.99. At what frequency are the magnitudes of $i_C(t)$ and $i_L(t)$ equal?

Figure P8.99

8.100 Using the PSPICE Schematics editor, draw the circuit in Figure P8.100. At what frequency are the phases of $i_1(t)$ and $v_X(t)$ equal?

Figure P8.100

Steady-State Power Analysis

9

Do you understand the electrical wiring circuit in your home that is used to power all the appliances and lights? Do you know the hazards associated with working with electrical power circuits? These are just two of the topics that will be examined in this chapter.

We have been primarily concerned in the preceding chapters with determining the voltage or current at some point in a network. Of equal importance to us in many situations is the power that is supplied or absorbed by some element. Typically, electrical and electronic devices have peak power or maximum instantaneous power ratings that cannot be exceeded without damaging the devices.

In electrical and electronic systems, power comes in all sizes. The power absorbed by some device on an integrated circuit chip may be in picowatts, whereas the power supplied by a large generating station may be in gigawatts. Note that the range between these two examples is phenomenally large (10^{21}).

In our previous work we defined instantaneous power to be the product of voltage and current. Average power, obtained by averaging the instantaneous power, is the average rate at which energy is absorbed or supplied. In the dc case, where both current and voltage are constant, the instantaneous power is equal to the average power. However, as we will demonstrate, this is not the case when the currents and voltages are sinusoidal functions of time.

In this chapter we explore the many ramifications of power in ac circuits. We examine instantaneous power, average power, maximum power transfer, average power for periodic nonsinusoidal waveforms, the power factor, complex power, and power measurement.

Some very practical safety considerations will be introduced and discussed through a number of examples.

Finally, a wide variety of application-oriented examples is presented. The origin of many of these examples is the common household wiring circuit.

1 INSTANTANEOUS POWER

By employing the sign convention adopted in the earlier chapters, we can compute the instantaneous power supplied or absorbed by any device as the product of the instantaneous voltage across the device and the instantaneous current through it.

Consider the circuit shown in Fig. 9.1. In general, the steady-state voltage and current for the network can be written as

$$v(t) = V_M \cos(\omega t + \theta_v) \qquad\qquad 9.1$$
$$i(t) = I_M \cos(\omega t + \theta_i) \qquad\qquad 9.2$$

The instantaneous power is then

$$
\begin{aligned}
p(t) &= v(t)i(t) \\
&= V_M I_M \cos(\omega t + \theta_v)\cos(\omega t + \theta_i) \qquad\qquad 9.3
\end{aligned}
$$

Employing the following trigonometric identity,

$$\cos \phi_1 \cos \phi_2 = \tfrac{1}{2}\left[\cos(\phi_1 - \phi_2) + \cos(\phi_1 + \phi_2)\right] \qquad\qquad 9.4$$

we find that the instantaneous power can be written as

$$p(t) = \frac{V_M I_M}{2}\left[\cos(\theta_v - \theta_i) + \cos(2\omega t + \theta_v + \theta_i)\right] \qquad\qquad 9.5$$

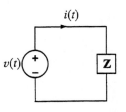

Figure 9.1 Simple ac network.

Note that the instantaneous power consists of two terms. The first term is a constant (i.e., it is time independent), and the second term is a cosine wave of twice the excitation frequency. We will examine this equation in more detail in Section 9.2.

EXAMPLE 9.1

The circuit in Fig. 9.1 has the following parameters: $v(t) = 4\cos(\omega t + 60°)$ V and $\mathbf{Z} = 2\,\underline{/30°}\,\Omega$. We wish to determine equations for the current and the instantaneous power as a function of time and plot these functions with the voltage on a single graph for comparison.

SOLUTION Since

$$\mathbf{I} = \frac{4\,\underline{/60°}}{2\,\underline{/30°}}$$
$$= 2\,\underline{/30°}\ \text{A}$$

then

$$i(t) = 2\cos(\omega t + 30°)\ \text{A}$$

From Eq. (9.5),

$$
\begin{aligned}
p(t) &= 4[\cos(30°) + \cos(2\omega t + 90°)] \\
&= 3.46 + 4\cos(2\omega t + 90°)\ \text{W}
\end{aligned}
$$

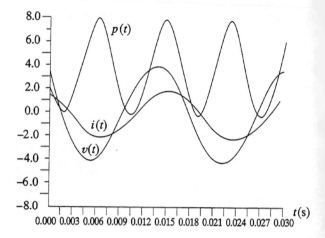

Note that $p(t)$ contains a dc term and a cosine wave with twice the frequency of $v(t)$ and $i(t)$.

Figure 9.2 Plots of $v(t)$, $i(t)$, and $p(t)$ for the circuit in Example 9.1 using $f = 60$ Hz.

A plot of this function, together with plots of the voltage and current, is shown in Fig. 9.2. As can be seen in the figure, the instantaneous power has an average value, and the frequency is twice that of the voltage or current. ☐

9.2 AVERAGE POWER

The average value of any periodic waveform (e.g., a sinusoidal function) can be computed by integrating the function over a complete period and dividing this result by the period. Therefore, if the voltage and current are given by Eqs. (9.1) and (9.2), respectively, the average power is

$$P = \frac{1}{T} \int_{t_0}^{t_0+T} p(t)\, dt$$

$$= \frac{1}{T} \int_{t_0}^{t_0+T} V_M I_M \cos(\omega t + \theta_v) \cos(\omega t + \theta_i)\, dt \qquad 9.6$$

where t_0 is arbitrary, $T = 2\pi/\omega$ is the period of the voltage or current, and P is measured in watts. Actually, we may average the waveform over any integral number of periods so that Eq. (9.6) can also be written as

$$P = \frac{1}{nT} \int_{t_0}^{t_0+nT} V_M I_M \cos(\omega t + \theta_v) \cos(\omega t + \theta_i)\, dt \qquad 9.7$$

where n is a positive integer.

Employing Eq. (9.5) for the expression in (9.6), we obtain

$$P = \frac{1}{T} \int_{t_0}^{t_0+T} \frac{V_M I_M}{2} \left[\cos(\theta_v - \theta_i) + \cos(2\omega t + \theta_v + \theta_i) \right] dt \qquad 9.8$$

We could, of course, plod through the indicated integration; however, with a little forethought we can determine the result by inspection. The first term is independent of t, and

erefore a constant in the integration. Integrating the constant over the period and di-
ding by the period simply results in the original constant. The second term is a cosine
ave. It is well known that the average value of a cosine wave over one complete period
r an integral number of periods is zero, and therefore the second term in Eq. (9.8) van-
hes. In view of this discussion, Eq. (9.8) reduces to

$$P = \tfrac{1}{2} V_M I_M \cos(\theta_v - \theta_i) \qquad\qquad \textbf{9.9}$$

Jote that since $\cos(-\theta) = \cos(\theta)$, the argument for the cosine function can be either
$\theta_v - \theta_i$ or $\theta_i - \theta_v$. In addition, note that $\theta_v - \theta_i$ is the angle of the circuit impedance as
shown in Fig. 9.1. Therefore, *for a purely resistive circuit,*

A frequently used equation for calculating the average power

$$P = \tfrac{1}{2} V_M I_M \qquad\qquad \textbf{9.10}$$

and *for a purely reactive circuit,*

$$P = \tfrac{1}{2} V_M I_M \cos(90°)$$
$$= 0$$

Because purely reactive impedances absorb no average power, they are often called *loss-less elements.* The purely reactive network operates in a mode in which it stores energy over one part of the period and releases it over another.

EXAMPLE 9.2

We wish to determine the average power absorbed by the impedance shown in Fig. 9.3.

SOLUTION From the figure we note that

$$\mathbf{I} = \frac{\mathbf{V}}{\mathbf{Z}} = \frac{V_M\,\angle\theta_v}{2 + j2} = \frac{10\,\angle 60°}{2.83\,\angle 45°} = 3.53\,\angle 15°\ \text{A}$$

Therefore,

$$I_M = 3.53\ \text{A} \qquad \text{and} \qquad \theta_i = 15°$$

Hence,

$$P = \tfrac{1}{2} V_M I_M \cos(\theta_v - \theta_i)$$
$$= \tfrac{1}{2}(10)(3.53)\cos(60° - 15°)$$
$$= 12.5\ \text{W}$$

Since the inductor absorbs no power, we can employ Eq. (9.10) provided that V_M in that equation is the voltage across the resistor. Using voltage division, we obtain

$$\mathbf{V}_R = \frac{(10\,\angle 60°)(2)}{2 + j2} = 7.07\,\angle 15°\ \text{V}$$

DRILL

D9.1 Find the average power for the following **V** and **I**:

(1) $\mathbf{V} = 2\,\angle 30°$ V,
$\mathbf{I} = 2\,\angle 30°$ A
(2) $\mathbf{V} = 2\,\angle 30°$ V,
$\mathbf{I} = 2\,\angle 120°$ A
(3) $\mathbf{V} = 2\,\angle 30°$ V,
$\mathbf{I} = 2\,\angle 60°$ A

ANSWER: (1) 2 W
(2) 0 W
(3) 1.732 W

Figure 9.3 Example *RL* circuit.

and therefore,

$$P = \tfrac{1}{2}(7.07)(3.53)$$
$$= 12.5 \text{ W}$$

In addition, using Ohm's law, we could also employ the expressions

$$P = \frac{1}{2}\frac{V_M^2}{R}$$

or

$$P = \tfrac{1}{2}I_M^2 R$$

where once again we must be careful that the V_M and I_M in these equations refer to the voltage across the resistor and the current through it, respectively. ❑

EXAMPLE 9.3

For the circuit shown in Fig. 9.4, we wish to determine both the total average power absorbed and the total average power supplied.

Figure 9.4 Example circuit for illustrating a power balance.

SOLUTION   From the figure we note that

$$\mathbf{I}_1 = \frac{12 \underline{/45°}}{4} = 3 \underline{/45°} \text{ A}$$

$$\mathbf{I}_2 = \frac{12 \underline{/45°}}{2 - j1} = \frac{12 \underline{/45°}}{2.24 \underline{/-26.57°}} = 5.36 \underline{/71.57°} \text{ A}$$

and therefore,

$$\mathbf{I} = \mathbf{I}_1 + \mathbf{I}_2$$
$$= 3 \underline{/45°} + 5.36 \underline{/71.57°}$$
$$= 8.15 \underline{/62.10°} \text{ A}$$

The average power absorbed in the 4-Ω resistor is

$$P_4 = \tfrac{1}{2}V_M I_M = \tfrac{1}{2}(12)(3) = 18 \text{ W}$$

The average power absorbed in the 2-Ω resistor is

$$P_2 = \tfrac{1}{2}I_M^2 R = \tfrac{1}{2}(5.34)^2(2) = 28.7 \text{ W}$$

Therefore, the total average power absorbed is

$$P_A = 18 + 28.7 = 46.7 \text{ W}$$

Note that we could have calculated the power absorbed in the 2-Ω resistor using $\tfrac{1}{2}V_M^2/R$ if we had first calculated the voltage across the 2-Ω resistor.
  The total average power supplied by the source is

$$
\begin{aligned}
P_S &= \tfrac{1}{2}V_M I_M \cos(\theta_v - \theta_i) \\
&= \tfrac{1}{2}(12)(8.15)\cos(45° - 62.10°) \\
&= 46.7 \text{ W}
\end{aligned}
$$

Thus, the total average power supplied is, of course, equal to the total average power absorbed. ❏

EXTENSION EXERCISES

E9.1   Find the average power adsorbed by each resistor in the network in Fig. E9.1.

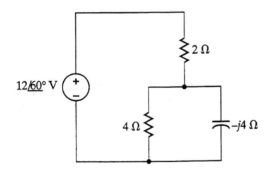

Figure E9.1

ANSWER: $P_{2\Omega} = 7.20 \text{ W}, P_{4\Omega} = 7.20 \text{ W}.$

E9.2   Given the network in Fig. E9.2, find the average power absorbed by each passive circuit element and the total average power supplied by the current source.

ANSWER: $P_{3\Omega} = 56.60 \text{ W}, P_{4\Omega} = 33.96 \text{ W}, P_L = 0, P_{cs} = 90.50 \text{ W}.$

Figure E9.2

When determining average power, if more than one source is present in a network, we can use any of our network analysis techniques to find the necessary voltage and/or current to compute the power. However, we must remember that in general we cannot apply

Superposition is not applicable to power. Why?

superposition to power. We should, however, mention that there is a special case in whi
superposition does apply to power. If the current is of the form

$$i(t) = I_1 \cos(\omega_1 t + \theta_1) + I_2 \cos(\omega_2 t + \theta_2)$$

and if $n\omega_1 = m\omega_2$, where n and m are two different integers, then $n \cdot 2\pi/T_1 = m \cdot 2\pi/$
so that $nT_2 = mT_1 = T$, where $\cos \omega_1 t$ has m full periods in time T and $\cos \omega_2 t$ has n fu
periods in time T. These sinusoids are said to be harmonically related. This subject will b
discussed in detail in Chapter 16.

The average power that is absorbed by resistor R, in a time interval T, is given by

$$P = \frac{1}{T} \int_0^T [I_1 \cos(\omega_1 t + \theta_1) + I_2 \cos(\omega_2 t + \theta_2)]^2 R \, dt$$

where both ω_1 and ω_2 are periodic in T. The preceding equation can be written as

$$
\begin{aligned}
P &= \frac{1}{T} \int_0^T [I_1^2 \cos^2(\omega_1 t + \theta_1) + I_2^2 \cos^2(\omega_2 t + \theta_2) \\
&\quad + 2I_1 I_2 \cos(\omega_1 t + \theta_1) \cos(\omega_2 t + \theta_2)] R \, dt \\
&= \frac{I_1^2}{2} R + \frac{I_2^2}{2} R + \frac{1}{T} \int_0^T \{ I_1 I_2 \cos[(\omega_1 + \omega_2)t + \theta_1 + \theta_2] \\
&\quad + I_1 I_2 \cos[(\omega_1 - \omega_2)t + \theta_1 - \theta_2] \} R \, dt
\end{aligned}
$$

Since $(\omega_1 + \omega_2) = (m + n)(2\pi/T)$ and $(\omega_1 - \omega_2) = (m - n)(2\pi/T)$, the integral will
be zero. Therefore,

$$P = \frac{I_1^2}{2} R + \frac{I_2^2}{2} R$$

Superposition also applies if one, and only one, of the sources is dc.

EXAMPLE 9.4

Consider the network shown in Fig. 9.5. We wish to determine the total average
power absorbed and supplied by each element.

Figure 9.5 Example *RL* circuit
with two sources.

SOLUTION From the figure we note that

$$\mathbf{I}_2 = \frac{12 \,\underline{/30°}}{2} = 6 \,\underline{/30°} \text{ A}$$

and

$$\mathbf{I}_3 = \frac{12 \,\underline{/30°} - 6 \,\underline{/0°}}{j1} = \frac{4.39 + j6}{j1} = 7.44 \,\underline{/-36.21°} \text{ A}$$

The power absorbed by the 2-Ω resistor is

$$P_2 = \tfrac{1}{2} V_M I_M = \tfrac{1}{2}(12)(6) = 36 \text{ W}$$

According to the direction of $\mathbf{I}_3$ the 6 $\underline{/0°}$-V source is absorbing power. The power it absorbs is given by

$$\begin{aligned}
P_{6\,\underline{/0°}} &= \tfrac{1}{2} V_M I_M \cos(\theta_v - \theta_i) \\
&= \tfrac{1}{2}(6)(7.44) \cos[0° - (-36.21°)] \\
&= 18 \text{ W}
\end{aligned}$$

At this point an obvious question arises: How do we know if the 6 $\underline{/0°}$-V source is supplying power to the remainder of the network or absorbing it? The answer to this question is actually straightforward. If we employ our passive sign convention that was adopted in the earlier chapters—that is, if the current reference direction enters the positive terminal of the source and the answer is positive—the source is absorbing power. If the answer is negative, the source is supplying power to the remainder of the circuit. A generator sign convention could have been used, and under this condition the interpretation of the sign of the answer would be reversed. Note that once the sign convention is adopted and used, the sign for average power will be negative only if the angle difference is greater than 90° (i.e., $|\theta_v - \theta_i| > 90°$).

Under the following condition

If $P = IV > 0$, power is being absorbed.
If $P = IV < 0$, power is being generated.

To obtain the power supplied to the network, we compute $\mathbf{I}_1$ as

$$\begin{aligned}
\mathbf{I}_1 &= \mathbf{I}_2 + \mathbf{I}_3 \\
&= 6 \,\underline{/30°} + 7.44 \,\underline{/-36.21°} \\
&= 11.29 \,\underline{/-7.10°} \text{ A}
\end{aligned}$$

Therefore, the power supplied by the 12 $\underline{/30°}$-V source using the generator sign convention is

$$\begin{aligned}
P_S &= \tfrac{1}{2}(12)(11.29) \cos(30° + 7.10°) \\
&= 54 \text{ W}
\end{aligned}$$

and hence the power absorbed is equal to the power supplied. ❏

EXAMPLE 9.5

The current in a 2-Ω resistor (12 $\underline{/30°}$-V source) is of the form

$$i(t) = 4 \cos(377t + 30°) + 2 \cos(754t + 60°)$$

We wish to find the average power absorbed by the resistor.

SOLUTION   Since $\omega_1 = 377$ and $\omega_2 = 754$ (i.e., $\omega_2 = 2\omega_1$), then

$$P = \tfrac{1}{2}(4^2 + 2^2)2 = 20 \text{ W}$$ ❏

EXTENSION EXERCISES

E9.3 Determine the total average power absorbed and supplied by each element in the network in Fig. E9.3.

Figure E9.3

ANSWER: $P_{cs} = -69.4\,\text{W}, P_{VS} = 19.8\,\text{W}, P_{4\Omega} = 49.6\,\text{W}, P_C = 0.$

E9.4 Given the network in Fig. E9.4, determine the total average power absorbed or supplied by each element.

Figure E9.4

ANSWER: $P_{24\,\underline{/0^\circ}} = -55.4\,\text{W}, P_{12\,\underline{/0^\circ}} = 5.5\,\text{W}, P_{2\Omega} = 22.2\,\text{W}, P_{4\Omega} = 27.7\,\text{W}, P_L = 0.$

This impedance matching concept is an important issue in the design of high-speed computer chips and motherboards. For today's high-speed chips with internal clocks running at about 1 GHz and motherboards with a bus speed above 100 MHz, impedance matching is necessary in order to obtain the required speed for signal propagation. Although this high-speed transmission line is based on a distributed circuit (discussed later in electrical engineering courses), the impedance matching technique for the transmission line is the same as that of the lumped parameter circuit for maximum average power transfer.

9.3 MAXIMUM AVERAGE POWER TRANSFER

In our study of resistive networks we addressed the problem of maximum power transfer to a resistive load. We showed that if the network excluding the load was represented by a Thévenin equivalent circuit, maximum power transfer would result if the value of the load resistor was equal to the Thévenin equivalent resistance (i.e., $R_L = R_{Th}$). We will now reexamine this issue within the present context to determine the load impedance for the network shown in Fig. 9.6 that will result in maximum average power being absorbed by the load impedance $\mathbf{Z}_L$.

The equation for average power at the load is

$$P_L = \tfrac{1}{2} V_L I_L \cos(\theta_{V_L} - \theta_{i_L}) \qquad \textbf{9.11}$$

gure 9.6 Circuit used to
amine maximum power
ansfer.

he phasor current and voltage at the load are given by the expressions

$$\mathbf{I}_L = \frac{\mathbf{V}_{oc}}{\mathbf{Z}_{Th} + \mathbf{Z}_L} \qquad \text{9.12}$$

$$\mathbf{V}_L = \frac{\mathbf{V}_{oc}\mathbf{Z}_L}{\mathbf{Z}_{Th} + \mathbf{Z}_L} \qquad \text{9.13}$$

vhere

$$\mathbf{Z}_{Th} = R_{Th} + jX_{Th} \qquad \text{9.14}$$

and

$$\mathbf{Z}_L = R_L + jX_L \qquad \text{9.15}$$

The magnitude of the phasor current and voltage are given by the expressions

$$I_L = \frac{V_{oc}}{\left[(R_{Th} + R_L)^2 + (X_{Th} + X_L)^2\right]^{1/2}} \qquad \text{9.16}$$

$$V_L = \frac{V_{oc}(R_L^2 + X_L^2)^{1/2}}{\left[(R_{Th} + R_L)^2 + (X_{Th} + X_L)^2\right]^{1/2}} \qquad \text{9.17}$$

The phase angles for the phasor current and voltage are contained in the quantity $(\theta_{V_L} - \theta_{i_L})$. Note also that $\theta_{V_L} - \theta_{i_L} = \theta_{\mathbf{Z}_L}$ and, in addition,

$$\cos\theta_{\mathbf{Z}_L} = \frac{R_L}{(R_L^2 + X_L^2)^{1/2}} \qquad \text{9.18}$$

Substituting Eqs. (9.16) to (9.18) into Eq. (9.11) yields

$$P_L = \frac{1}{2}\frac{V_{oc}^2 R_L}{(R_{Th} + R_L)^2 + (X_{Th} + X_L)^2} \qquad \text{9.19}$$

which could, of course, be obtained directly from Eq. (9.16) using $P_L = \frac{1}{2}I_L^2 R_L$. Once again, a little forethought will save us some work. From the standpoint of maximizing P_L, V_{oc} is a constant. The quantity $(X_{Th} + X_L)$ absorbs no power, and therefore any nonzero value of this quantity only serves to reduce P_L. Hence, we can eliminate this term by selecting $X_L = -X_{Th}$. Our problem then reduces to maximizing

$$P_L = \frac{1}{2}\frac{V_{oc}^2 R_L}{(R_L + R_{Th})^2} \qquad \text{9.20}$$

DRILL

D9.2 Determine $\mathbf{Z}_L$ for maximum average power transfer to the load and the maximum average power transferred to the load.

ANSWER: $\mathbf{Z}_L = 1 - j1\,\Omega$ Maximum average power transfer $= \frac{1}{2}$ W.

However, this is the same quantity we maximized in the purely resistive case by selectir $R_L = R_{Th}$. Therefore, for maximum average power transfer to the load shown in Fig.(ffe 9.6, Z_L should be chosen so that

$$\mathbf{Z}_L = R_L + jX_L = R_{Th} - jX_{Th} = \mathbf{Z}_{Th}^*$$ 9.2

Finally, if the load impedance is purely resistive (i.e., $X_L = 0$), the condition for max imum average power transfer can be derived via the expression

$$\frac{dP_L}{dR_L} = 0$$

where P_L is the expression in Eq. (9.19) with $X_L = 0$. *The value of R_L that maximizes P_L under the condition $X_L = 0$ is*

$$R_L = \sqrt{R_{Th}^2 + X_{Th}^2}$$ 9.22

Problem-solving Strategy for Maximum Average Power Transfer

1. Remove the load $\mathbf{Z}_L$ and find the Thévenin equivalent for the remainder of the circuit.
2. Construct the circuit shown in Fig. 9.6.
3. Select $\mathbf{Z}_L = \mathbf{Z}_{Th}^* = R_{Th} - jX_{Th}$, and then $\mathbf{I}_L = \mathbf{V}_{oc}/2 R_{Th}$ and the maximum av-erage power transfer $= \dfrac{1}{2}\mathbf{I}_L^2 R_{Th} = \mathbf{V}_{oc}^2/8 R_{Th}$.

EXAMPLE 9.6

Given the circuit in Fig. 9.7a, we wish to find the value of $\mathbf{Z}_L$ for maximum average power transfer. In addition, we wish to find the value of the maximum average power delivered to the load.

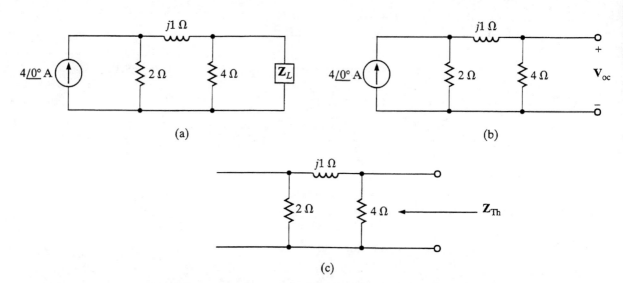

(a)

(b)

(c)

Figure 9.7 Circuits for illustrating maximum average power transfer.

SOLUTION   In order to solve the problem, we form a Thévenin equivalent at the load. The circuit in Fig. 9.7b is used to compute the open-circuit voltage

$$\mathbf{V}_{oc} = \frac{4\ \underline{/0°}\ (2)}{6 + j1}\ (4) = 5.26\ \underline{/-9.46°}\ \text{V}$$

The Thévenin equivalent impedance can be derived from the circuit in Fig. 9.7c. As shown in the figure,

$$\mathbf{Z}_{Th} = \frac{4(2 + j1)}{6 + j1} = 1.41 + j0.43\ \Omega$$

Therefore, $\mathbf{Z}_L$ for maximum average power transfer is

$$\mathbf{Z}_L = 1.41 - j0.43\ \Omega$$

With $\mathbf{Z}_L$ as given previously, the current in the load is

$$\mathbf{I} = \frac{5.26\ \underline{/-9.46°}}{2.82} = 1.87\ \underline{/-9.46°}\ \text{A}$$

Therefore, the maximum average power transferred to the load is

$$P_L = \tfrac{1}{2}I_M^2 R_L = \tfrac{1}{2}(1.87)^2(1.41) = 2.47\ \text{W}$$ ❑

In this Thévenin analysis,

1 Remove $\mathbf{Z}_L$ and find the voltage across the open terminals, $\mathbf{V}_{oc}$.

2 Determine the impedance $\mathbf{Z}_{Th}$ at the open terminals with all independent sources made zero.

3 Construct the following circuit and determine $\mathbf{I}$ and P_L.

EXAMPLE 9.7

For the circuit shown in Fig. 9.8a, we wish to find the value of $\mathbf{Z}_L$ for maximum average power transfer. In addition, let us determine the value of the maximum average power delivered to the load.

(a)   (b)   (c)

Figure 9.8 Circuits for illustrating maximum average power transfer.

SOLUTION   We will first reduce the circuit, with the exception of the load, to a Thévenin equivalent circuit. The open-circuit voltage can be computed from Fig. 9.8b. The equations for the circuit are

$$\mathbf{V}'_x + 4 = (2 + j4)\mathbf{I}_1$$
$$\mathbf{V}'_x = -2\mathbf{I}_1$$

When there is a dependent source, both $\mathbf{V}_{oc}$ and $\mathbf{I}_{sc}$ must be found and $\mathbf{Z}_{Th}$ computed from the equation

$$\mathbf{Z}_{Th} = \frac{\mathbf{V}_{oc}}{\mathbf{I}_{sc}}$$

Solving for $\mathbf{I}_1$, we obtain

$$\mathbf{I}_1 = \frac{1 \underline{/-45°}}{\sqrt{2}}$$

The open-circuit voltage is then

$$\begin{aligned}
\mathbf{V}_{oc} &= 2\mathbf{I}_1 - 4 \underline{/0°} \\
&= \sqrt{2} \underline{/-45°} - 4 \underline{/0°} \\
&= -3 - j1 \\
&= -3.16 \underline{/-161.57°} \text{ V}
\end{aligned}$$

The short-circuit current can be derived from Fig. 9.8c. The equations for this circuit are

$$\begin{aligned}
\mathbf{V}_x'' + 4 &= (2 + j4)\mathbf{I} - 2\mathbf{I}_{sc} \\
-4 &= -2\mathbf{I} + (2 - j2)\mathbf{I}_{sc} \\
\mathbf{V}_x'' &= -2(\mathbf{I} - \mathbf{I}_{sc})
\end{aligned}$$

Solving these equations for $\mathbf{I}_{sc}$ yields

$$\mathbf{I}_{sc} = -(1 + j2) \text{ A}$$

The Thévenin equivalent impedance is then

$$\mathbf{Z}_{Th} = \frac{\mathbf{V}_{oc}}{\mathbf{I}_{sc}} = \frac{3 + j1}{1 + j2} = 1 - j1 \ \Omega$$

Therefore, for maximum average power transfer the load impedance should be

$$\mathbf{Z}_L = 1 + j1 \ \Omega$$

The current in this load $\mathbf{Z}_L$ is then

$$\mathbf{I}_L = \frac{\mathbf{V}_{oc}}{\mathbf{Z}_{Th} + \mathbf{Z}_L} = \frac{-3 - j1}{2} = 1.58 \underline{/-161.57°} \text{ A}$$

Hence, the maximum average power transferred to the load is

$$\begin{aligned}
P_L &= \tfrac{1}{2}(1.58)^2(1) \\
&= 1.25 \text{ W}
\end{aligned}$$

EXTENSION EXERCISES

E9.5   Given the network in Fig. E9.5, find $\mathbf{Z}_L$ for maximum average power transfer and the maximum average power transferred to the load.

ANSWER: $\mathbf{Z}_L = 1 + j1 \ \Omega, P_L = 45 \text{ W.}$

E9.6   Find $\mathbf{Z}_L$ for maximum average power transfer and the maximum average power transferred to the load in the network in Fig. E9.6.

Figure E9.5

Figure E9.6

ANSWER: $\mathbf{Z}_L = 2 - j2 \,\Omega, P_L = 45 \,\text{W}.$

9.4 EFFECTIVE OR RMS VALUES

In the preceding sections of this chapter we have shown that the average power absorbed by a resistive load is directly dependent on the type, or types, of sources that are delivering power to the load. For example, if the source was dc, the average power absorbed was $I^2 R$, and if the source was sinusoidal, the average power was $\frac{1}{2} I_M^2 R$. Although these two types of waveforms are extremely important, they are by no means the only waveforms we will encounter in circuit analysis. Therefore, a technique by which we can compare the *effectiveness* of different sources in delivering power to a resistive load would be quite useful.

In order to accomplish this comparison, we define what is called the *effective value of a periodic waveform*, representing either voltage or current. Although either quantity could be used, we will employ current in the definition. Hence, we define the effective value of a periodic current as a constant or dc value, which as current would deliver the same average power to a resistor R. Let us call the constant current I_{eff}. Then the average power delivered to a resistor as a result of this current is

$$P = I_{\text{eff}}^2 R$$

Similarly, the average power delivered to a resistor by a periodic current $i(t)$ is

$$P = \frac{1}{T} \int_{t_0}^{t_0+T} i^2(t) R \, dt$$

Equating these two expressions, we find that

$$I_{\text{eff}} = \sqrt{\frac{1}{T} \int_{t_0}^{t_0+T} i^2(t) \, dt}$$

9.2

Note that this effective value is found by first determining the *square* of the current, the computing the average or *mean* value, and finally taking the square *root*. Thus in "reading" the mathematical Eq. (9.23), we are determining the root mean square, which we abbreviate as rms, and therefore I_{eff} is called I_{rms}.

Since dc is a constant, the rms value of dc is simply the constant value. Let us now determine the rms value of other waveforms. The most important waveform is the sinusoid and, therefore, we address this particular one in the following example.

EXAMPLE 9.8

We wish to compute the rms value of the waveform $i(t) = I_M \cos(\omega t - \theta)$, which has a period of $T = 2\pi/\omega$.

SOLUTION   Substituting these expressions into Eq. (9.23) yields

$$I_{\text{rms}} = \left[\frac{1}{T} \int_0^T I_M^2 \cos^2(\omega t - \theta) \, dt \right]^{1/2}$$

Using the trigonometric identity

$$\cos^2 \phi = \tfrac{1}{2} + \tfrac{1}{2} \cos 2\phi$$

we find that the preceding equation can be expressed as

$$I_{\text{rms}} = I_M \left\{ \frac{\omega}{2\pi} \int_0^{2\pi/\omega} \left[\tfrac{1}{2} + \tfrac{1}{2} \cos(2\omega t - 2\theta) \right] dt \right\}^{1/2}$$

Since we know that the average or mean value of a cosine wave is zero,

$$I_{\text{rms}} = I_M \left(\frac{\omega}{2\pi} \int_0^{2\pi/\omega} \frac{1}{2} \, dt \right)^{1/2}$$

$$= I_M \left[\frac{\omega}{2\pi} \left(\frac{t}{2} \right) \Big|_0^{2\pi/\omega} \right]^{1/2} = \frac{I_M}{\sqrt{2}}$$

9.24

Therefore, the rms value of a sinusoid is equal to the maximum value divided by the $\sqrt{2}$. Hence, a sinusoidal current with a maximum value of I_M delivers the same average power to a resistor R as a dc current with a value of $I_M/\sqrt{2}$. Recall that earlier a phasor **X** was defined as $X_M \underline{/\theta}$ for a sinusoidal wave of the form $X_m \cos(\omega t + \theta)$. This phasor can also be represented as $X_m/\sqrt{2} \underline{/\theta}$ if the units are given in rms. For example, $120 \underline{/30°}$ V rms is equivalent to $170 \underline{/30°}$ V.   ❏

Upon using the rms values for voltage and current, the average power can be written, in general, as

$$P = V_{\text{rms}} I_{\text{rms}} \cos(\theta_v - \theta_i)$$

9.25

power absorbed by a resistor R is

$$P = I_{rms}^2 R = \frac{V_{rms}^2}{R}$$

9.26

In dealing with voltages and currents in numerous electrical applications, it is impor-
t to know whether the values quoted are maximum, average, rms, or what. For exam-
, the normal 120-V ac electrical outlets have an rms value of 120 V, an average value
0 V, and a maximum value of $120\sqrt{2}$ V.

Finally, if the current in a resistor R is composed of a sum of sinusoidal waves of dif-
rent frequencies, the power absorbed by the resistor can be expressed as

$$P = \left(I_{1rms}^2 + I_{2rms}^2 + \cdots + I_{nrms}^2\right)R$$

9.27

here the rms value of the total current is

$$I_{rms} = \sqrt{I_{1rms}^2 + I_{2rms}^2 + \cdots + I_{nrms}^2}$$

9.28

Each component represents a current of different frequency.

EXAMPLE 9.9

We wish to compute the rms value of the voltage waveform shown in Fig. 9.9.

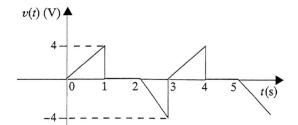

Figure 9.9 Waveform used to
illustrate rms values.

SOLUTION   The waveform is periodic with period $T = 3$ s. The equation for the
voltage in the time frame $0 \le t \le 3$ s is

$$v(t) = \begin{cases} 4t \text{ V} & 0 < t \le 1 \text{ s} \\ 0 \text{ V} & 1 < t \le 2 \text{ s} \\ -4t + 8 \text{ V} & 2 < t \le 3 \text{ s} \end{cases}$$

The rms value is

$$V_{rms} = \left\{ \frac{1}{3}\left[\int_0^1 (4t)^2 \, dt + \int_1^2 (0)^2 \, dt + \int_2^3 (8 - 4t)^2 \, dt \right] \right\}^{1/2}$$

$$= \left[\frac{1}{3}\left(\frac{16t^3}{3}\Big|_0^1 + \left(64t - \frac{64t^2}{2} + \frac{16t^3}{3} \right)\Big|_2^3 \right) \right]^{1/2}$$

$$= 1.89 \text{ V}$$

□

EXAMPLE 9.10

Determine the rms value of the current waveform in Fig. 9.10 and use this value to
compute the average power delivered to a 2-Ω resistor through which this current
is flowing.

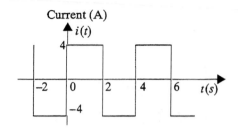

Figure 9.10 Waveform used to illustrate rms values.

SOLUTION The current waveform is periodic with a period of $T = 4$ s. The rms value is

$$I_{rms} = \left\{ \frac{1}{4} \left[\int_0^2 (4)^2 \, dt + \int_2^4 (-4)^2 \, dt \right] \right\}^{1/2}$$

$$= \left[\frac{1}{4} \left(16t \Big|_0^2 + 16t \Big|_2^4 \right) \right]^{1/2}$$

$$= 4 \text{ A}$$

The average power delivered to a 2-Ω resistor with this current is

$$P = I_{rms}^2 R = (4)^2 (2) = 32 \text{ W}$$

EXAMPLE 9.11

We wish to find the rms value of the current

$$i(t) = 12 \sin 377t + 6 \sin(754t + 30°) \text{ A}$$

SOLUTION Since the frequencies of the two sinusoidal waves are different,

$$I_{rms}^2 = I_{1rms}^2 + I_{2rms}^2$$

$$= \left(\frac{12}{\sqrt{2}} \right)^2 + \left(\frac{6}{\sqrt{2}} \right)^2$$

Therefore,

$$I_{rms} = 9.49 \text{ A}$$

EXTENSION EXERCISES

E9.7 Compute the rms value of voltage waveform shown in Fig. E9.7.

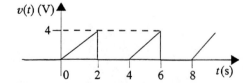

Figure E9.7

ANSWER: $V_{rms} = 1.633$ V.

E9.8   The current waveform in Fig. E9.8 is flowing through a 4-Ω resistor. Compute the average power delivered to the resistor.

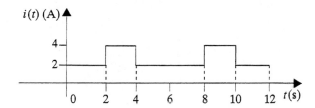

Figure E9.8

ANSWER: $P = 32$ W.

E9.9   The current waveform in Fig. E9.9 is flowing through a 10-Ω resistor. Determine the average power delivered to the resistor.

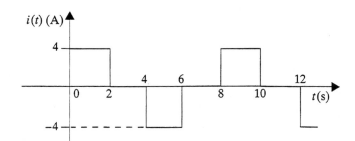

Figure E9.9

ANSWER: $P = 80$ W.

E9.10   The current $i(t)$ is flowing in a 4-Ω resistor. Determine the power absorbed by the resistor if

$$i(t) = 4 \sin(377t - 30°) + 6 \cos(754t - 45°) \text{ A}$$

ANSWER: $P = 104$ W.

9.5 THE POWER FACTOR

The power factor is a very important quantity. Its importance stems in part from the economic impact it has on industrial users of large amounts of power. In this section we carefully define this term and then illustrate its significance via some practical examples.

In Section 9.4 we showed that a load operating in the ac steady state is delivered an average power of

$$P = V_{rms} I_{rms} \cos(\theta_v - \theta_i)$$

We will now further define the terms in this important equation. The product $V_{rms} I_{rms}$ is referred to as the *apparent power*. Although the term $\cos(\theta_v - \theta_i)$ is a dimensionless quantity, and the units of P are watts, apparent power is normally stated in volt-ampe̶r̶ (VA) or kilovolt-amperes (kVA) in order to distinguish it from average power.

We now define the *power factor* (pf) as the ratio of the average power to the appar̶ ent power; that is,

$$\text{pf} = \frac{P}{V_{\text{rms}} I_{\text{rms}}} = \cos(\theta_v - \theta_i) \qquad 9.\text{:}$$

where

$$\cos(\theta_v - \theta_i) = \cos \theta_{\mathbf{Z}_L} \qquad 9.3$$

DRILL

D9.3 If a load impedance is 10 $\underline{/30°}$ Ω, find the power factor.

ANSWER:
pf = cos 30° = $\sqrt{3}/2$.

The angle $\theta_v - \theta_i = \theta_{\mathbf{Z}_L}$ is the phase angle of the load impedance and is often referre̶ to as the *power factor angle*. The two extreme positions for this angle correspond to ̶ purely resistive load where $\theta_{\mathbf{Z}_L} = 0$ and the pf is 1, and the purely reactive load wher̶ $\theta_{\mathbf{Z}_L} = \pm 90°$ and the pf is 0. It is, of course, possible to have a unity pf for a load contain̶ ing R, L, and C elements if the values of the circuit elements are such that a zero phase angle is obtained at the particular operating frequency.

There is, of course, a whole range of power factor angles between $\pm 90°$ and 0°. If the load is an equivalent RC combination, then the pf angle lies between the limits $-90° < \theta_{\mathbf{Z}_L} < 0°$. On the other hand, if the load is an equivalent RL combination, then the pf angle lies between the limits $0 < \theta_{\mathbf{Z}_L} < 90°$. Obviously, confusion in identifying the type of load could result, due to the fact that $\cos \theta_{\mathbf{Z}_L} = \cos(-\theta_{\mathbf{Z}_L})$. To circumvent this problem, the pf is said to be either *leading* or *lagging*, where these two terms *refer to the phase of the current with respect to the voltage*. Since the current leads the voltage in an RC load, the load has a leading pf. In a similar manner, an RL load has a lagging pf; therefore, load impedances of $\mathbf{Z}_L = 1 - j1$ and $\mathbf{Z}_L = 2 + j1$ have power factors of cos $(-45°) = 0.707$ leading and $\cos(26.57°) = 0.894$ lagging, respectively.

EXAMPLE 9.12

An industrial load consumes 88 kW at a pf of 0.707 lagging from a 480-V rms line. The transmission line resistance from the power company's transformer to the plant is 0.08 Ω. Let us determine the power that must be supplied by the power company (a) under present conditions and (b) if the pf is somehow changed to 0.90 lagging. (It is economically advantageous to have a power factor as close to one as possible.)

SOLUTION (a) The equivalent circuit for these conditions is shown in Fig. 9.11. Using Eq. (9.29), we obtain the magnitude of the rms current into the plant:

Technique

1 Given P_L, pf and V rms, determine I rms.

2 Then
$P_s = P_L + I^2 \text{ rms } R_{\text{line}}$, where R_{line} is the line resistance.

$$I_{\text{rms}} = \frac{P_L}{(\text{pf})(V_{\text{rms}})}$$

Figure 9.11 Example circuit for examining changes in power factor.

$$= \frac{(88)(10^3)}{(0.707)(480)}$$

$$= 259.3 \text{ A rms}$$

The power that must be supplied by the power company is

$$P_S = P_L + (0.08)I_{rms}^2$$
$$= 88{,}000 + (0.08)(259.3)^2$$
$$= 93.38 \text{ kW}$$

(b) Suppose now that the pf is somehow changed to 0.90 lagging but the voltage remains constant at 480 V. The rms load current for this condition is

$$I_{rms} = \frac{P_L}{(pf)(V_{rms})}$$

$$= \frac{(88)(10^3)}{(0.90)(480)}$$

$$= 203.7 \text{ A rms}$$

Under these conditions the power company must generate

$$P_S = P_L + (0.08)I_{rms}^2$$
$$= 88{,}000 + (0.08)(203.7)^2$$
$$= 91.32 \text{ kW}$$

Note carefully the difference between the two cases. A simple change in the pf of the load from 0.707 lagging to 0.90 lagging has had an interesting effect. Note that in the first case the power company must generate 93.38 kW in order to supply the plant with 88 kW of power because the low power factor means that the line losses will be high—5.38 kW. However, in the second case the power company need only generate 91.32 kW in order to supply the plant with its required power, and the corresponding line losses are only 3.32 kW.   ❑

The example clearly indicates the economic impact of the load's power factor. A low power factor at the load means that the utility generators must be capable of carrying more current at constant voltage, and they must also supply power for higher $I_{rms}^2 R$ line losses than would be required if the load's power factor were high. Since line losses represent energy expended in heat and benefit no one, the utility will insist that a plant maintain a high pf, typically 0.90 lagging, and adjust its rate schedule to penalize plants that do not conform to this requirement. In the next section we will demonstrate a simple and economical technique for achieving this power factor correction.

EXTENSION EXERCISE

E9.11   An industrial load consumes 100 kW at 0.707 pf lagging. The 60-Hz line voltage at the load is 480 $\underline{/0°}$ V rms. The transmission-line resistance between the

power company transformer and the load is 0.1 Ω. Determine the power savings th⸺
could be obtained if the pf is changed to 0.94 lagging.

ANSWER: Power saved is 3.771 kW.

9.6 COMPLEX POWER

In our study of ac steady-state power, it is convenient to introduce another quantity, whic⸺
is commonly called *complex power*. To develop the relationship between this quantit⸺
and others we have presented in the preceding sections, consider the circuit shown i⸺
Fig. 9.12.

The complex power is defined to be

$$\mathbf{S} = \mathbf{V}_{rms} \mathbf{I}_{rms}^* \qquad \text{9.3}⸺$$

Figure 9.12 Circuit used to explain power relationships.

where $\mathbf{I}_{rms}^*$ refers to the complex conjugate of $\mathbf{I}_{rms}$; that is, if $\mathbf{I}_{rms} = I_{rms} \underline{/\theta_i} = I_R + jI_I$,
then $\mathbf{I}_{rms}^* = I_{rms} \underline{/-\theta_i} = I_R - jI_I$. Complex power is then

$$\mathbf{S} = V_{rms} \underline{/\theta_v}\, I_{rms} \underline{/-\theta_i} = V_{rms} I_{rms} \underline{/\theta_v - \theta_i} \qquad \text{9.32}$$

or

$$\mathbf{S} = V_{rms} I_{rms} \cos(\theta_v - \theta_i) + jV_{rms} I_{rms} \sin(\theta_v - \theta_i) \qquad \text{9.33}$$

where, of course, $\theta_v - \theta_i = \theta_{\mathbf{Z}}$. We note from Eq. (9.33) that the real part of the complex
power is simply the *real* of *average power*. The imaginary part of $\mathbf{S}$ we call the *reactive* or
quadrature power. Therefore, complex power can be expressed in the form

$$\mathbf{S} = P + jQ \qquad \text{9.34}$$

where

$$P = \text{Re}(\mathbf{S}) = V_{rms} I_{rms} \cos(\theta_v - \theta_i) \qquad \text{9.35}$$
$$Q = \text{Im}(\mathbf{S}) = V_{rms} I_{rms} \sin(\theta_v - \theta_i) \qquad \text{9.36}$$

As shown in Eq. (9.33), the magnitude of the complex power is what we have called the
apparent power, and the phase angle for complex power is simply the power factor angle.
Complex power, like apparent power, is measured in volt-amperes, real power is measured in watts, and in order to distinguish Q from the other quantities, which in fact have
the same dimensions, it is measured in volt-amperes reactive, or var.

In addition to the relationships expressed previously, we note that since

$$\cos(\theta_v - \theta_i) = \frac{\text{Re}(\mathbf{Z})}{Z}$$

$$\sin(\theta_v - \theta_i) = \frac{\text{Im}(\mathbf{Z})}{Z}$$

$$I_{rms} = \frac{V_{rms}}{Z}$$

. (9.35) and (9.36) can be written as

$$P = I_{rms}^2 \operatorname{Re}(\mathbf{Z}) \qquad \qquad \textbf{9.37}$$

$$Q = I_{rms}^2 \operatorname{Im}(\mathbf{Z}) \qquad \qquad \textbf{9.38}$$

d therefore, Eq. (9.34) can be expressed as

$$\mathbf{S} = I_{rms}^2 \mathbf{Z} \qquad \qquad \textbf{9.39}$$

D9.4 If $\mathbf{I}_{rms} = 10\ \underline{/30°}$ A and $\mathbf{Z} = 10\ \underline{/45°}\ \Omega$, as shown in Fig. 9.12, find $\mathbf{S}$ and Q.

ANSWER:
$\mathbf{S} = 1000\ \underline{/45°}$ VA
$= (500\sqrt{2} + j500\sqrt{2})$ VA
$Q = 500\sqrt{2}$ var

The diagrams in Fig. 9.13 serve to explain further the relationships among the various antities of power. As shown in Fig. 9.13a, the phasor current can be split into two components: one that is in phase with $\mathbf{V}_{rms}$ and one that is 90° out of phase with $\mathbf{V}_{rms}$. Equations (9.35) and (9.36) illustrate that the in-phase component produces the real power, and e 90° component, called the *quadrature component*, produces the reactive or quadrature ower. In addition, Eqs. (9.34) to (9.36) indicate that

$$\tan(\theta_v - \theta_i) = \frac{Q}{P} \qquad \qquad \textbf{9.40}$$

which relates the pf angle to P and Q in what is called the *power triangle*.

The relationships among $\mathbf{S}$, P, and Q can be expressed via the diagrams shown in Figs. 9.13b and c. In Fig. 9.13b we note the following conditions. If Q is positive, the load is inductive, the power factor is lagging, and the complex number $\mathbf{S}$ lies in the first quadrant. If Q is negative, the load is capacitive, the power factor is leading, and the complex number $\mathbf{S}$ lies in the fourth quadrant. If Q is zero, the load is resistive, the power factor is unity, and the complex number $\mathbf{S}$ lies along the positive real axis. Figure 9.13c illustrates the relationships expressed by Eqs. (9.37) to (9.39) for an inductive load.

Finally, it is important to state that complex power, like energy, is conserved; that is, the total complex power delivered to any number of individual loads is equal to the sum of the complex powers delivered to each individual load, regardless of how the loads are interconnected.

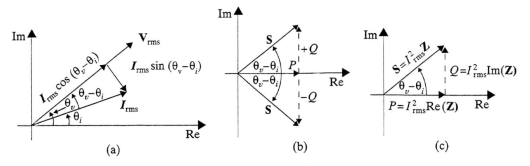

Figure 9.13 Diagram for illustrating power relationships.

Problem-solving strategies involving P or S

If $v(t)$ and $i(t)$ are known and we wish to find P given an impedance $\mathbf{Z}\,\underline{/\theta} = R + jX$, t viable approaches are as follows:

1   Determine $\mathbf{V}$ and $\mathbf{I}$ and then calculate

$$P = VI \cos \theta \quad \text{or} \quad P = VI \cos(\underline{/\mathbf{V}} - \underline{/\mathbf{I}})$$

2   Use $\mathbf{I}$ to calculate the real part of $\mathbf{S}$; that is,

$$P = R_e(\mathbf{S}) = I^2 R$$

The latter method may be easier to calculate than the former. However, if the imaginar part of the impedance, X, is not zero, then

$$P \neq \frac{V^2}{R}$$

which is a common mistake. Furthermore, the P and Q portions of $\mathbf{S}$ are directly relat ed to $Z\,\underline{/\theta}$ and provide a convenient way in which to relate power, current, and imped ance. That is,

$$\tan \theta = \frac{Q}{P}$$

$$\mathbf{S} = I^2 \mathbf{Z}$$

The following example illustrates the usefulness of $\mathbf{S}$.

EXAMPLE 9.13

A load operates at 20 kW, 0.8 pf lagging. The load voltage is $220\,\underline{/0°}$ V rms at 60 Hz. The impedance of the line is $0.09 + j0.3\ \Omega$. We wish to determine the voltage and power factor at the input to the line.

SOLUTION   The circuit diagram for this problem is shown in Fig. 9.14. As illus trated in Fig. 9.13,

$$S = \frac{P}{\cos \theta} = \frac{P}{\text{pf}} = \frac{20{,}000}{0.8} = 25{,}000 \text{ VA}$$

Technique

1   Use the given P_L, $\cos \theta$, and V_L rms to obtain S_L and I_L based on Eqs. (9.35) and (10.31), respectively.

2   Use I_L and Z_{line} to obtain S_{line} using Eq. (9.39).

3   Use $\mathbf{S}_s = \mathbf{S}_{\text{line}} + \mathbf{S}_L$.

4   $\mathbf{V}_s = \mathbf{S}_s/\mathbf{I}_L^*$ yields V_s and θ_v Since $\mathbf{V}_s = V_s\,\underline{/\theta_v}$ and θ_v is the phase of I_L, pf $= \cos(\theta_v - \theta_i)$.

Therefore, at the load

$$\mathbf{S}_L = 25{,}000\,\underline{/\theta} = 25{,}000\,\underline{/36.87°} = 20{,}000 + j15{,}000 \text{ VA}$$

Figure 9.14   Example circuit for power analysis.

Since $\mathbf{S}_L = \mathbf{V}_L \mathbf{I}_L^*$

$$\mathbf{I}_L = \left[\frac{25{,}000 \, \underline{/36.87^\circ}}{220 \, \underline{/0^\circ}} \right]^*$$

$$= 113.64 \, \underline{/36.87^\circ} \text{ A rms}$$

The complex power losses in the line are

$$\mathbf{S}_{\text{line}} = I_L^2 \mathbf{Z}_{\text{line}}$$

$$= (113.64)^2 (0.09 + j03)$$

$$= 1162.26 + j3874.21 \text{ VA}$$

As stated earlier, complex power is conserved and, therefore, the complex power at the generator is

$$\mathbf{S}_S = \mathbf{S}_L + \mathbf{S}_{\text{line}}$$

$$= 21{,}162.26 + j18{,}874.21$$

$$= 28{,}356.25 \, \underline{/41.73^\circ} \text{ VA}$$

Hence, the generator voltage is

$$V_L = \frac{|\mathbf{S}_L|}{I_L} = \frac{28{,}356.25}{113.64}$$

$$= 249.53 \text{ V rms}$$

and the generator power factor is

$$\cos(41.73^\circ) = 0.75 \text{ lagging}$$

We could have solved this problem using KVL. For example, we calculated the load current as

$$\mathbf{I}_L = 113.64 \, \underline{/-36.87^\circ} \text{ A rms}$$

Hence, the voltage drop in the transmission line is

$$\mathbf{V}_{\text{line}} = \left(113.64 \, \underline{/-36.87^\circ} \right)(0.09 + j0.3)$$

$$= 35.59 \, \underline{/36.43^\circ} \text{ V rms}$$

Therefore, the generator voltage is

$$\mathbf{V}_S = 220 \, \underline{/0^\circ} + 35.59 \, \underline{/36.43^\circ}$$

$$= 249.53 \, \underline{/4.86^\circ} \text{ V rms}$$

Hence, the generator voltage is 249.53 V rms. In addition,

$$\theta_v - \theta_i = 4.86^\circ - (-36.87^\circ) = 41.73^\circ$$

and therefore,

$$\text{pf} = \cos(41.73^\circ) = 0.75 \text{ lagging} \qquad \square$$

EXAMPLE 9.14

Two networks A and B are connected by two conductors having a net impedance of $\mathbf{Z} = 0 + j1\Omega$, as shown in Fig. 9.15. The voltages at the terminals of the networks

Figure 9.15 Network used in Example 9.14.

are $\mathbf{V}_A = 120\ \underline{/30°}$ V rms and $\mathbf{V}_B = 120\ \underline{/0°}$ V rms. We wish to determine the average power flow between the networks and identify which is the source and which is the load.

SOLUTION As shown in Fig. 9.15,

$$\mathbf{I} = \frac{\mathbf{V}_A - \mathbf{V}_B}{\mathbf{Z}}$$

$$= \frac{120\ \underline{/30°} - 120\ \underline{/0°}}{j1}$$

$$= 62.12\ \underline{/15°}\ \text{A rms}$$

The power delivered by network A is

$$P_A = |\mathbf{V}_A||\mathbf{I}|\cos(\theta_{\mathbf{V}_A} - \theta_{\mathbf{I}})$$
$$= (120)(62.12)\cos(30° - 15°)$$
$$= 7200.4\ \text{W}$$

The power absorbed by network B is

$$P_B = |\mathbf{V}_B||\mathbf{I}|\cos(\theta_{\mathbf{V}_B} - \theta_{\mathbf{I}})$$
$$= (120)(62.12)\cos(0° - 15°)$$
$$= 7200.4\ \text{W}$$

If the power flow had actually been from network B to network A, the resultant signs on P_A and P_B would have been negative. ❏

EXTENSION EXERCISES

E9.12 An industrial load requires 40 kW at 0.84 pf lagging. The load voltage is $220\ \underline{/0°}$ V rms at 60 Hz. The transmission-line impedance is $0.1 + j0.25\ \Omega$. Determine the real and reactive power losses in the line and the real reactive power required at the input to the transmission line.

ANSWER: $P_{\text{line}} = 4.685$ kW, $Q_{\text{line}} = 11.713$ kvar, $P_S = 44.685$ kW, $Q_S = 45.31$ kvar.

E9.13 A load requires 60 kW at 0.85 pf lagging. The 60-Hz line voltage at the load is $220\ \underline{/0°}$ V rms. If the transmission-line impedance is $0.12 + j0.18\ \Omega$, determine the line voltage and power factor at the input.

ANSWER: $\mathbf{V}_{\text{in}} = 284.6\ \underline{/5.8°}$ V rms, $\text{pf}_{\text{in}} = 0.792$ lagging.

POWER FACTOR CORRECTION

~~ustrial~~ plants that require large amounts of power have a wide variety of loads. How-
~~r,~~ by nature the loads normally have a lagging power factor. In view of the results ob-
~~ned~~ in Example 9.12, we are naturally led to ask if there is any convenient technique
~~for~~ raising the power factor of a load. Since a typical load may be a bank of induction
~~tors~~ or other expensive machinery, the technique for raising the pf should be an eco-
~~mical~~ one in order to be feasible.

To answer the question we pose, let us examine the circuit shown in Fig. 9.16. The cir-
~~it~~ illustrates a typical industrial load. In parallel with this load we have placed a capac-
~~r.~~ The original complex power for the load $\mathbf{Z}_L$, which we will denote as $\mathbf{S}_{old}$, is

$$\mathbf{S}_{old} = P_{old} + jQ_{old} = |\mathbf{S}_{old}| \,\underline{/\theta_{old}}$$

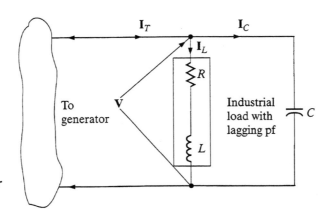

Figure 9.16 Circuit for power factor correction.

The complex power for the capacitor is

$$\mathbf{S}_{cap} = 0 + jQ_{cap} = |\mathbf{S}_{cap}| \,\underline{/-90°}$$

The new complex power that results from adding a capacitor is

$$\mathbf{S}_{old} + \mathbf{S}_{cap} = \mathbf{S}_{new} = P_{old} + jQ_{new} = |\mathbf{S}_{new}| \,\underline{/\theta_{new}} \qquad \textbf{9.41}$$

where θ_{new} is specified by the required power factor. The difference between the new and
old complex powers is caused by the addition of the capacitor and since the capacitor is
purely reactive,

$$\mathbf{S}_{cap} = +jQ_{cap} = -j\omega C V_{rms}^2 \qquad \textbf{9.42}$$

Equation (9.42) can be used to find the required value of C in order to achieve the new
specified power factor. The procedure we have described is illustrated in Fig. 9.17, where
$\theta_{old} = \theta_{v_L} - \theta_{i_L}$ and $\theta_{new} = \theta_{v_T} - \theta_{i_T}$. Hence, we can obtain a particular power factor for
the total load simply by a judicious selection of a capacitor and placing it in parallel with

the original load. In general, we want the power factor to be large, and therefore the po[
factor angle must be small (i.e., the larger the desired power factor, the smaller the ar
$\theta_{v_T} - \theta_{i_T}$).

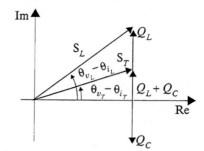

Figure 9.17 Illustration of the technique for power factor correction.

EXAMPLE 9.15

Plastic kayaks are manufactured using a process called rotomolding, which is dia
grammed in Fig. 9.18. Molten plastic is injected into a mold, which is then spun or
the long axis of the kayak until the plastic cools, resulting in a hollow one-piec[
craft. Suppose that the induction motors used to spin the molds consume 50 kW a[
a pf of 0.8 lagging from a 220 $\underline{/0°}$-V rms, 60-Hz line. We wish to raise the pf to 0.95
lagging by placing a bank of capacitors in parallel with the load.

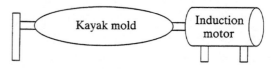

Figure 9.18 Rotomolding manufacturing process.

Technique

1 Find Q_{old} from P_L and θ_{old}, or the equivalent pf_{old}.

2 Find θ_{new} from the desired pf_{new}.

3 Determine $Q_{new} = P_{old} \tan \theta_{new}$.

4 $Q_{new} - Q_{old} = Q_{cap} = -\omega C V^2$ rms.

SOLUTION The circuit diagram for this problem is shown in Fig. 9.19. $P_L = 50\,kW$ and since $\cos^{-1} 0.8 = 36.87°$, $\theta_{old} = 36.87°$. Therefore,

$$Q_{old} = P_{old} \tan \theta_{old} = (50)(10^3)(0.75) = 37.5 \text{ kvar}$$

Hence,

$$\mathbf{S}_{old} = P_{old} + jQ_{old} = 50,000 + j37,500$$

and

$$\mathbf{S}_{cap} = 0 + jQ_{cap}$$

Since the required power factor is 0.95.

Figure 9.19 Example circuit for power factor correction.

Furthermore,

$$\theta_{new} = \cos^{-1}(pf_{new}) = \cos^{-1}(0.95)$$
$$= 18.19°$$

Then

$$Q_{new} = P_{old} \tan \theta_{new}$$
$$= 50,000 \tan(18.19°)$$
$$= 16,430 \text{ var}$$

Hence

$$Q_{new} - Q_{old} = Q_{cap} = -\omega CV^2 \text{ rms}$$
$$16,430 - 37,500 = -\omega CV^2 \text{ rms}$$

Solving the equation for C yields

$$C = \frac{21,070}{(377)(220)^2}$$
$$= 1155 \text{ } \mu F$$

By using a capacitor of this magnitude in parallel with the industrial load, we create, from the utility's perspective, a load pf of 0.95 lagging. However, the parameters of the actual load remain unchanged. Under these conditions, the current supplied by the utility to the kayak manufacturer is less and therefore they can use smaller conductors for the same amount of power. Or, if the conductor size is fixed, the line losses will be less since these losses are a function of the square of the current.   ❑

EXTENSION EXERCISE

E9.14   Compute the value of the capacitor necessary to change the power factor in Extension Exercise E9.11 to 0.95 lagging.

ANSWER: $C = 773 \text{ } \mu F$.

9.8 POWER MEASUREMENTS

In the preceding sections we have illustrated techniques for computing power. We will now show how power is actually measured in an electrical network. An instrument used to measure average power is the wattmeter. This instrument contains a low-impedance current coil (which ideally has zero impedance) that is connected in series with the load, and a high-impedance voltage coil (which ideally has infinite impedance) that is connected

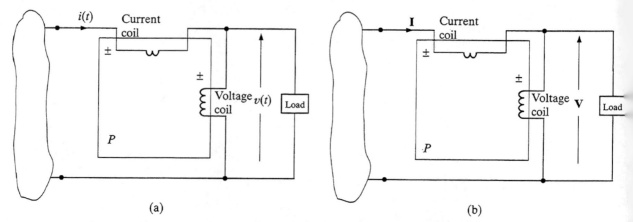

Figure 9.20 (a) and (b) Wattmeter connections for power measurement.

across the load. If the voltage and current are periodic and the wattmeter is connected as shown in Fig. 9.20a, it will read

$$P = \frac{1}{T} \int_0^T v(t)i(t) \, dt$$

where $v(t)$ and $i(t)$ are defined on the figure. Note that $i(t)$ is referenced entering the $\pm$ terminal of the current coil and $v(t)$ is referenced positive at the $\pm$ terminal of the voltage coil.

In addition, the $\pm$ terminal of the current coil in Fig. 9.20a is usually connected to the source of electric power and the $\pm$ terminal of the voltage coil is normally connected to the line that links the current terminal with no sign. When the current and voltage coils are connected in the configuration shown, the wattmeter will provide an upscale reading.

In the frequency domain the equivalent circuit is shown in Fig. 9.20b, where the current and voltage are referenced the same as in the time domain and the wattmeter reading will be

$$P = \text{Re}(\mathbf{V}\mathbf{I}^*) = VI \cos(\theta_v - \theta_i)$$

If $v(t)$ and $i(t)$ or $\mathbf{V}$ and $\mathbf{I}$ are correctly chosen, the reading will be average power. In Fig. 9.20 the connections will produce a reading of power delivered to the load. Since the two coils are completely isolated from one another within the meter, they could be connected anywhere in the circuit and the reading may or may not have meaning.

Note that if one of the coils on the wattmeter is reversed, the equations for the power are the negative of what they were before the coil reversal due to the change in the variable reference as related to the $\pm$ terminal. Due to the physical construction of wattmeters, the $\pm$ terminal of the potential coil should always be connected to the same line as the current coil, as shown in Fig. 9.21a. If it becomes necessary to reverse a winding to produce an upscale reading, either coil can be reversed. For example, reversing the current coil results in the network shown in Fig. 9.21b. Note that if the $\pm$ terminal of the

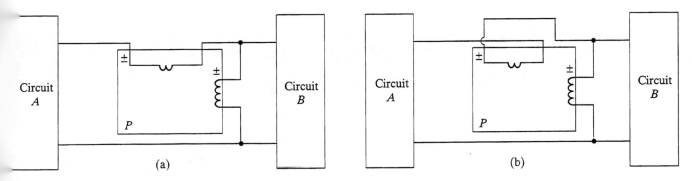

Figure 9.21 Wattmeter connections.

potential coil is connected to the line containing the current coil and the meter is reading upscale, the power, P, is flowing through the wattmeter from circuit A to circuit B.

EXAMPLE 9.16

Given the network shown in Fig. 9.22, we wish to determine the wattmeter reading.

Figure 9.22 Use of the wattmeter in power measurement.

SOLUTION   Using any one of a number of techniques (e.g., Thévenin's theorem), we can show that

$$\mathbf{I} = 2.68 \ \underline{/153.43°} \ \text{A rms}$$

$$\mathbf{V} = 11.99 \ \underline{/216.86°} \ \text{V rms}$$

and therefore the wattmeter reads

$$P = \text{Re}(\mathbf{VI}^*) = |\mathbf{V}||\mathbf{I}| \cos(\theta_v - \theta_i)$$

$$= (11.99)(2.68) \cos(216.86° - 153.43°)$$

$$= 14.37 \ \text{W}$$

the average power delivered to the impedance $(2 + j4) \ \Omega$.   ❑

EXTENSION EXERCISE

E9.15 Determine the wattmeter reading in the network in Fig. E9.15.

Figure E9.15

ANSWER: $P = 1440$ W.

Modern electronics technology has reduced the size and enhanced the capabilities of all types of measurement equipment. For example, the handheld power meter shown in Fig. 9.23 can be used to measure and display voltage, current, and power.

9.9 SINGLE-PHASE THREE-WIRE CIRCUITS

The single-phase three-wire ac circuit shown in Figure 9.24 is an important topic because it is the typical ac power network found in households. Note that the voltage sources are equal; that is, $\mathbf{V}_{an} = \mathbf{V}_{nb} = \mathbf{V}$. Thus, the magnitudes are equal and the phases are equal (single phase). The line-to-line voltage $\mathbf{V}_{ab} = 2\mathbf{V}_{an} = 2\mathbf{V}_{nb} = 2\mathbf{V}$. Within a household, lights and small appliances are connected from one line to *neutral n*, and large appliances such as hot water heaters and air conditioners are connected line to line. Lights operate at about 120 V rms and large appliances operate at approximately 240 V rms.

Let us now attach two identical loads to the single-phase three-wire voltage system using perfect conductors as shown in Fig. 9.24b. From the figure we note that

$$\mathbf{I}_{aA} = \frac{\mathbf{V}}{\mathbf{Z}_L}$$

and

$$\mathbf{I}_{bB} = -\frac{\mathbf{V}}{\mathbf{Z}_L}$$

Figure 9.23 A Handheld Power Meter (Courtesy of Fluke Corporation).

D9.5   In Fig. 9.24b, if
$\mathbf{V} = 120\ \underline{/0°}$ V and
$\mathbf{Z}_L = 100 + j1\ \Omega$, find $\mathbf{I}_{nN}$.

ANSWER: $\mathbf{I}_{nN} = 0$.

D9.6   In Fig. 9.24d, if

$$\mathbf{V} = 120\ \underline{/0°}\,\text{V}$$
$$\mathbf{Z}_L = 100 + j1\ \Omega$$
$$\mathbf{Z}_{LL} = 10 + j0.1\ \Omega$$
$$\mathbf{Z}_{\text{line}} = 0$$
$$\mathbf{Z}_n = 0$$

find $\mathbf{I}_a$.

ANSWER: $13.2\ \underline{/0.57°}$ A.

Figure 9.24   Single-phase three-wire system.

KCL at point N is

$$\mathbf{I}_{nN} = -\left(\mathbf{I}_{aA} + \mathbf{I}_{bB}\right)$$
$$= -\left(\frac{\mathbf{V}}{\mathbf{Z}_L} - \frac{\mathbf{V}}{\mathbf{Z}_L}\right)$$
$$= 0$$

Note that there is no current in the neutral wire, and therefore it could be removed without affecting the remainder of the system; that is, all the voltages and currents would be unchanged. One is naturally led to wonder just how far the simplicity exhibited by this system will extend. For example, what would happen if each line had a line impedance, if the neutral conductor had an impedance associated with it, and if there were a load tied from line to line? To explore these questions, consider the circuit in Fig. 9.24c. Although we could examine this circuit using many of the techniques we have employed in previous chapters, the symmetry of the network suggests that perhaps superposition may lead us to some conclusions without having to resort to a brute-force assault. Employing superposition, we consider the two circuits in Figs. 9.24d and e. The currents in Fig. 9.24d are labeled arbitrarily. Because of the symmetrical relationship between Figs. 9.24d and e, the currents in Fig. 9.24e correspond directly to those in Fig. 9.24d. If we add the two *phasor* rents in each branch, we find that the neutral current is again zero. A neutral current zero is a direct result of the symmetrical nature of the network. If either the line impedances $\mathbf{Z}_{line}$ or the load impedances $\mathbf{Z}_L$ are unequal, the neutral current will be nonze-. We will make direct use of these concepts when we study three-phase networks in hapter 10.

EXAMPLE 9.17

A three-wire single-phase household circuit is shown in Fig. 9.25a. The use of the lights, stereo, and range for a 24-hour period is demonstrated in Fig. 9.25b. Let us calculate the energy use over the 24 hours in kilowatt-hours. Assuming that this represents a typical day and that our utility rate is \$0.08/kWh, let us also estimate the power bill for a 30-day month.

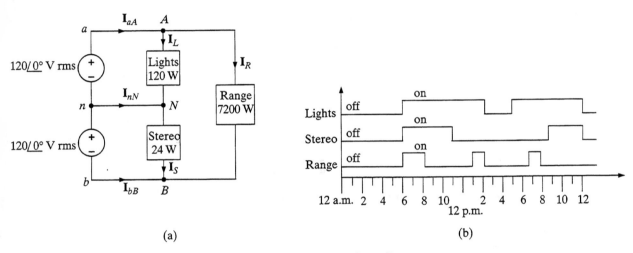

(a)   (b)

Figure 9.25 Household three-wire network and appliance usage.

SOLUTION   Applying nodal analysis to Fig. 9.25a yields

$$\mathbf{I}_{aA} = \mathbf{I}_L + \mathbf{I}_R$$

$$\mathbf{I}_{bB} = -\mathbf{I}_S - \mathbf{I}_R$$

$$\mathbf{I}_{nN} = \mathbf{I}_S - \mathbf{I}_L$$

The current magnitudes for each load can be found from the corresponding power levels as follows.

$$I_L = \frac{P_L}{V_{an}} = \frac{120}{120} = 1 \text{ A rms}$$

$$I_S = \frac{P_S}{V_{nb}} = \frac{24}{120} = 0.2 \text{ A rms}$$

$$I_R = \frac{P_R}{V_{ab}} = \frac{7200}{240} = 30 \text{ A rms}$$

The energy used is simply the integral of the power delivered by the two sources ove[r] the 24-hour period. Since the voltage magnitudes are constants, we can express th[e] energy delivered by the sources as

$$E_{an} = V_{an} \int I_{aA} \, dt$$

$$E_{nb} = V_{nb} \int -I_{bB} \, dt$$

The integrals of I_{aA} and I_{bB} can be determined graphically from Fig. 9.25b.

$$\int_{12a.m.}^{12a.m.} I_{aA} \, dt = 4I_R + 15I_L = 135$$

$$\int_{12a.m.}^{12a.m.} -I_{bB} \, dt = 8I_S + 4I_R = 121.6$$

Therefore, the daily energy for each source and the total energy is

$$E_{an} = 16.2 \text{ kWh}$$

$$E_{nb} = 14.6 \text{ kWh}$$

$$E_{total} = 30.8 \text{ kWh}$$

Over a 30-day month, a \$0.08/kWh utility rate results in a power bill of

$$\text{Cost} = (30.8)(30)(0.08) = \$73.92$$

9.10 SAFETY CONSIDERATIONS

Although this book is concerned primarily with the theory of circuit analysis, we recognize that most students, by this point in their study, will have begun to relate the theory to the electrical devices and systems that they encounter in the world around them. Thus, it seems advisable to depart briefly from the theoretical and spend some time discussing the very practical and important subject of safety. Electrical safety is a very broad and diverse topic that would require several volumes for a comprehensive treatment. Instead, we will limit our discussion to a few introductory concepts and illustrate them with examples.

It would be difficult to imagine that anyone in our society could have reached adolescence without having experienced some form of electrical shock. Whether that shock was from a harmless electrostatic discharge or from accidental contact with an energized electrical circuit, the response was probably the same—an immediate and involuntary muscular reaction. In either case, the cause of the reaction is current flowing through the body. The severity of the shock depends on several factors, the most important of which are the magnitude, the duration, and the pathway of the current through the body.

The effect of electrical shock varies widely from person to person. Figure 9.26 shows the general reactions that occur as a result of 60-Hz ac current flow through the body from hand to hand, with the heart in the conduction pathway. Observe that there is an intermediate range of current, from about 0.1 to 0.2 A, which is most likely to be fatal. Current levels in this range are apt to produce ventricular fibrillation, a disruption of the

erly contractions of the heart muscle. Recovery of the heartbeat generally does not
ur without immediate medical intervention. Current levels above that fatal range tend
ause the heart muscle to contract severely, and if the shock is removed soon enough,
heart may resume beating on its own.

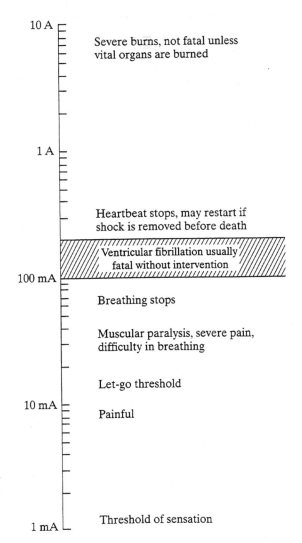

Figure 9.26 Effects of electrical shock. (From C. F. Dalziel and W. R. Lee, "Lethal Electric Currents," *IEEE Spectrum*, February 1969, pp. 44–50, and C. F. Dalziel, "Electric Shock Hazard," *IEEE Spectrum*, February 1972, pp. 41–50.)

The voltage required to produce a given current depends on the quality of the contact to the body and the impedance of the body between the points of contact. The electrostatic voltage such as might be produced by sliding across a car seat on a dry winter day may be on the order of 20,000 to 40,000 V, and the current surge upon touching the door handle, on the order of 40 A. However, the pathway for the current flow is mainly over the body surface, and its duration is for only a few microseconds. Although that shock could be disastrous for some electronic components, it causes nothing more than mild discomfort and aggravation to a human being.

Electrical appliances found about the home typically require 120 or 240 V rms for operation. Although the voltage level is small compared with that of the electrostatic shock, the potential for harm to the individual and to property is much greater. Accidental ctact is more apt to result in current flow either from hand to hand or from hand to fooeither of which will subject the heart to shock. Moreover, the relatively slowly chang(low-frequency) 60-Hz current tends to penetrate more deeply into the body as oppoto remaining on the surface as a rapidly changing (high-frequency) current would tenddo. In addition, the energy source has the capability of sustaining a current flow withodepletion. Thus, subsequent discussion will concentrate primarily on hazards associatwith the 60-Hz ac power system.

The single-phase three-wire system introduced earlier is commonly, although nexclusively, used for electrical power distribution in residences. Two important aspecof this, or any, system that relate to safety were not mentioned earlier: circuit fusinand grounding.

Each branch circuit, regardless of the type of load it serves, is protected from excesive current flow by circuit breakers or fuses. Receptacle circuits are generally limiteto 20 amps and lighting circuits to 15 amps. Clearly, these cannot protect persons frolethal shock. The primary purpose of these current-limiting devices is to protect equipment

The neutral conductor of the power system is connected to ground (earth) at a multitude of points throughout the system and, in particular, at the service entrance to the residence. The connection to earth may be by way of a driven ground rod or by contact to acold water pipe of a buried metallic water system. The 120-V branch circuits radiatingfrom the distribution panel (fuse box) generally consist of three conductors rather thanonly two, as was shown in Fig. 9.24. The third conductor is the ground wire, as shown inFig. 9.27.

The ground conductor may appear to be redundant, since it plays no role in the normal operation of a load that might be connected to the receptacle. Its role is illustratedby the following example.

Figure 9.27 A household receptacle.

EXAMPLE 9.18

Joe has a workshop in his basement where he uses a variety of power tools such as drills, saws, and sanders. The basement floor is concrete, and being below ground level, it is usually damp. Damp concrete is a relatively good conductor. Unknown to Joe, the insulation on a wire in his electric drill has been nicked and the wire is in contact with (or shorted to) the metal case of the drill, as shown in Fig. 9.28. Is Joe in any danger when using the drill?

Figure 9.28 Faulty circuit, when the tool does not have the case grounded through the power cord.

SOLUTION   Without the ground conductor connected to the metal case of the tool, Joe would receive a severe, perhaps fatal, shock when he attempts to use the drill. The voltage between his hand and his feet would be 120 V, and the current through his body would be limited by the resistance of his body and of the concrete floor. Typically, the circuit breakers would not operate. However, if the ground conductor is present and properly connected to the drill case, the case will remain at ground potential, the 120-V source becomes shorted to ground, the circuit breaker operates, and Joe lives to drill another hole.   ❏

It was mentioned earlier that the circuit breaker or fuse cannot provide effective protection against shock. There is, however, a special type of device called a ground-fault interrupter (GFI) that can provide protection for personnel. This device detects current flow outside the normal circuit. Consider the circuit of Fig. 9.28. In the normal safe operating condition, the current in the neutral conductor must be the same as that in the line conductor. If at anytime the current in the line does not equal the current in the neutral, then a secondary path has somehow been established, creating an unsafe condition. This secondary path is called a fault. For example, the fault path in Fig. 9.28 is through Joe and the concrete floor. The GFI detects this fault and opens the circuit in response. Its principle of operation is illustrated by the following example.

EXAMPLE 9.19

Let us describe the operation of a GFI.

SOLUTION   Consider the action of the magnetic circuit in Fig. 9.29. Under normal operating conditions, i_1 and i_2 are equal, and if the coils in the neutral and line conductors are identical, as we learned in basic physics, the magnetic flux in the core will be zero. Consequently, no voltage will be induced in the sensing coil.

Figure 9.29 Ground-fault interrupter circuit.

If a fault should occur at the load, current will flow in the ground conductor and perhaps in the earth; thus, i_1 and i_2 will no longer be equal, the magnetic flux will not be zero, and a voltage will be induced in the sensing coil. That voltage can be used to activate a circuit breaker. This is the essence of the GFI device. ❏

Ground-fault interrupters are available in the form of circuit breakers and also as receptacles. They are now required in branch circuits that serve outlets in areas such as bathrooms, basements, garages, and outdoor sites. The devices will operate at ground-fault currents on the order of a few milliamperes. Unfortunately, the GFI is a relatively new device and electrical code requirements are generally not retroactive. Thus few older residences have them.

Requirements for the installation and maintenance of electrical systems are meticulously defined by various codes that have been established to provide protection of personnel and property. Installation, alteration, or repair of electrical devices and systems should be undertaken only by qualified persons. The subject matter that we study in circuit analysis does not provide that qualification.

The following examples illustrate the potential hazards that can be encountered in a variety of everyday situations. We begin by revisiting a situation described in a previous example.

EXAMPLE 9.20

Suppose that a man is working on the roof of a mobile home with a hand drill. It is early in the day, the man is barefoot, and dew covers the mobile home. The ground prong on the electrical plug of the drill has been removed. Will the man be shocked if the "hot" electrical line shorts to the case of the drill?

SOLUTION To analyze this problem, we must construct a model that adequately represents the situation described. In his book *Medical Instrumentation* (Boston: Houghton Mifflin Company, 1978), John G. Webster suggests the following values for resistance of the human body: R_{skin} (dry) = 15 kΩ, R_{skin} (wet) = 150 Ω, R_{limb} (arm or leg) = 100 Ω, and R_{trunk} = 200 Ω.

The network model is shown in Fig. 9.30. Note that since the ground line is open circuited, a closed path exists from the hot wire through the short, the human body, the mobile home, and the ground. For the conditions stated previously, we assume that the surface contact resistances R_{sc_1} and R_{sc_2} are 150 Ω each. The body resistance, R_{body}, consisting of arm, trunk, and leg, is 400 Ω. The mobile home resistance is assumed to be zero, and the ground resistance, R_{gnd}, from the mobile home ground to the actual source ground is assumed to be 1 Ω. Therefore, the magnitude of the current through the body from hand to foot would be

$$\mathbf{I}_{body} = \frac{120}{R_{sc_1} + R_{body} + R_{sc_2} + R_{gnd}}$$

$$= \frac{120}{701}$$

$$= 171 \text{ mA}$$

Figure 9.30 Model for Example 9.20.

A current of this magnitude can easily cause heart failure.

It is important to note that additional protection would be provided if the circuit breaker were a ground-fault interrupter. ❑

EXAMPLE 9.21

Two boys are playing basketball in their backyard. In order to cool off, they decide to jump into their pool. The pool has a vinyl lining, so the water is electrically insulated from the earth. Unknown to the boys, there is a ground fault in one of the pool lights. One boy jumps in and while standing in the pool with water up to his

(a)

(b)

Figure 9.31 Diagrams used in Example 9.21.

chest, reaches up to pull in the other boy, who is holding onto a grounded hand rail, as shown in Fig. 9.31a. What is the impact of this action?

SOLUTION The action in Fig. 9.31a is modeled as shown in Fig. 9.31b. Note that since a ground fault has occurred, there exists a current path through the two boys. Assuming that the fault, pool, and railing resistances are approximately zero, the magnitude of the current through the two boys would be

$$\mathbf{I} = \frac{120}{(3R_{\text{arm}}) + 3(3R_{\text{wet contact}}) + R_{\text{trunk}}}$$

$$= \frac{120}{950}$$

$$= 126 \text{ mA}$$

This current level would cause severe shock in both boys. The boy outside the pool could experience heart failure. ❏

EXAMPLE 9.22

A patient in a medical laboratory has a muscle stimulator attached to her left fore-arm. Her heart rate is being monitored by an EKG machine with two differential electrodes over the heart and the ground electrode attached to her right ankle. This activity is illustrated in Fig. 9.32a. The stimulator acts as a current source that drives 150 mA through the muscle from the active electrode to the passive electrode. If the laboratory technician mistakenly decides to connect the passive electrode of the stimulator to the ground electrode of the EKG system to achieve a common ground, is there any risk?

(a)   (b)

Figure 9.32 Diagrams used in Example 9.22.

SOLUTION   When the passive electrode of the stimulator is connected to the ground electrode of the EKG system, the equivalent network in Fig. 9.32b illustrates the two paths for the stimulator current: one through half an arm and the other through half an arm and the body. Using current division, the body current is

$$I_{body} = \frac{(150)(10^{-3})(50)}{50 + 50 + 200 + 100}$$

$$= 19 \text{ mA}$$

Therefore, a dangerously high level of current will flow from the stimulator through the body to the EKG ground.   ❏

EXAMPLE 9.23

A cardiac care patient with a pacing electrode has ignored the hospital rules and is listening to a cheap stereo. The stereo has an amplified 60-Hz hum that is very an-noying. The patient decides to dismantle the stereo partially in an attempt to elim-inate the hum. In the process, while he is holding one of the speaker wires, the other touches the pacing electrode. What are the risks in this situation?

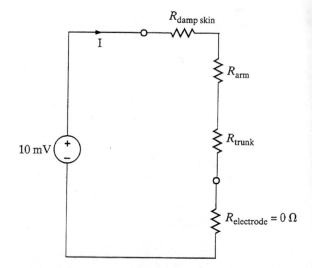

Figure 9.33 Circuit model
for Example 9.23.

SOLUTION   Let us suppose that the patient's skin is damp and that the 60-Hz voltage across the speaker wires is only 10 mV. Then the circuit model in this case would be shown in Fig. 9.33. The current through the heart would be

$$I = \frac{(10)(10^{-3})}{150 + 100 + 200}$$

$$= 22.2 \ \mu A$$

It is known that 10 μA delivered directly to the heart is potentially lethal.   ❏

EXAMPLE 9.24

While maneuvering in a muddy area, a crane operator accidentally touched a high-voltage line with the boom of the crane, as illustrated in Fig. 9.34a. The line potential was 7200 V. The neutral conductor was grounded at the pole. When the crane operator realized what had happened, he jumped from the crane and ran in the direction of the pole, which was approximately 10 m away. He was electrocuted as he ran. Can we explain this very tragic accident?

SOLUTION   The conditions depicted in Fig. 9.34a can be modeled as shown in Fig. 9.34b. The crane was at 7200 V with respect to earth. Therefore, a gradient of 720 V/m existed along the earth between the crane and the power pole. This earth between the crane and the pole is modeled as a resistance. If the man's stride was about 1 m, the difference in potential between his feet was approximately 720 V. A man standing in the same area with his feet together was unharmed.   ❏

The examples of this section have been provided in an attempt to illustrate some of the potential dangers that exist when working or playing around electric power. In the worst case, failure to prevent an electrical accident can result in death. However, even

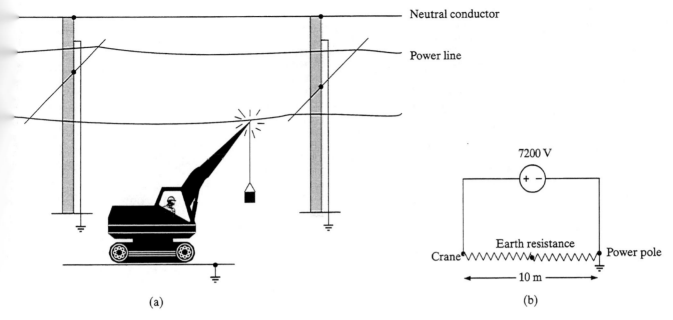

Figure 9.34 Illustrations used in Example 9.24.

nonlethal electrical contacts can cause such things as burns or falls. Therefore, we must always be alert to ensure not only our own safety, but also that of others who work and play with us.

The following guidelines will help us minimize the chances of injury.

SAFETY Guidelines

1. Avoid working on energized electrical systems.

2. Always assume that an electrical system is energized unless you can absolutely verify that it is not.

3. Never make repairs or alterations that are not in compliance with the provisions of the prevailing code.

4. Do not work on potentially hazardous electrical systems alone.

5. If another person is "frozen" to an energized electrical circuit, deenergize the circuit, if possible. If that cannot be done, use nonconductive material such as dry wooden boards, sticks, belts, and articles of clothing to separate the body from the contact. Act quickly but take care to protect yourself.

6. When handling long metallic equipment, such as ladders, antennas, and so on, outdoors, be continuously aware of overhead power lines and avoid any possibility of contact with them.

EXTENSION EXERCISE

E9.16   A woman is driving her car in a violent rainstorm. While she is waiting at an intersection, a power line falls on her car and makes contact. The power line voltage is 7200 V.

(a) Assuming that the resistance of the car is negligible, what is the potential current-through her body if, while holding the door handle with a dry hand, she steps out onto the wet ground?

(b) If she remainded in the car, what would happen?

ANSWER: (a) I = 463 mA, extremely dangerous; (b) she should be safe.

Safety when working with electric power must always be a primary consideration. Regardless of how efficient or expedient an electrical network is for a particular application, it is worthless if it is also hazardous to human life.

The safety device shown in Fig. 9.35, which is also used for troubleshooting, is a proximity-type sensor that will indicate if a circuit is energized by simply touching the conductor on the outside of the insulation. This device is typically carried by all electricians and i helpful when working on electric circuits.

In addition to the numerous deaths that occur each year due to electrical accidents fire damage that results from improper use of electrical wiring and distribution equipment amounts to millions of dollars per year.

To prevent the loss of life and damage to property, very detailed procedures and specifications have been established for the construction and operation of electrical systems to ensure their safe operation. The *National Electrical Code* ANSI C1 (ANSI—American National Standards Institute) is the primary guide. There are other codes, however: for example, the national electric safety code, ANSI C2, which deals with safety requirements for public utilities. The Underwriters' Laboratory (UL) tests all types of devices and systems to ensure that they are safe for use by the general public. We find the UL label on all types of electrical equipment that is used in the home, such as appliances and extension cords.

Electric energy plays a very central role in our lives. It is extremely important to our general health and well-being. However, if not properly used, it can be lethal.

9.11 APPLICATIONS

The following application-oriented examples illustrate a practical use of the material studied in this chapter.

EXAMPLE 9.25

An electric lawn mower requires 12 A rms at 120 V rms but will operate down to 110 V rms without damage. At 110 V rms, the current draw is 13.1 A rms, as shown in Fig. 9.36. Give the maximum length extension cord that can be used with a 120-V rms power source if the extension cord is made from

1. 16-gauge wire (4 mΩ/ft)

2. 14-gauge wire (2.5 mΩ/ft)

SOLUTION The voltage drop across the extension cord is

$$V_{cord} = (2)(13.1)R_{cord} = 10 \text{ V rms}$$

Figure 9.35 A modern safety or troubleshooting device (courtesy of Fluke Corporation).

Figure 9.36 Circuit model in Example 9.25.

or

$$R_{\text{cord}} = 0.382 \ \Omega$$

If ℓ_{cord} is the length of the extension cord, then for 16-gauge wire we find

$$\ell_{\text{cord}} = \frac{R_{\text{cord}}}{0.004} = 95.5 \text{ feet}$$

and for 14-gauge wire

$$\ell_{\text{cord}} = \frac{R_{\text{cord}}}{0.0025} = 152.8 \text{ feet} \qquad \square$$

EXAMPLE 9.26

The dc power supply shown in Fig. 9.37 is designed to supply the power requi̶
ment for a computer. The dc supply has an efficiency of 92%, and at unity in̶
power factor provides 5 V at 40 A dc. Let us determine the current requirement̶
the ac wall plug.

Figure 9.37 Circuit model in Example 9.26.

SOLUTION   The output power of the dc supply is

$$P_{out} = (40)(5) = 200 \text{ W}$$

The input power to the supply is

$$P_{in} = \frac{P_{out}}{0.92} = 217.4 \text{ W}$$

The ac power requirement is

$$P_{ac} = P_{in} = V_{rms} I_{rms} \text{pf}$$

and hence,

$$I_{rms} = \frac{P_{in}}{V_{rms} \text{pf}} = 1.81 \text{ A rms}$$

❏

EXAMPLE 9.27

After stopping for coffee after dark, Joe starts his car and notices that his alternator light fails to turn off, indicating total alternator failure. Highway conditions permit driving at 55 mph after dark and Joe is 82.5 miles from home. He recalls that he
had purchased a 120 Ah 12-V car battery just last month, advertised to produce "600
cold cranking amps." Joe estimates that it had taken about 6 seconds to start the car.

Joe immediately turns off the radio and heater to conserve power. His two headlights each draw 50 watts on low beam. Assuming the rest of the car's electrical load
totals to 50 watts, will Joe make it home before the battery is completely discharged?

SOLUTION   If Joe assumes that the battery is completely charged, then
charge = 120 Ah. Since the exact current required to start the engine is unknown,

Joe decides to use the worst case value of 600 A. Thus the charge required to start the engine is estimated to be

$$\text{charge}_{\text{start}} = (600 \text{ A}) \frac{(6 \text{ s})}{3600 \text{ s/h}}$$

$$= 1 \text{ Ah}$$

The electrical load on the battery caused by the two lights and other elements within the car is

$$I = \frac{150 \text{ W}}{12 \text{ V}}$$

$$= 12.5 \text{ A}$$

The time required to drive home at 55 mph is

$$t = \frac{82.5}{55}$$

$$= 1.5 \text{ h}$$

The total charge drain on the battery including starting is

$$\text{charge}_{\text{total}} = 1 + 12.5(1.5) = 19.75 \text{ Ah}$$

Therefore, Joe should make it home with charge to spare. ❑

EXAMPLE 9.28

The circuit in Fig. 9.38a models an electric field probe amplifier including capacitive effects. As drawn, the average power delivered to the load is not optimized. However, the compensating impedance $\mathbf{Z}_C$ can be used to modify the output power. Let us design $\mathbf{Z}_C$ for maximum power transfer if the input signal $\mathbf{V}_S$ is sinusoidal at 10^6 rad/s.

SOLUTION   For maximum power transfer, the impedance of $\mathbf{Z}_C$ in parallel with R_L must be the complex conjugate of the Thévenin equivalent impedance at nodes A-B, looking back toward the source. Therefore, let us find the Thévenin impedance for the network as redrawn in Fig. 9.38b. Nodal analysis yields

$$(\mathbf{V}_S - \mathbf{V}_{oc})j\omega C_2 = g_m \mathbf{V}_S + \frac{\mathbf{V}_{oc}}{R_2}$$

or

$$\mathbf{V}_{oc} = \mathbf{V}_S(j\omega C_2 - g_m)\mathbf{Z}_2$$

where

$$\mathbf{Z}_2 = R_2 // \frac{1}{j\omega C_2}$$

To find $\mathbf{I}_{sc}$ we short nodes A and B as shown in Fig. 9.38c. All of the dependent source current flows through the short; therefore,

$$\mathbf{I}_{sc} = -g_m \mathbf{V}_S$$

Use the problem-solving strategy for maximum average power transfer delineated in Section 9.3.

1 Remove the load $\mathbf{Z}_L$ and find the Thévenin equivalent for the remainder of the circuit.

2 Construct a circuit consisting of the Thévenin equivalent circuit and the load.

3 Select a load $\mathbf{Z}_L$ such that $\mathbf{Z}_L = \mathbf{Z}_{\text{Th}}^*$ and determine the values of the load elements.

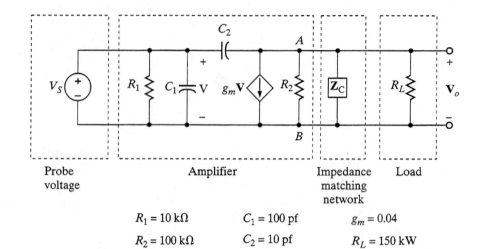

$R_1 = 10 \text{ k}\Omega$ $C_1 = 100 \text{ pf}$ $g_m = 0.04$

$R_2 = 100 \text{ k}\Omega$ $C_2 = 10 \text{ pf}$ $R_L = 150 \text{ kW}$

(a)

(b)   (c)

(d)

Figure 9.38 Circuits used in Example 9.28.

and $\mathbf{Z}_{Th}$ is

$$\mathbf{Z}_{Th} = \frac{\mathbf{V}_{oc}}{\mathbf{I}_{sc}} = \mathbf{Z}_2 \left(1 - \frac{j\omega C_2}{g_m} \right)$$

Using element values from Fig. 9.38a yields

$$\mathbf{Z}_{Th} = \frac{10^5}{1 + j1} \left(1 - j2.5 \times 10^{-4} \right) \approx \frac{10^5}{1 + j1}$$

or

$$\mathbf{Z}_{Th} = 70.7 \; \underline{/-45°} \text{ k}\Omega$$

Maximum power transfer requires

$$\mathbf{Z}_C // R_L = \mathbf{Z}_{Th}^*$$

Since $\mathbf{Z}_{Th}$ is capacitive, it is sensible that $\mathbf{Z}_C$ should be inductive. Let us use the network in Fig. 9.38d for the compensating impedance and find the values of L_C and R_C for maximum average power transfer. The equivalent parallel impedance is

$$\mathbf{Z}_{eq} = R_C // R_L // j\omega L_C$$

or

$$\mathbf{Z}_{eq} = \frac{R_C // R_L}{1 - j\dfrac{R_C // R_L}{\omega L_C}}$$

Based on the requirements for maximum power transfer, i.e.

$$\frac{R_C // R_L}{1 - j\dfrac{R_C // R_L}{\omega L_C}} = \left(\frac{10^5}{1 - j1}\right)^*$$

we find that

$$R_C // R_L = 100 \text{ k}\Omega$$

or

$$R_C = 300 \text{ k}\Omega$$

and

$$\frac{R_C // R_L}{\omega L_C} = 1$$

or

$$L_C = 100 \text{ mH}$$

These values ensure that the amplifier sees an equivalent load, $\mathbf{Z}_{eq}$, which produces maximum average power transfer.   ❑

EXAMPLE 9.29

For safety reasons the National Electrical Code restricts the circuit-breaker rating in a 120-V household lighting branch circuit to no more than 20 A. Furthermore, the code also requires a 25% safety margin for continuous-lighting loads. Under these conditions let us determine the number of 100-W lighting fixtures that can be placed in one branch circuit.

SOLUTION   The model for the branch circuit is shown in Fig. 9.39. The current drawn by each 100-W bulb is

$$I_{bulb} = 100/120 = 0.833 \text{ A rms}$$

Using the safety margin recommendation, the estimated current drawn by each bulb is 25% higher or

$$I_{bulb} = (1.25)(0.83) = 1.04 \text{ A rms}$$

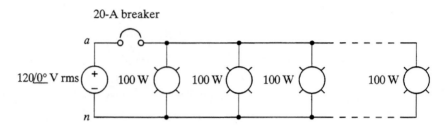

Figure 9.39 20-A branch circuit for household lighting.

Therefore, the maximum number of fixtures on one circuit breaker is

$$n = 20/1.04 = 19 \text{ fixtures}$$

EXAMPLE 9.30

While sitting in a house reading a book, we notice that every time the air conditioner comes on, the lights momentarily dim. Let us investigate this phenomenon using
the single-phase three-wire circuit shown in Fig. 9.40a and some typical current requirements for a 10,000-Btu/h air conditioner, assuming a line resistance of 0.5 Ω.

Figure 9.40 Networks used in Example 9.30.

SOLUTION   The 60-W light bulb can be roughly modeled by its equivalent resistance:

$$P_{bulb} = \frac{V_{an}^2}{R_{bulb}}$$

or

$$R_{bulb} = 240 \ \Omega$$

When the air conditioner unit first turns on, the current requirement is 40 A, as shown in Fig. 9.40b. As the compressor motor comes up to speed, the current requirement drops quickly to a steady-state value of 10 A, as shown in Fig. 9.40c. We will compare the voltage across the light fixture, V_{AN}, both at turn on and in steady state.

Using superposition, let us first find that portion of V_{AN} caused by the voltage sources. The appropriate circuit is shown in Fig. 9.40d. Using voltage division, we find that

$$V_{AN1} = V_{AN}\left(\frac{R_{bulb}}{R_{bulb} + 2R_L}\right)$$

or

$$V_{AN1} = 119.50 \ \text{V rms}$$

Figure 9.40e will yield the contribution to V_{AN} caused by the 10-A steady-state current. Using current division to calculate the current through the light bulb, we find that

$$V_{AN2} = -\left\{I_{AB}\left(\frac{R_L}{R_{bulb} + 2R_L}\right)\right\}R_{bulb}$$

or

$$V_{AN2} = -4.98 \ \text{V rms}$$

Therefore, the steady-state value of V_{AN} is

$$V_{AN} = V_{AN1} + V_{AN2} = 114.52 \ \text{V rms}$$

At start-up, our expression for V_{AN2} can be used with $I_{AB} = 40$ A, which yields $V_{AN2} = -19.92$ V rms. The resulting value for V_{AN} is

$$V_{AN} = V_{AN1} + V_{AN2} = 119.50 - 19.92 = 99.58 \ \text{V rms}$$

The voltage delivered to the light fixture at start-up is 13% lower than the steady-state value, resulting in a momentary dimming of the lights.

9.12 CIRCUIT DESIGN

The following design example is presented in the context of a single-phase three-wire household circuit and serves as a good introduction to the material in the next chapter.

Technique

1 Find the resistance of the light bulb.

2 Use a large current source to represent the transient current of the air conditioner and a small current source to represent the steady-state current.

3 Find the voltage drop across the light bulb during both the transient and steady-state operations.

EXAMPLE 9.31

A light-duty commercial single-phase three-wire 60-Hz circuit serves lighting, heating, and motor loads, as shown in Fig. 9.41a. Lighting and heating loads are essentially pure resistance and, hence, unity power factor (pf), whereas motor loads have lagging pf.

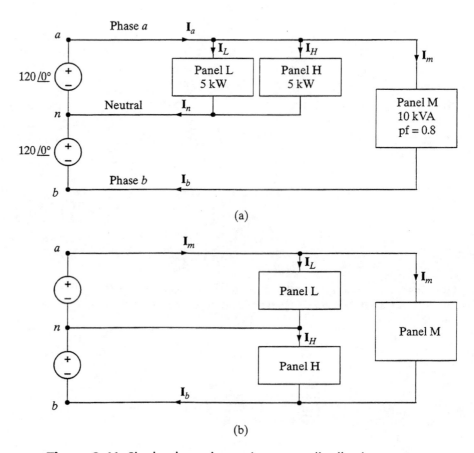

(a)

(b)

Figure 9.41 Single-phase three-wire power distribution system.

We wish to design a balanced configuration for the network and determine its economic viability using the following procedure.

(a) Compute the phase and neutral currents, and the complex power and pf for each source.

(b) Now move the heating load (panel H) to phase b, as shown in Fig. 9.41b. This is called "balancing" the load. Repeat the analysis of (a).

(c) Assume that the phase and neutral conductor resistances are each 0.05 Ω and have negligible effect on the results of (a). Evaluate the system line losses for (a) and (b). If the loads in question operate 24 hours per day and 365 days per year, at $0.08 kWh, how much energy (and money) is saved by operating in the balanced mode?

SOLUTION   (a) The magnitudes of the rms currents are

$$I_L = I_H = \frac{P}{V} = \frac{5000}{120} = 41.67 \text{ A}$$

and

$$I_m = \frac{10,000}{240} = 41.67 \text{ A}$$

In addition,

$$\theta_m = \cos^{-1}(0.8) = -36.9°$$

Therefore,

$$\mathbf{I}_a = \mathbf{I}_L + \mathbf{I}_H + \mathbf{I}_m$$
$$= 41.67 \ \underline{/0°} + 41.67 \ \underline{/0°} + 41.67 \ \underline{/-36.9°}$$
$$= 119.4 \ \underline{/-12.1°} \text{ A}$$

The currents in the neutral and phase b lines are

$$\mathbf{I}_n = \mathbf{I}_L + \mathbf{I}_H = 83.34 \ \underline{/0°} \text{ A}$$
$$\mathbf{I}_b = 41.67 \ \underline{/-36.9°} \text{ A}$$

The complex power and power factor for each source are

$$\mathbf{S}_a = \mathbf{V}_{an} \mathbf{I}_a^* = (120 \ \underline{/0°})(119.4 \ \underline{/+12.1°}) = 14 + j3 \text{ kVA}$$
$$\text{pf}_a = \cos(12.1°) = 0.9778 \text{ lagging}$$

and in a similar manner

$$\mathbf{S}_b = \mathbf{V}_{bn} \mathbf{I}_b^* = 4 + j3 \text{ kVA}$$
$$\text{pf}_b = 0.8 \text{ lagging}$$

(b) Under the balanced condition

$$\mathbf{I}_a = \mathbf{I}_L + \mathbf{I}_m = 41.67 \ \underline{/0°} + 41.67 \ \underline{/-36.9°}$$
$$= 79.06 \ \underline{/-18.4°} \text{ A}$$

and

$$\mathbf{I}_b = 79.06 \ \underline{/-18.4°} \text{ A}$$
$$\mathbf{I}_n = 0$$

Therefore,

$$\mathbf{S}_a = \mathbf{V}_{na} \mathbf{I}_a^* = 9 + j3 \text{ kVA}$$
$$\text{pf}_a = 0.9487 \text{ lagging}$$

and

$$\mathbf{S}_b = \mathbf{V}_{bn} \mathbf{I}_b^* = 9 + j3 \text{ kVA}$$
$$\text{pf}_b = 0.9487 \text{ lagging}$$

(c) The power loss in the lines in kW is

$$P_{loss} = R_a I_a^2 + R_b I_b^2 + R_n I_c^2$$
$$= 0.05(I_a^2 + I_b^2 + I_n^2)/1000$$

The total energy loss for a year is

$$W_{loss} = (24)(365)P_{loss} = 8760 \, P_{loss}$$

and the annual cost is

$$\text{Cost} = \$0.08 \, W_{loss}$$

A comparison of the unbalanced and balanced cases is shown in the following table.

	Unbalanced Case	Balanced Case
P_{loss} (kW)	1.147	0.625
W_{loss} (kW-hr)	10,034	5,475
Cost ($)	804	438

Therefore, the annual savings obtained using the balanced configuration is

Annual energy savings = 10,048 − 5,475 = 4,573 kWh

Annual cost savings = 804 − 438 = $366 ❏

9.13 SUMMARY

Instantaneous power   If the current and voltage are sinusoidal functions of time, the instantaneous power is equal to a time-independent average value plus a sinusoidal term that has a frequency twice that of the voltage or current.

Average power $P = \frac{1}{2}VI \cos(\theta_v - \theta_i) = \frac{1}{2}VI \cos \theta$, where θ is the phase of the impedance.

Resistive load $P = \frac{1}{2}I^2 R = \frac{1}{2}VI$ since $\mathbf{V}$ and $\mathbf{I}$ are in phase.

Reactive load $P = \frac{1}{2}VI \cos(\pm 90°) = 0$

Superposition   When multiple sources are present in a network, superposition cannot be used in computing power unless each source is of a different frequency.

Maximum average power transfer   To obtain the maximum average power transfer to a load, the load impedance should be chosen equal to the conjugate of the Thévenin equivalent impedance representing the remainder of the network.

rms or effective value of a periodic waveform   The effective, or rms, value of a periodic waveform was introduced as a means of measuring the effectiveness of a source in delivering power to a resistive load. The effective value of a periodic waveform is found by determining the root-mean-square value of the waveform.

The rms value of a sinusoidal function is equal to the maximum value of the sinusoid divided by $\sqrt{2}$.

ower factor   Apparent power is defined as the product $V_{rms} I_{rms}$. The power factor is defined as the ratio of the average power to the apparent power and is said to be leading when the phase of the current leads the voltage, and lagging when the phase of the current lags the voltage. The power factor of a load with a lagging power factor can be corrected by placing a capacitor in parallel with the load.

omplex power   The complex power, **S**, is defined as the product $\mathbf{V}_{rms} \mathbf{I}_{rms}^{*}$. The complex power **S** can be written as $\mathbf{S} = P + jQ$, where P is the real or average power and Q is the imaginary or quadrature power.

$$\mathbf{S} = I^2\mathbf{Z} = I^2 R + jI^2 X$$

he single-phase three-wire circuit   The single-phase three-wire circuit is the one commonly used in households. Large appliances are connected line to line and small appliances and lights are connected line to neutral.

Safety   Safety must be a primary concern in the design and use of any electrical circuit. The National Electric Code is the primary guide for the construction and operation of electrical systems.

PROBLEMS

Section 9.2

9.1   The voltage and current at the input of a circuit are given by the expressions

$$v(t) = 170 \cos(\omega t + 30°) \text{ V}$$
$$i(t) = 5 \cos(\omega t + 45°) \text{ A}$$

$P = \frac{1}{2}VI \cos\theta$

Determine the average power absorbed by the circuit.

9.2   The voltage and current at the input of a network are given by the expressions

$$v(t) = 6 \cos \omega t \text{ V}$$
$$i(t) = 4 \sin \omega t \text{ A}$$

Similar to P9.1.

Determine the average power absorbed by the network.

9.3   Find the average power absorbed by the resistor in the circuit shown in Fig. P9.3 if $v_1(t) = 10 \cos(377t + 60°)$ V and $v_2(t) = 20 \cos(377t + 120°)$ V.

$P = \frac{1}{2}\dfrac{V^2}{R}$

Figure P9.3

9.4 Find the average power absorbed by the resistor in the circuit shown in Fig. P9.4.
Let $i_1(t) = 4\cos(377t + 60°)$ A, $i_2(t) = 6\cos(754t + 10°)$ A, and $i_3(t) = 4\cos(377t - 30°)$ A.

$P = \frac{1}{2}I^2 R$

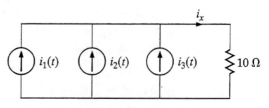

Figure P9.4

9.5 Compute the average power absorbed by each of the elements to the right of the dashed line in the circuit shown in Fig. P9.5.

Only the resistor absorbs
average power.

Figure P9.5

9.6 Compute the average power absorbed by the elements to the right of the dashed line in the network shown in Fig. P9.6.

Similar to P9.5.

Figure P9.6

9.7 Determine the average power supplied by each source in the network shown in Fig. P9.7.

KCL can be used to find all
voltages and currents needed
to determine the unknowns.

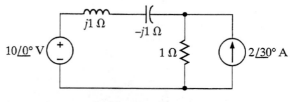

Figure P9.6

Given the network in Fig. P9.8, show that the power supplied by the sources is equal to the power absorbed by the passive elements.

Figure P9.8

Similar to P9.7.

.9   Find the average power absorbed by the network shown in Fig. P9.9.

Figure P9.9

Similar to P9.7.

9.10   Calculate the average power absorbed by the 1-Ω resistor in the network shown in Fig. P9.10.

Figure P9.10

Use the voltage division rule to find the voltage across the 1-Ω resistor. Then employ $P = \frac{1}{2}V^2/R$.

9.11   Find the average power supplied and/or absorbed by each element in Fig. P9.11.

Figure P9.11

Use nodal analysis to find voltages. Find the current in the voltage source. Then use $P = \frac{1}{2}VI \cos\theta$ for the sources and $P = \frac{1}{2}V^2/R$ for the resistor.

9.12 Given the circuit in Fig. P9.12, find the average power supplied and the average power absorbed by each element.

The current division rule will yield the currents, and they, in turn, will produce the source voltage.

Figure P9.12

9.13 Determine the average power absorbed by the 4-Ω resistor in the network shown in Fig. P9.13.

Using nodal analysis, find the voltage across the 4-Ω resistor.

Figure P9.13

9.14 Find the average power absorbed by each element in the network shown in Fig. P9.14.

Nodal analysis will yield the node voltages and then the element currents.

Figure P9.14

9.15 Given the network in Fig. P9.15, find the average power supplied to the circuit.

Use nodal analysis to find the voltage across the current source. Then all currents can be calculated.

Figure P9.15

6 Determine the average power absorbed by the 4-Ω in the network shown in Fig. P9.16.

Figure P9.16

Similar to P9.13.

.17 Determine the average power supplied to the network shown in Fig. P9.17.

Figure P9.17

Use the supernode technique to find the node voltages. Then all currents can be obtained immediately.

9.18 Given the circuit in Fig. P9.18, determine the amount of average power supplied to the network.

Figure P9.18

Similar to P9.17.

9.19 Determine the average power supplied to the network in Fig. P9.19.

Figure P9.19

Similar to P9.17.

Section 9.3

Find the Thévenin equivalent circuit for the network with the load Z_L removed. Then let $Z_L = Z_{Th}^*$ and find the maximum power.

9.20 Determine the impedance Z_L for maximum average power transfer and the value of the maximum power transferred to the Z_L for the circuit shown in Fig. P9.20.

Figure P9.20

Find the current for maximum average power transfer. Then use the average power equation.

9.21 Show that the maximum average power transferred to a load is given by the following expression, where V_{oc} and R_{Th} are defined in the chapter.

$$P_L = \frac{V_{oc}^2}{8R_{Th}}$$

Similar to P9.20.

9.22 Determine the impedance Z_L for maximum average power transfer and the value of the maximum average power absorbed by the load in the network shown in Fig. P9.22.

Figure P9.22

Similar to P9.20.

9.23 Determine the impedance Z_L for maximum average power transfer and the value of the maximum average power transferred to Z_L for the circuit shown in Fig. P9.23.

Figure P9.23

24 Repeat Problem 9.23 for the network in Fig. P9.24.

Figure P9.24

.25 Repeat Problem 9.23 for the network in Fig. P9.25.

Figure P9.25

9.26 Repeat Problem 9.23 for the network in Fig. P9.26.

Figure P9.26

9.27 Repeat Problem 9.23 for the network in Fig. P9.27.

Figure P9.27

Similar to P9.20.

9.28 Repeat Problem 9.23 for the network in Fig. P9.28.

Similar to P9.20.

Figure P9.28

9.29 Repeat Problem 9.23 for the network in Fig. P9.29.

Similar to P9.20.

Figure P9.29

Section 9.4

$$V_{rms} = \sqrt{\frac{1}{T}\int_0^T v^2(t)\,dt}$$

9.30 Compute the rms value of the voltage given by the expression $v(t) = 10 + 20\cos(377t + 30°)$ V.

9.31 Compute the rms value of the voltage given by the waveform shown in Fig. P9.31.

Similar to P9.30; carefully select the bounds of the integral.

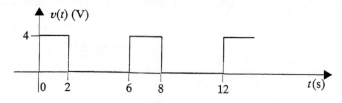

Figure P9.31

9.32 Calculate the rms value of the waveform shown in Fig. P9.32.

Similar to P9.31.

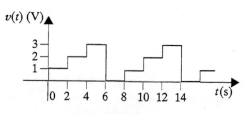

Figure P9.32

3 Calculate the rms value of the waveform in Fig. P9.33.

Figure P9.33

The equation of the voltage between 0 and 2 s is 3t.

.34 Calculate the rms value of the waveform in Fig. P9.34.

Figure P9.34

Similar to P9.33.

9.35 Calculate the rms value of the waveform in Fig. P9.35.

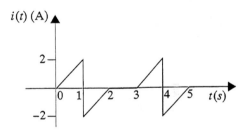

Figure P9.35

Find the equation of the functions in the two intervals $0 \leq t < 1$ s, and $0 \leq t < 2$ s. Integrate to find rms value.

9.36 Find the rms value of the exponential waveform shown in Fig. P9.36.

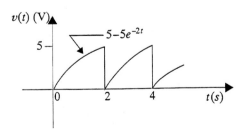

Figure P9.36

Use $T = 2$ seconds.

9.37 Determine the rms value of the rectified sine wave shown in Fig. P9.37. Use the fact that $\sin^2 \theta = \frac{1}{2} - \frac{1}{2} \cos 2\theta$.

The integral's upper bound = 1, the lower bound = 0, $T = 2$, and the function is $\sin \pi t$.

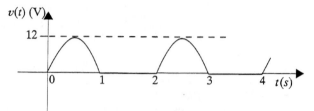

Figure P9.37

See Example 9.11.

9.38 Find the rms value of the current defined by the expression

$$i(t) = 10 \cos (377t) + 12 \sin (754t + 80°) \text{ A}$$

Shortcut: Note that $\cos (t + 240°) + \cos t + \cos (t + 120°) = 0$.

9.39 Find the rms value of the voltage defined by the expression

$$v(t) = \cos t + \cos (t + 120°).$$

9.40 Write the phasor with rms magnitude in polar form for the following waveforms.

(a) $v_1(t) = 170 \cos (377t + 30°)$ V

(b) $i_2(t) = 10 \cos (377t + 45°)$ A

(c) $i_3(t) = 20 \cos (377t + 120°)$ A

(d) $v_4(t) = 32 \sqrt{2} \cos (377t - 120°)$ V

(e) $v_5(t) = 20 \cos (377t + 30°)$ V

Section 9.5

Find θ from pf = $\cos \theta$.

9.41 A plant consumes 20 kW of power from a 240-V rms line. If the load power factor is 0.9, what is the angle by which the load voltage leads the load current? What is the load current phasor if the line voltage has a phasor of 240 $\underline{/0°}$ V rms?

Find V using $P = VI \cos \theta$.

9.42 A plant consumes 100 kW of power at 0.9 pf lagging. If the load current is 200 A rms, find the load voltage.

pf = $\cos \theta$

9.43 A plant draws 250 A rms from a 240-V rms line to supply a load with 100 kW. What is the power factor of the load?

44 An industrial load consumes 100 kW at 0.8 pf lagging. If an ammeter in the transmission line indicates that the load current is 200 A rms, find the load voltage.

$P = VI \cos \theta.$

---

45 An industrial load that consumes 40 kW is supplied by the power company through a transmission line with 0.1 Ω resistance, with 44 kW. If the voltage at the load is 240 V rms, find the power factor at the load.

$P_s = P_{line} + P_{Load}$ will yield the line current.

.46 The power company supplies 40 kW to an industrial load. The load draws 200 A rms from the transmission line. If the load voltage is 240 V rms and the load power factor is 0.8 lagging, find the losses in the transmission line.

Similar to P9.45.

Section 9.6

9.47 Determine the real power, the reactive power, the complex power, and the power factor for a load having the following characteristics.

(a) $\mathbf{I} = 2\ \underline{/40°}$ A rms, $\mathbf{V} = 450\ \underline{/70°}$ V rms.

(b) $\mathbf{I} = 1.5\ \underline{/-20°}$ A rms, $\mathbf{Z} = 5000\ \underline{/15°}$ Ω.

(c) $\mathbf{V} = 200\ \underline{/+35°}$ V rms, $\mathbf{Z} = 1500\ \underline{/-15°}$ Ω.

The complex power can be expressed as

$$\mathbf{S} = \mathbf{VI}^* = I^2 \mathbf{Z} = \left(\frac{V}{Z}\right)^2 \mathbf{Z}$$

9.48 An industrial load operates at 30 kW, 0.8 pf lagging. The load voltage is $240\ \underline{/0°}$ V rms. The real and reactive power losses in the transmission-line feeder are 1.8 kW and 2.4 kvar, respectively. Find the impedance of the trasmission line and the input voltage to the line.

Recall that

$$R_{Line} = \frac{P_{Line}}{I_L^2}$$

$$X_{Line} = \frac{Q_{Line}}{I_L^2}$$

9.49 A transmission line with impedance $0.08 + j0.25$ Ω is used to deliver power to a load. The load is inductive and the load voltage is $240\ \underline{/0°}$ V rms at 60 Hz. If the load requires 12 kW and the real power loss in the line is 560 W, determine the power factor angle of the load.

The line losses can be used to find the line current.

9.50 Calculate the voltage $\mathbf{V}_S$ that must be supplied to obtain 2 kW, $240\ \underline{/0°}$ V rms, and a power factor of 0.9 lagging at the load $\mathbf{Z}_L$ in the network in Fig. P9.50.

Figure P9.50

I_L can be found using the expression $P = VI \cos \theta.$

9.51 Use Kirchhoff's laws to compute the source voltage of the network shown in Fig. P9.51.

$\mathbf{V}_S = \mathbf{I}_S \mathbf{Z}_{\text{Line}} + 240 \underline{/0°}$,
where $\mathbf{I}_S$ is the source current.

Figure P9.51

9.52 Given the network in Fig. P9.52, determine the input voltage $\mathbf{V}_S$.

Similar to P9.51.

Figure P9.52

9.53 Given the network in Fig. P9.53, determine the input voltage $\mathbf{V}_S$.

Similar to P9.51.

Figure P9.53

9.54 Find the input source voltage and the power factor of the source for the network shown in Fig. P9.54.

Find the current in the 24-kW load. Then find the voltage across the 12-kW load. Find the current in the 12-kW load. Then determine $\mathbf{V}_S$.

Figure P9.54

·ction 9.7

·5 A plant consumes 60 kW at a power factor of 0.75 lagging from a 240-V rms 60-Hz line. Determine the value of the capacitor that when placed in parallel with the load will change the load power factor to 0.9 lagging.

$Q_{new} - Q_{old} = -\omega C V_{rms}^2$

56 A particular load has a pf of 0.8 lagging. The power delivered to the load is 40 kW from a 270-V rms 60-Hz line. What value of capacitance placed in parallel with the load will raise the pf to 0.9 lagging?

Similar to P9.55.

·57 A small plant has a bank of induction motors that consume 64 kW at a pf of 0.68 lagging. The 60-Hz line voltage across the motors is $240 \underline{/0°}$ V rms. The local power company has told the plant to raise the pf to 0.92 lagging. What value of capacitance is required?

Similar to P9.55.

9.58 The 60-Hz line voltage for a 60-kW, 0.76-pf lagging industrial load is $240 \underline{/0°}$ V rms. Find the value of capacitance that when placed in parallel with the load will raise the power factor to 0.9 lagging.

Similar to P9.55.

9.59 An industrial load consumes 44 kW at 0.82 pf lagging from a $240 \underline{/0°}$-V-rms 60-Hz line. A bank of capacitors totaling 600 μF is available. If these capacitors are placed in parallel with the load, what is the new power factor of the total load?

Similar to P9.55.

9.60 A bank of induction motors consumes 36 kW at 0.78 pf lagging from a 60-Hz $240 \underline{/0°}$-V-rms line. If 200 μF of capacitors are placed in parallel with the load, what is the new power factor of the total load?

Similar to P9.55.

Section 9.8

9.61 Given the network in Fig. P9.61, determine the wattmeter reading

Figure P9.61

Assume the current coil is replaced with a short circuit and the voltage coil is open circuited. Find the current and voltage of the watt meter. $P = \frac{1}{2} IV \cos \theta$.

9.62 Given the network in Fig. P9.62, determine the wattmeter reading.

Similar to P9.61.

Figure P9.62

Section 9.9

9.63 A single-phase three-wire 60-Hz circuit serves three loads, as shown in Fig. P9.63. Determine $\mathbf{I}_{aA}, \mathbf{I}_{nN}, \mathbf{I}_c$, and the energy use over a 24-hour period in kilowatt-hours.

See Example 9.17.

Figure P9.63

Section 9.10

9.64 A 5.1-kW household range is designed to operate on a 240-V rms sinusoidal voltage, as shown in Fig. P9.64a. However, the electrician has mistakenly connected the range to 120 V rms, as shown in Fig. P9.64b. What is the effect of this error?

Find the power consumed by the range by assuming that I_L is the maximum current allowed in both connections. Check to see if the power causes any damage.

Figure P9.64

5 A man and his son are flying a kite. The kite becomes entangled in a 7200-V power line close to a power pole. The man crawls up the pole to remove the kite. While trying to remove the kite, the man accidentally touches the 7200-V line. Assuming the power pole is well grounded, what is the potential current through the man's body?

Use the resistances of an arm or leg, trunk, and skin defined in Problem 9.20 to find the current.

66 A number of 120-V household fixtures are to be used to provide lighting for a large room. The total lighting load is 8 kW. The National Electric Code requires that no circuit breaker be larger than 20 A with a 25% safety margin. Determine the number of identical branch circuits needed for this requirement.

Find I_{load} in the 8 kW load. $I_{breaker} = (1.25)I_{load}$, $I_{breaker}/(20 \text{ A})$ is used to determine the number of circuits.

.67 In order to test a light socket, a woman, while standing on cushions that insulate her from the ground, sticks her finger into the socket, as shown in Fig. P9.67. The tip of her finger makes contact with one side of the line, and the side of her finger makes contact with the other side of the line. Assuming that any portion of a limb has a resistance of 95 Ω, is there any current in the body? Is there any current in the vicinity of the heart?

Construct an equivalent circuit similar to that in P9.65. Determine if the current will go through the heart.

Figure P9.67

9.68 An inexperienced mechanic is installing a 12-V battery in a car. The negative terminal has been connected. He is currently tightening the bolts on the positive terminal. With a tight grip on the wrench, he turns it so that the gold ring on his finger makes contact with the frame of the car. This

Figure P9.68

situation is modeled in Fig. P9.68, where we assume that the resistance of the wrench is negligible and the resistance of the contact is as follows:

Find the current in the circuit. Then find the power.

$$R_1 = R_{\text{bolt to wrench}} = 0.012 \ \Omega$$
$$R_2 = R_{\text{wrench to ring}} = 0.012 \ \Omega$$
$$R_3 = R_{\text{ring}} = 0.012 \ \Omega$$
$$R_4 = R_{\text{ring to frame}} = 0.012 \ \Omega$$

What power is quickly dissipated in the gold ring, and what is the impact of this power dissipation?

Advanced Problems

9.69 Find the average power absorbed by the 2-Ω resistor in the circuit shown in Fig P9.69.

Figure P9.69

9.70 Determine the average power absorbed by a 2-Ω resistor connected at the output terminals of the network shown in Fig. P9.70.

Figure P9.70

9.71 Determine the average power absorbed by the 2-kΩ output resistor in Fig. P9.71.

Figure P9.71

2 Determine the average power absorbed by the 4-kΩ resistor in Fig. P9.72.

Figure P9.72

.73 Given the network in Fig. P9.73, determine which elements are supplying power, which ones are absorbing power, and how much power is being supplied and absorbed.

Figure P9.73

9.74 Given the network in Fig. P9.74, find the total average power supplied and the average power absorbed in the 4-Ω resistor.

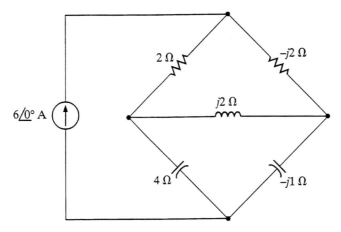

Figure P9.74

9.75 Given the network in Fig. P9.75, find the average power supplied and the total average power absorbed.

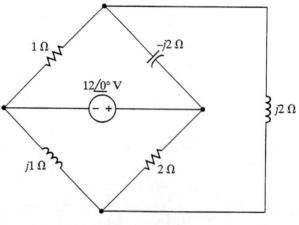

Figure P9.75

9.76 Find the impedance $\mathbf{Z}_L$ for maximum average power transfer and the value of the maximum average power transferred to $\mathbf{Z}_L$ for the circuit shown in Fig. P9.76.

Figure P9.76

9.77 Determine $\mathbf{Z}_L$ for maximum average power transfer and the value of the maximum average power transferred for the network shown in Fig. P9.77.

Figure P9.77

8 Find the impedance $\mathbf{Z}_L$ for maximum average power transfer and the value of the maximum average power transferred to $\mathbf{Z}_L$ for the circuit shown in Fig. P9.78.

Figure P9.78

9.79 Repeat Problem 9.78 for the network in Fig. P9.79.

Figure P9.79

9.80 Given the network in Fig. P9.80, find $\mathbf{Z}_L$ for maximum average power transfer and the maximum average power transferred.

Figure P9.80

9.81 Determine the impedance $\mathbf{Z}_L$ for maximum average power transfer and the value of the maximum average power absorbed by the load in the network shown in Fig. P9.81.

Figure P9.81

9.82 Determine the impedance $\mathbf{Z}_L$ for maximum average power transfer and the value of the maximum average power absorbed by the load in the network shown in Fig. P9.82.

Figure P9.82

9.83 Determine the voltage $\mathbf{V}_S$ that must be supplied to obtain 1.5 kW, 0.75 $\underline{/0°}$ A rms, and a power factor of 0.85 leading at the load $\mathbf{Z}_L$ in the network in Fig. P9.83.

Figure P9.83

9.84 Given the network in Fig. P9.84, compute the wattmeter reading.

Figure P9.84

5 For the network in Fig. P9.85, find the impedance $\mathbf{Z} = R + jX$ that is required for maximum average power transfer.

Figure P9.85